6th International Conference of Global Network for Innovative Technology (IGNITE 2019)

IOP Conference Series: Materials Science and Engineering
Volume 920

Penang, Malaysia
2 - 3 December 2019

ISBN: 978-1-7138-3189-1
ISSN: 1757-8981

Printed from e-media with permission by:

Curran Associates, Inc.
57 Morehouse Lane
Red Hook, NY 12571

Some format issues inherent in the e-media version may also appear in this print version.

Institute of Physics
Dirac House, Temple Back
Bristol BS1 6BE UK

Phone: 44 1 17 929 7481
Fax: 44 1 17 920 0979

techtracking@iop.org

Additional copies of this publication are available from:

Curran Associates, Inc.
57 Morehouse Lane
Red Hook, NY 12571 USA
Phone: 845-758-0400
Fax: 845-758-2633
Email: curran@proceedings.com
Web: www.proceedings.com

TABLE OF CONTENTS

PREFACE (IGNITE 2019)

Dear Colleagues,

I would like to congratulate all speakers, paper presenters and participants both from abroad and local for presenting and submitting their research papers to the 6th International Conference of Global Network for Innovative Technology (IGNITE 2019) held on 2 – 3 December 2019, in Parkroyal Penang Resort, Penang. The conference is hosted by the School of Mechanical Engineering, Universiti Sains Malaysia (USM) and jointly organized by the School of Aerospace Engineering, USM and Toyohashi University of Technology (TUT). The IGNITE 2019 conference provides a platform for researchers and experts to present their recent innovative works pertaining to mechanical, manufacturing, and civil engineering. The conference is also aimed to foster new researches and cultivate collaborations in the near future across different countries and institutions. This year, a total of 37 participants have submitted their paper to the IGNITE 2019 conference and all papers have gone through rigorous double-blind review before being accepted. This is a golden opportunity for the researchers and practitioners to be exposed to the advancement of new innovations and researches with the aim of identifying ways to contribute back to the society. I would like to thank the committees of the IGNITE 2019 for their continuous efforts in organizing this conference. Special thanks also to our partner from TUT and other collaborators for their continuous support to the IGNITE conference.

Thank you and "Terima kasih"

Associate Professor Mohamad Aizat bin Abas

Chairman IGNITE 2019

School of Mechanical Engineering, Universiti Sains Malaysia

IOP Conf. Series: Materials Science and Engineering **920** (2020) 011002 doi:10.1088/1757-899X/920/1/011002

Peer review declaration

All papers published in this volume of *IOP Conference Series: Materials Science and Engineering* have been peer reviewed through processes administered by the Editors. Reviews were conducted by expert referees to the professional and scientific standards expected of a proceedings journal published by IOP Publishing.

- **Type of peer review:** Double-blind
- **Conference submission management system:**
  E-mail submission (ignite.mech.2019@gmail.com)
- **Number of submissions received:** 37 submissions
- **Number of submissions sent for review:** 37 submissions
- **Number of submissions accepted:** 37 submissions
- **Acceptance Rate (Number of Submissions Accepted / Number of Submissions Received X 100):** 100%
- **Average number of reviews per paper:** 2
- **Total number of reviewers involved:** 41
- **Any additional info on review process:** -
- **Contact person for queries:** Dr. Norwahida Yusoff (menorwahida@usm.my)

IOP Conf. Series: Materials Science and Engineering **920** (2020) 012001 doi:10.1088/1757-899X/920/1/012001

Design and Development of a Small-Scale Wind Tunnel for Flow Visualization

M S M Shamsuddin and Noorfazreena M. Kamaruddin*

School of Aerospace Engineering, Universiti Sains Malaysia, 14300 Nibong Tebal, Penang, Malaysia

*Corresponding author: fazreena@usm.my

Abstract. A small-scale wind tunnel model with a length of 100 cm was designed and manufactured for flow visualization. The flow was visualized by using a mini smoke generator developed specialized for this small-scale wind tunnel and used the same principal as the conventional fog generator. To capture the flow motion, several parameters such as honeycomb screens, illumination systems, smoke liquids and fan speeds were varied to illustrate the smoke lines that represent the flow characteristics. The qualitative results based on these various configurations were presented and discussed in this paper with clear visibility for the smoke line intensity at 6 m/s with an airfoil used as the model to effectively demonstrate the flow visualization.

1. Introduction

A wind tunnel is commonly used to study the flow motion of a particular test model. It offers the ability to visualize airflow to allow further understanding of the aerodynamic characteristics of the model [1]. There are various techniques established in the literature to visualize the flow. Batil and Muellerf [2] suggested that smoke flow visualization is the most effective technique, as it is capable of producing a qualitative macroscopic picture of the overall flow field, as compared to other techniques which are limited to measuring the flow conditions at discrete point within the flow field. In addition, Martin and Mark [1] recommended the usage of a fog machine as a smoke generator device to enhance flow visualization, with the fog produced by heating a water-based liquid that has superior dispersal properties. They also emphasized the usage of a smoke rake to channel the smoke from the fog machine into the wind tunnel test section to enhance multiple smoke lines formation for better illustration of the flow.

To capture the flow motion effectively, several parameters must be considered. Wind tunnel flow velocity is one of the most common parameters that affect smoke formation [3]. The airflow velocity for each wind tunnel configurations vary ranging from 2 m/s to 10 m/s and the smoke lines can be easily dispersed based on this range. It is very important to add a screen and a honeycomb to retain the smoke lines [4-5]. A screen and honeycomb can reduce the turbulence flow caused by the air intake, resulting in smoother smoke lines. Martin and Mark [1] also studied the effect of various smoke liquid solutions for the smoke generator. They suggested a water-based solution as the best liquid for smoke generation, because the smoke is less toxic and denser compared to other liquids. In the current study, Safex oil has been used to illustrate the smoke pattern clearly.

The aim of this work is to investigate the effect of honeycomb screens, fan speeds, smoke liquids and illuminations systems as these parameters will significantly influence the quality of the flow visualization. The flow behaviour produced by the most effective configurations was discussed based on the smoke line pattern which represents the model's aerodynamic characteristics. The final small-scale wind tunnel model produced from this work will be used for an in-house demonstration as it is very practical due to its capability to be dismantled easily, small in terms of size, portable and reasonably low cost for its capability to produce a good quality of flow visualization.

2. Methodology

A small-scale wind tunnel was designed and integrated with a smoke system to allow for flow visualization demonstration. An overview of the methodology for this work is shown in Figure 1.

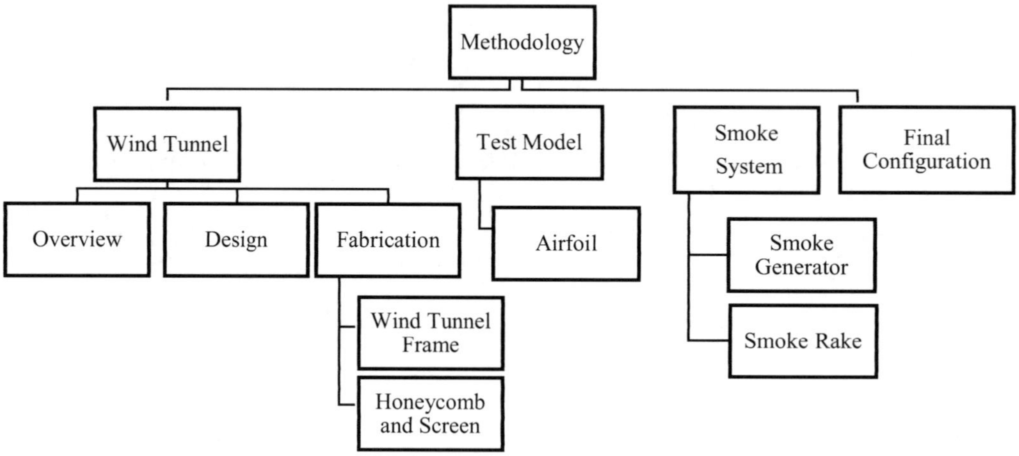

Figure 1. An overview of the methodology.

2.1. Wind Tunnel

A small-scale wind tunnel was designed in CATIA R21 software and fabricated by using wood, aluminium and Acrylic Perspex. The design was based on the basic configurations established in the literature [6-8]. A contraction nozzle was placed inside the wind tunnel to eliminate the need for any attachment between the nozzle and the test section. Two rectangular supports were added to level the test section height. The wind tunnel has an overall length of 1 m, a contraction ratio of two, and a test section length of 0.56 m as shown in Figure 2.

A honeycomb made of a 6 mm circular cell with a length of 30 mm width, combined with a screen, was added to straighten the incoming flow in the test section. The size of the honeycomb was the same as the opening of the contraction nozzle. For the screen, a wire gauze of 1 mm square cell was used, stretched and sewed to a rectangular metal frame, as shown in Figure 3a and 3b.The blockage effect was less than 10% and not significant to influence the qualitative results.

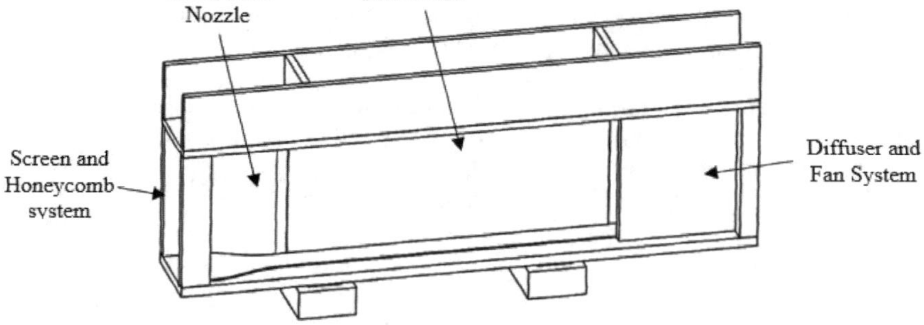

Figure 2. Three-dimensional modelling of Wind Tunnel using CATIA software.

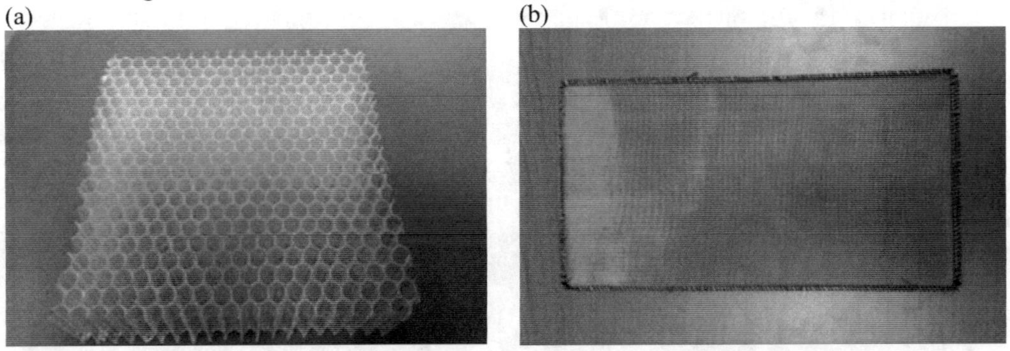

Figure 3. The image of (a) honeycomb and (b) screen used in the current work

2.2. Test Model

An NACA 0018 airfoil was used as a test model for the flow visualization study due to its good stall condition and minimum effect of roughness. The length was set at 15 cm to satisfy Barlow's [5] recommendation that a test model should be lower than 0.7 times the height of the test section. The airfoil profile points were exported from software (Airfoiltools.com) and configured into CATIA. The points were connected and linked until a solid line of the airfoil shape was formed. The airfoil model was then printed on a paper based on the real scale size. The actual model was fabricated using a polystyrene reinforced glass fibre composite via the wet layup process. The final product was then painted in black to provide contrast and better illumination, as shown in Figure 4.

2.3. Smoke System

A smoke was generated by heating a liquid fuel using a double coil 0.18 mm Nichrome wire, as shown in Figure 5a. A Nichrome wire was used as the heating element, as it has a very high electrical resistivity. The length of the wire was set at 100 mm, and a voltage of 12V was applied to heat the wire. The oil was carried to the heating coiled wire with a lamp wick. The wick acted as a mechanism to feed the oil to the heated wire through the effect of capillary action. The smoke produced by the heated oil was then stored in a closed jar to prevent it from dispersing into the surrounding environment. However, a mechanism to regulate the pressure inside the jar was required to channel the smoke into the smoke rake. A B-10200 model air pump was used to create a pressure difference and force the smoke into the smoke rake. The system works similarly to a conventional fog generator. The configuration of the smoke system and the smoke rake are shown in Figure 5a and 5b.

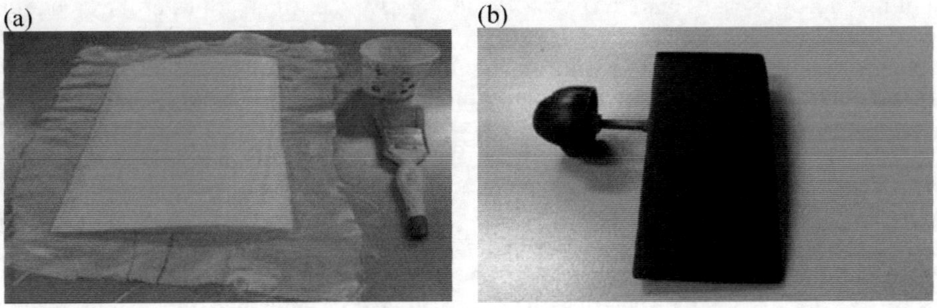

Figure 4. Airfoil model (a) during wet layup process, and (b) after painting and attachment to a knob.

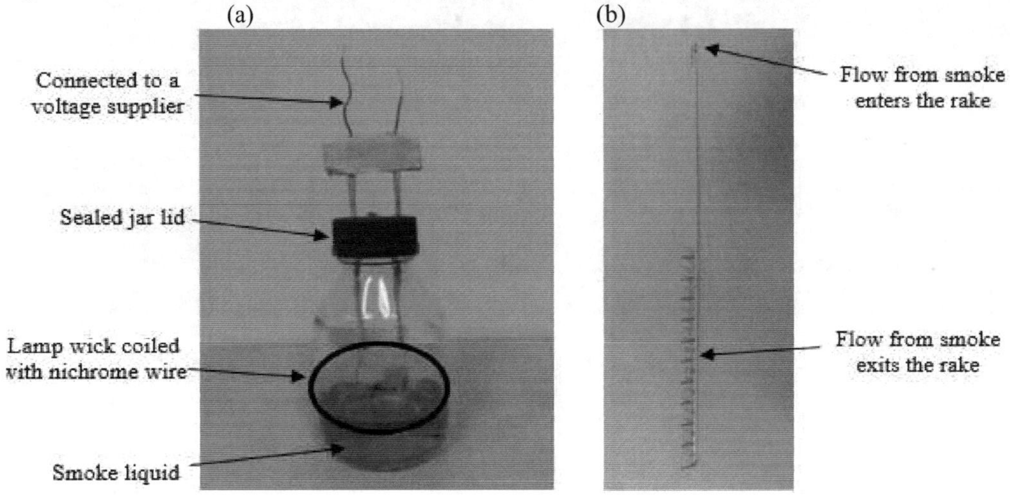

Figure 5. (a) Smoke generator system, and (b) Smoke rake.

2.4. Image Configurations

To capture the effect of various parameters of the smoke line characteristics in the wind tunnel, a D3500 Nikon camera at 60 fps was mounted on a tripod and positioned directly in front of the test section. The camera was set to 'automatic' mode with 'High Definition Image' to clarify the generated images. Black cloth was used to cover the whole test section areas to reduce any reflections caused by the Perspex and the illumination systems.

3. Result and Discussion

3.1. Effect of honeycomb and screen configurations

The results for the smoke flow visualization system for the lab-small scale wind tunnel are shown in Figure 6. The smoke was set to be in a single line with the operating fan speed at 4 m/s, while the illumination system employed was a fluorescent lamp. The initial configuration for the small-scale wind tunnel consisted of only one honeycomb. The smoke generated for the initial configuration (Figure 6a) showed irregular and faded smoke patterns. Since the smoke entered directly from the honeycomb, a solid line was only present if the tube rake was aligned with the honeycomb cell. Otherwise, the smoke would break into several small streamlines and be dispersed rapidly later. The addition of two honeycombs (Figure 6b) appeared to improve the irregularities of the smoke line due to the synchronized alignment. However, the flow fluctuation was very significant in this case. The addition of a screen before the honeycomb (Figure 6c) resolved the fluctuation rate of the smoke. The flow became more coherent and concentrated, and the smoke line showed straighter and more streamline pattern. These results agree well with Groth [4], who showed that the usage of a screen will reduce the turbulence level of the flow and improve its streamline quality. The screen was use to break up the large-scale turbulent eddies that came along with the channelled smoke into a number of small scale eddies. The small scale eddies will subsequently have decayed, and then produced a more uniform and stable flow pattern. The further incorporation of another honeycomb (Figure 6d) however, results in a very severe dispersed flow that is not favourable. The line was observed to fade and disperse rapidly even before entering the test section. The effective smoke flow visualization configuration for the current study has been identified as the one with the screen positioned just before the honeycomb

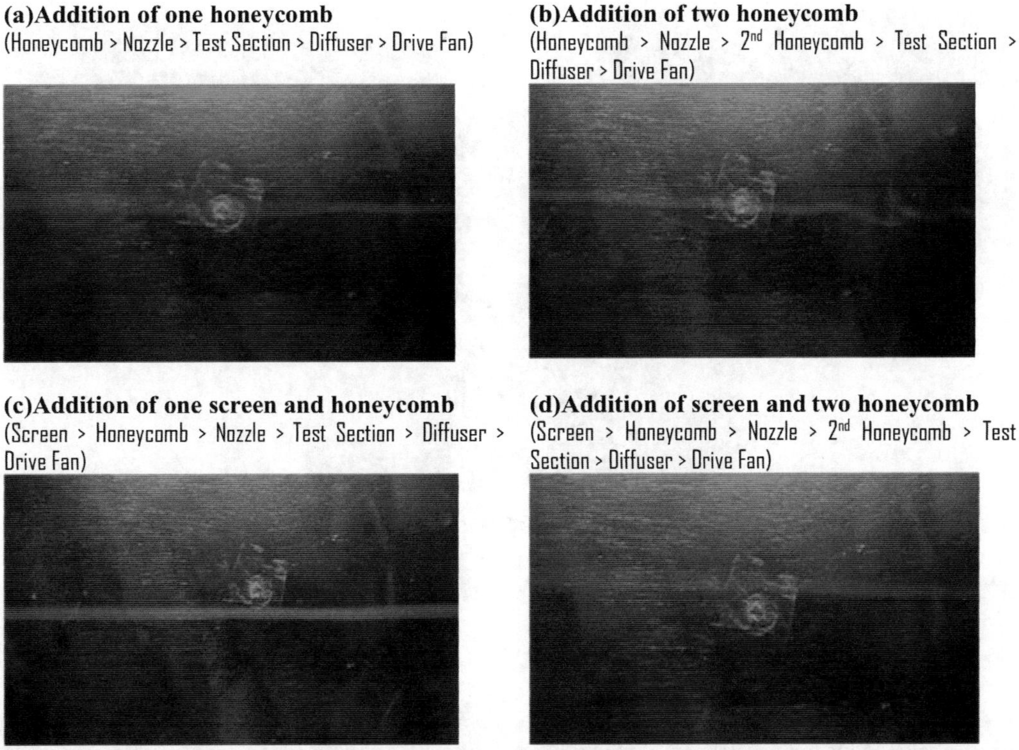

Figure 6. Smoke line observation at various honeycomb and screen configurations

3.2. Effect of illumination systems

It was essential to illuminate the test section of the wind tunnel so that the smoke line could be easily viewed and recorded. Comparison were made with three illumination systems: a fluorescent lamp, LED lamp, and halogen lamp. The test was conducted at a constant fan speed of 4 m/s, and multiple smoke lines were generated by using a smoke rake were used to clearly distinguish the effects of the illumination system as shown in Figure 7. The results showed the smoke lines were dim and fairly visible when a fluorescent lamp was used (Figure 7a). The fluorescent system was not able to provide adequate illumination, however, as the smoke lines could only be seen at the middle region of the test section. When using an LED lamp, the illumination of the test section was relatively better, as almost all parts of the test section were brighter. However, the smoke lines were still difficult to see clearly. Flare existed at the top region near the location of the lamp, which in turn lowering the visibility of the smoke line (Figure 7b). The final illumination system was the halogen lamp (Figure 7c). The smoke lines were visible in all parts of the test section. Signs of flare were still evident, but their intensity was reduced as compared to LED lamp. The results revealed that the halogen lamp generated the most visible smoke lines pattern, similar to the results of previous studies [9]. The lightings from the halogen system produced less flare, less haze effect, and more detailed smoke stream line patterns. These effects are important to better capture the flow around a model. The shadow from the test model will prevent an accurate observation of the smoke pattern, requiring the use of a more reliable source

IOP Conf. Series: Materials Science and Engineering **920** (2020) 012001 doi:10.1088/1757-899X/920/1/012001

of illumination. In addition, the halogen bulb also came in a small and compact size, making modification and arrangement easier.

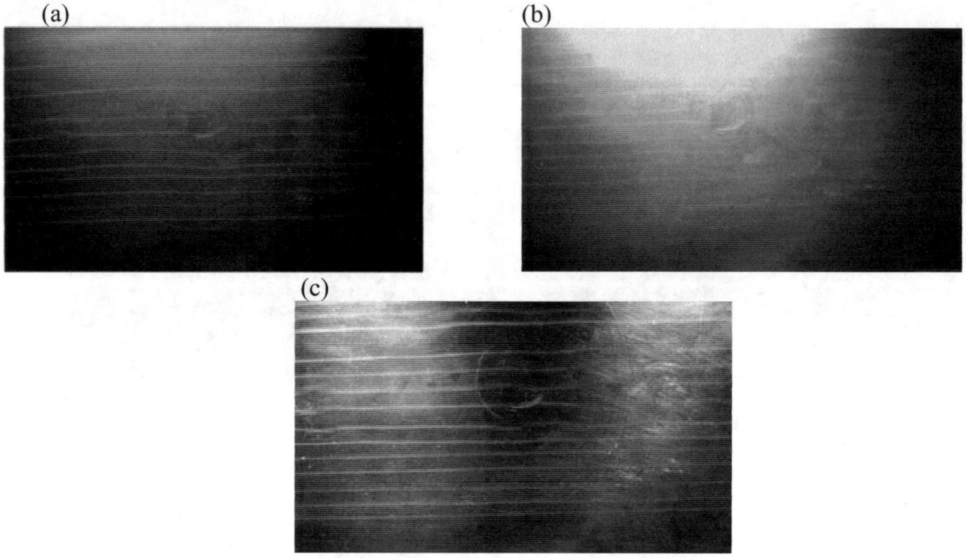

Figure 7. Comparison between (a) Fluorescent, (b) LED, and (c) Halogen illumination system

3.3. Effect of smoke liquid solutions

In general, it is important for the smoke lines to feature a good dispersion rate. In this study, the smoke was generated by heating a 0.18mm Nichrome wire using 12 Volt DC power supply. The wind tunnel operated at a fan speed of 4 m/s, and the lighting system was changed to a halogen lamp based on the previous discussion. The comparison of the smoke liquid includes; kerosene, corn oil, Safex (fog liquid), and the results are shown in Figure 8. All smoke liquids were capable of generating smoke. Kerosene (Figure 8a) smoke had a tendency to disperse rapidly as it progressed through the test section. Corn oil (Figure 8b) smoke allowed for good uniformity of streak lines. However, the intensity of the smoke was less dense as compared to the kerosene. The Safex produced the most coherent and dense smoke, as shown in Figure 8c. The smoke lines were clearly visible throughout the test section. The Safex smoke also had good disposal value, as it did not contaminate the test section with excessive leftover oil, as it, unlike kerosene and corn oil, is water-based. This resulted in a non-irritant, non-flammable, and non-hazardous fog cloud and considered suitable especially for a closed-environment and indoor usage.

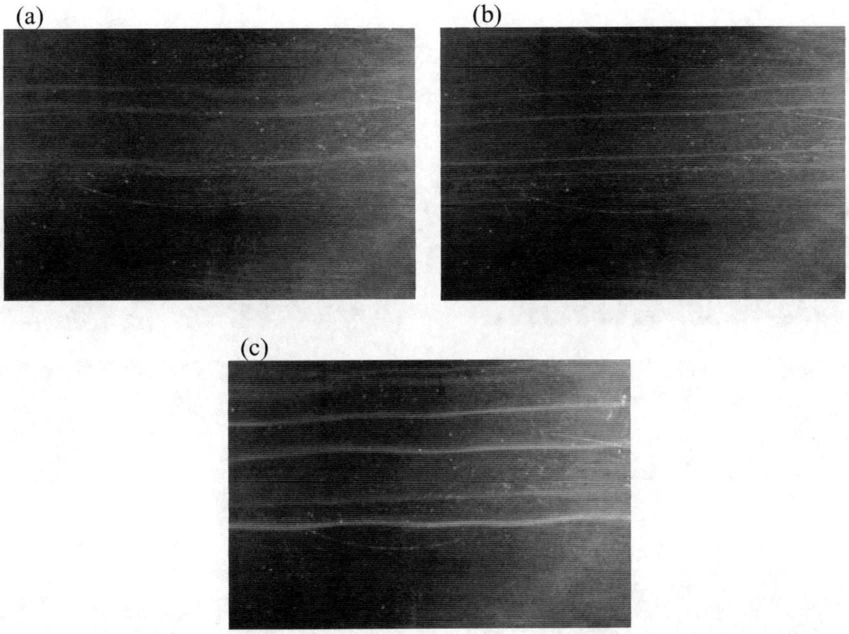

Figure 8. Comparison between (a) kerosene, (b) corn oil, and (c) Safex for smoke liquid application.

3.4. Effect of fan speeds

The flow qualities of the smoke visualization system at fan speeds of 2, 4, 6, and 8 m/s were compared. The smoke was generated using Safex fog liquid, and the lighting system was fixed as halogen lamp. The smoke lines generated at 2 m/s as shown in Figure 9a indicated a dense smoke intensity, but the line was not uniform and fluctuates severely. An improved smoke line can be seen at 4 m/s (Figure 9b), where the flow showed a better streamline, although fluctuations due to the turbulence of the rake opening could still be seen throughout the test section. At 6 m/s (Figure 9c), the smoke was less dense as compared to the previous one. The lines showed no fluctuation, indicating a uniform streamline flow pattern. At higher speed of 8 m/s (Figure 9d) the line was relatively straighter. Again, no fluctuation could be seen in this case, but the smoke losses most of its coherency. Only a plume of smoke was visible at the end of the wind tunnel test section. It is important to note that in order to find the optimum velocity for this smoke visualization technique, a trade-off between the smoke intensity and the streamline pattern is necessary. Lower speeds will result in a denser and coherent smoke line, while higher speeds will result in a straighter and more uniform line. The optimum fan speed for the current studies was 6 m/s, as the smoke line could still remain thick and coherent, with no visible fluctuations throughout the test section area.

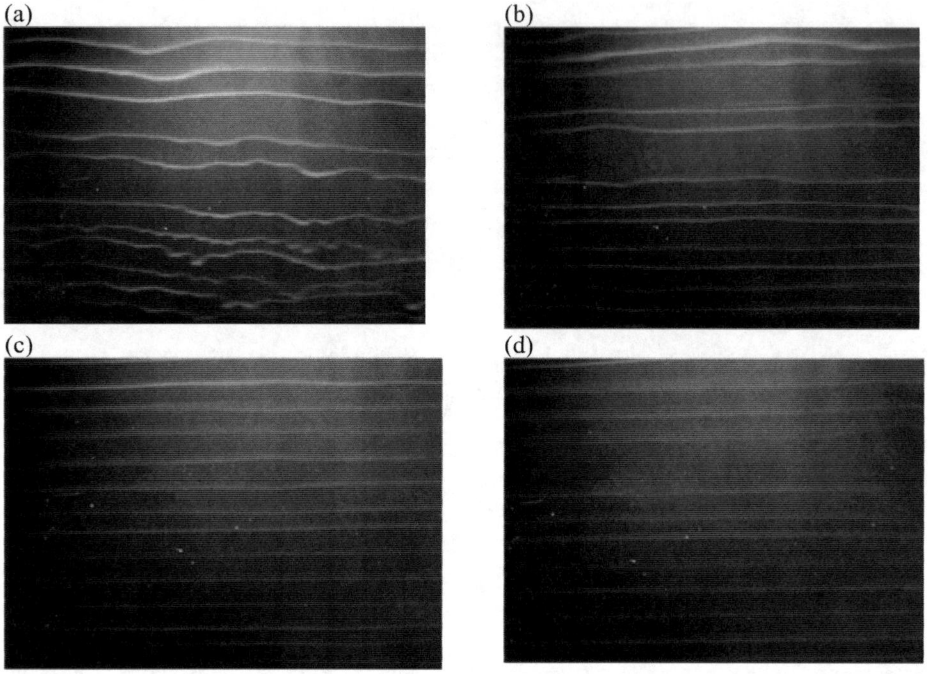

Figure 9. Comparison of smoke lines at fan speed of (a) **2** ms^{-1}, (b) **4** ms^{-1}. (c) **6** ms^{-1} and (d) **8** ms^{-1}.

3.5. Effective Smoke Flow Visualization Configurations

The smoke flow visualization of an airfoil test model is shown in Figure 10. The airfoil was set to be at 15° angle of attack, α, to highlight its capability to capture the flow separation at the trailing edge. The configurations were set to have only a screen and a honeycomb with the illumination system consisted of a halogen lamp. The most effective setup was the Safex fog liquid solution with a fan speed of around 6 m/s. The smoke patterns were observed to be uniform around the lower half of the airfoil, as it did not encounter any disturbance. The smoke seemed to closely follow the shape of the airfoil and its streamline patterns appeared to pass through the end of the test section. The flow could be identified as laminar across the bottom part of the airfoil. However, the smoke patterns were distorted at the middle region on the upper half of the airfoil. The flow exhibited a trailing edge stall, in which the flow becomes separated from the upper surface downstream of the leading edge. The occurrence of this flow separation can be explained in terms of an adverse pressure gradient, a phenomenon in which static pressure increases in the direction of the flow, leading to a reduced kinetic energy and deceleration of the fluid motion. The detached flow then created a region of turbulent wake, taking form of eddies and swirling motion.

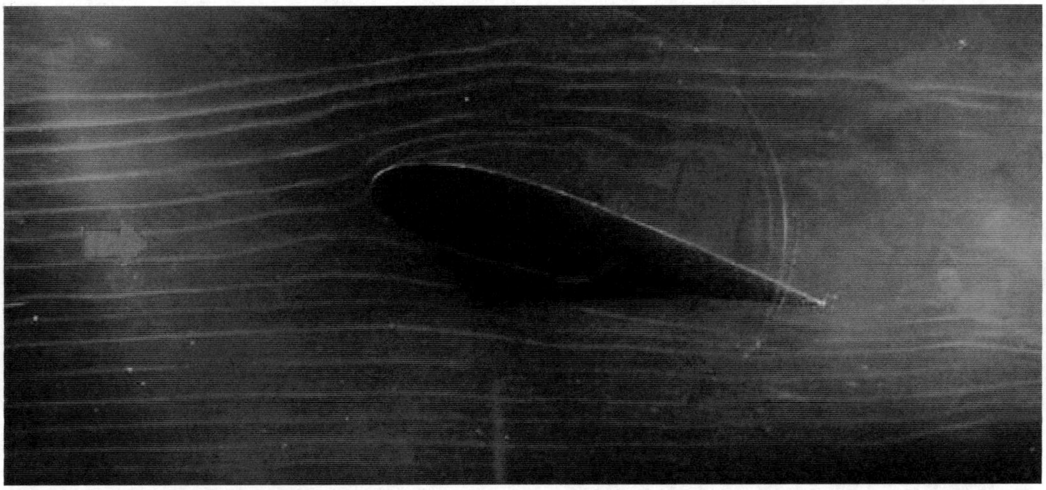

Figure 10. Smoke flow visualization using airfoil test model at 15° angle of attack. Red arrow indicates airflow direction

The smoke patterns visualized in this work are consistent with those found in previous studies [10-12]. The smoke technique developed in this study was able to illustrate the airflow around the airfoil. The optimized parameters were qualitatively compared based on the configuration of the current lab-scale smoke wind tunnel, and the flow visualization is considered acceptable for a lab-scale measurement. However, to fully analyse the effect of such parameters accurately, quantitative measurements would be required in future studies.

4. Conclusion

A smoke flow visualization system was developed in a small-scale wind tunnel system to demonstrate flow characteristics. It was found that the screen and honeycomb were essential, but excessive usage of these two components may result in a very thin smoke lines with a high dispersion rate. A halogen lamp can be considered to be very effective in illuminating the smoke lines. A Safex fog liquid (water-based solution) managed to produce the thickest and most stable smoke streak lines. In terms of fan speed, the smoke line intensity was clearly visible at 6 m/s. The developed smoke flow visualization system managed to successfully demonstrate smoke streak lines for an airfoil that represents classical aerodynamic behaviour.

5. Acknowledgement

The authors would like to acknowledge the financial support from the Ministry of Higher Education Malaysia under the Fundamental Research Grant Scheme (FRGS): 203.PAERO.6071433

6. References

[1] M. Trinder and M. Jabbal, "*Development of a smoke visualisation system for wind tunnel laboratory experiments*," Int. J. Mech. Eng. Educ., vol. 41, no. 1, pp. 27–43, 2013.
[2] S. M. Batill and T. J. Muellerf, "*Visualization of Transition in the Flow over an Airfoil Using the Smoke-Wire Technique*," vol. 19, no. 3, pp. 340–345.

IOP Conf. Series: Materials Science and Engineering **920** (2020) 012001 doi:10.1088/1757-899X/920/1/012001

[3] S. S. Dol, M. A. M. Nor, and M. Khairun, "*An Improved Smoke-Wire Flow Visualization Technique*," Proc. 4th WSEAS Int. Conf. Fluid Mech. Aerodyn. Elounda, Greece, August 21-23, Vol. 2006, pp. 231–236, (2006).
[4] B. J. Groth and A. R. N. E. V Johansson, "*Turbulence reduction by screens*," J. Fluid Mech. (1988), vol. 197. pp. 139-155, vol. 197, (1988).
[5] J. B. Barlow, W. H. R. A. E. Jr, A. Pope, G. L. Martin, W. Tunnel, and W. H. Rae, *Low Speed Wind Tunnel Testing*, vol. 7, no. 1. (2015).
[6] A. Welsh, "*Low Turbulence Wind Tunnel Design and Wind Turbine Wake Characterization*," The University of Wisconsin-Milwaukee, MSc Thesis Dissertation (2013).
[7] L. Cattafesta, C. Bahr, and J. Mathew, "*Fundamentals of Wind-Tunnel Design*," Encycl. Aerosp. Eng., pp. 1–10 (2010).
[8] P. Bradshaw and R. C. Pankhurst, "*The Design of Low Speed Wind Tunnel*," Prog Aerosp. Sci., vol. 5, pp. 1-69. (1964).
[9] N. B. Abdullah, "*Design and Fabrication of Portable Smoke Tunnel for Flow Visualization*," Universiti Teknologi Pertonas, Bachelor of Engineering Thesis Dissertation (2009).
[10] J. D. Anderson, B. Ridge, I. L. Dubuque, I. A. Madison, W. I. New, and Y. San, "*Fundamentals of Aerodynamics*". Tata McGraw-Hill Education (2010).
[11] Kalyankar, H., Choudhary, D., Melwanki, R., Chaudhari, D., Jethwa, S: Design and analysis of low speed water tunnel for flow visualization of bluff body. In: 2nd International Conference on Advances in Mechanical Engineering and its Interdisciplinary Areas (ICAMEI). India (2015).
[12] Tong P E , Design and Development of Water Tunnel for Flow Visualization Study on Micro-Hydrokinetice Turbine, Final Year Project, Universiti Sains Malaysia (2018).

IOP Conf. Series: Materials Science and Engineering **920** (2020) 012002 doi:10.1088/1757-899X/920/1/012002

Ankle joint rotator prototype designed for creating different angles of rotation at the ankle joint in performing various daily activities

T P Li and M S H Bhuiyan*

School of Mechanical Engineering, Engineering Campus,
University of Science Malaysia, 14300 Nibong Tebal, Pulau Pinang, Malaysia.

Email: sayem@usm.my; sayem_um@yahoo.com

Abstract. The prosthetic ankle joint in lower limb prosthesis replaces the missing ankle of the amputee. The prosthesis users want the prosthesis to enable them creating the desired angle of rotation at the ankle joint like that of a healthy biological ankle. The ankle joint prostheses currently available in the market are unable to produce sagittal plane rotations for squatting and kneeling postures. Absence of a proper ankle joint rotator and lack of their universal adaptability have limited the utility and the performance of the prosthetic ankle joint. An ankle joint rotator capable of enabling its users to squat and kneel with ease has been designed. A model of the knee joint rotator has been developed, and simulation has been carried to out to verify the functionality of the design before fabrication. The new ankle joint rotator was found capable of withstanding an external force of 3000N with a minimum safety factor of 5.18. The newly developed ankle joint rotator is able to facilitate the squatting and kneeling postures safely by enabling the prosthetic ankle joint to rotate in different angles.

1. Introduction

The prevalence of limb loss and amputation due to congenital effect, war effect, accident, natural disaster and so on has been increasing steadily around the world. Amputation or limb defect causes difficulties to the subjects in performing different daily activities. It creates the need of getting help from different assistive devices depending on the nature of the disability. The amputees have been aided by walking stick/cane, walking frame, crutch, wheelchair, motor wheelchair, stroller, motor stroller, and more recently prosthesis is being used to rehabilitate them [1, 2]. A prosthetic limb or joint is a kind of prosthesis that replaces a missing limb and joint respectively. In recent years, the design of the prosthesis has been significantly improved by incorporating some extra features as well as introducing new controlling methods. In advanced and more improved designs of prosthetic limb and joint, more control is given to the users by employing different control system, muscle of carbon fiber, mechanical linkages, motors, computer microprocessors, and innovative combinations of these technologies mentioned [3]. The essential factors like weight-force ratio, strength, durability, adaptability, wear-ability, degree of freedom, resistance to environment, functional capabilities, etc. are being addressed seriously to develop a more efficient prosthesis arrangement. The desired movement of prosthetic limb is obtained with the help of different types of joints, sensors, controllers and actuators.

The prosthetic joints of the prosthesis assist users to produce necessary movement in the prosthesis to yield different posture required in performing different daily activities. From literature and market study, it has been found that most of the ankle joint prosthesis can only rotate medially and laterally. Though numerous types of ankle joint prosthesis is available in the market, none of these are proving sagittal plane rotation flexibility to its users. This type of rotation flexibility is very much essential for producing, squatting and kneeling posture in performing different daily activities. In order to overcome the limitation of sagittal plane rotation, a universality adaptable ankle joint rotator that enabling squatting and kneeling movements for the lower-limb prosthesis user has to be developed.

2. Ankle joint prosthesis

Ankle joint plays an important role in the gait cycle during forward progression of the human body. It plays the most significant role in creating different body postures like squatting and kneeling. The foot and ankle act as a medium between the body and ground adjust themselves flexibly in order to achieve a harmonious coupling for successful movements [4, 5]. The ankle joint prosthesis is used to reproduce the same functions in the absence of the biological ankle. The main function of the ankle-foot design is to facilitate the terrain adaptability in maintaining natural gait patterns and stability for a person with single limb transtibial amputation [6]. Several factors like the functionality, ease of use, comfort, structural strength, mass, size, visual appearance, sound, passive support, energy, output power and torque delivery, lateral flexibility, shock tolerance, variability of use, durability, personalization, modularity and cost, etc. affect the design of the prosthetic devices [7, 8]. The stiffness is also considered as a variable in designing a functional prosthetic ankle joint [9]. The stability of prosthesis, necessary propulsion and body weight support in the gait cycle are influenced by the ground reaction force (GRF) [10]. The user needs have to be fulfilled, and the technical feasibility has to be achieved after taking all these factors into consideration. A prosthesis should be designed such a way that it is capable of mimicking the ankle movements during different activities [11]. A lower-limb prosthesis is usually equipped with a socket, a pylon, an SACH foot, and an ankle joint rotator. The foot and pylon are usually connected to the ankle joint rotator which merely permits the ankle joint to rotate in the sagittal plane. Therefore, it is indispensable to design such an ankle joint rotator that allows the prosthesis to rotate to some desired angle when bending is required during performing different daily activities.

3. Design of ankle joint rotator

The ankle joint rotator has been designed in such a way that it allows rotation in the sagittal plane (plantar- and dorsiflexion) rather than focusing on rotators in the lateral and medial planes like that of existing ankle joint rotator. The ankle joint rotator design has been performed considering two angles of rotation for the squatting and kneeling postures respectively. The design and simulation of the ankle joint rotator has been carried out using the SolidWorks 2019 software. The modelling and simulation have been run on the ankle joint rotator design to determine the maximum stress that can be withstood and the displacement that would occur in the components.

The main design parameters of the ankle joint rotator were rotation angle, weight and size. Two angles of rotations of 60^0 and 45^0 were chosen, lowest minimum weight and smallest in height to make sure fitting it within the space available between the shank, and foot were opted for the design. The designed ankle joint rotator was comprised of a female component, a male component, a rotator pin, and a spring where the first three components were designed and fabricated, and the spring was procured from the market.

The female component was designed to act as a connector to the lower end of the pylon and also to assist in angle adjustment. The inverted pyramid on the top section of the female component is to be connected to the pylon with screw from four sides of the pylon lower end. The size of the inverted pyramid was designed to be universally adaptable. The internal hemispheric shape provided room for rotational to the male component. The three holes of *6.00mm* diameter on the lateral side of the female component represents three respective angles for standing (0^0), squatting ($60°$) and kneeling ($45°$). The *10.20mm* diameter hole of the female component has to be mounted on the male component with a M10 nut-bolt in the rotator assemble.

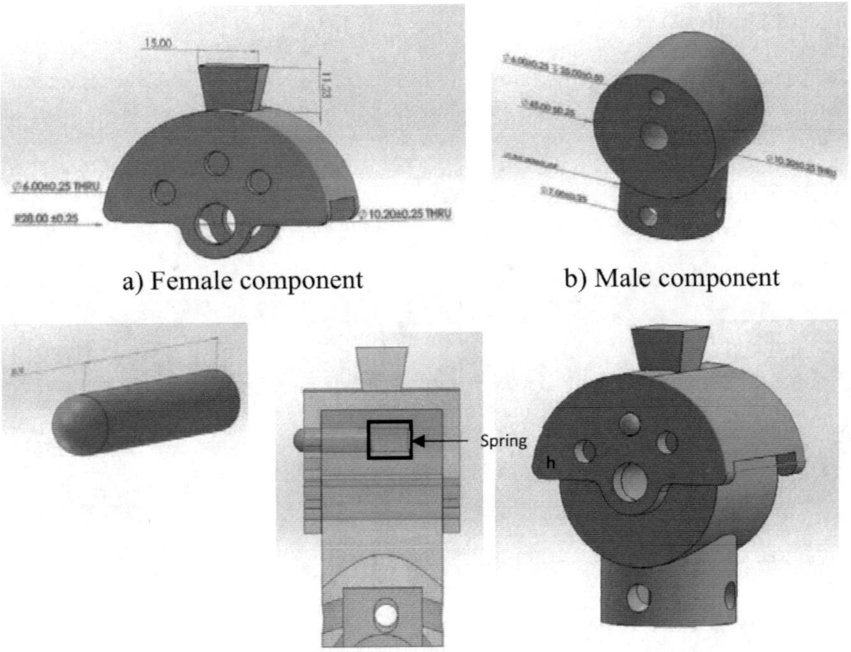

a) Female component b) Male component

c) Rotator pin d) Assembly of the ankle joint rotator

Figure 1. Ankle joint rotator design.

The male component was designed to be fitted with the female component. The cylindrical shape of the male component helped the rotator to create rotation when to adjust the angle of the ankle at different position. The *6.00mm* hole on the lateral side of the male component has to be aligned with any of the three holes in the female component for angle adjustment. The *10.20mm* hole at the center of the male component has to be aligned with the same diameter hole of the female component to assemble it with a M10 nut-bolt arrangement. The cross-body extension of the cylindrical section of the male component acted as a connecter to the prosthetic foot. The four *7.00mm* holes around the periphery of the extension were to connect the male component with the foot using screw. The holes are slightly slanted at *10°* to provide a stronger grip to the foot.

The rotator pin acted as a common locking means for the male and female components during adjustment of angle of rotation. After assembling and aligning the male and female components at the upright position (*0°*), the rotator pin attached with a spring connected to the base of the hole at *0°* of the male component. The spring is positioned in the spring housing of the male component market with a box in the design. Both parts will be locked in the position by the rotator pin. When rotation is required, the rotator pin has to be pushed into the hole of the female component to allow the male component to rotate. After the holes of the male and female parts are aligned at a desired angle, the rotator pin has to be released to lock both the male and female components again. Figure 1 shows the different components and the assembly of the ankle joint rotator design.

4. Modeling and Simulation results

The modeling and simulation of the ankle joint rotator design was carried out using SolidWorks 2019 software. The static model was developed for finite-element analysis of the ankle joint rotator assembly. Titanium alloy Ti-6Al-4V was selected as the material of the diffident components of the ankle joint rotator.

For the static studies, Fixed Geometry was selected as fixtures for the ankle joint rotator. Fixed Geometry was applied to those eight surfaces which would be connected to the pylon and foot with screws. The boundary condition setup and mesh creation are shown in Figure 2.

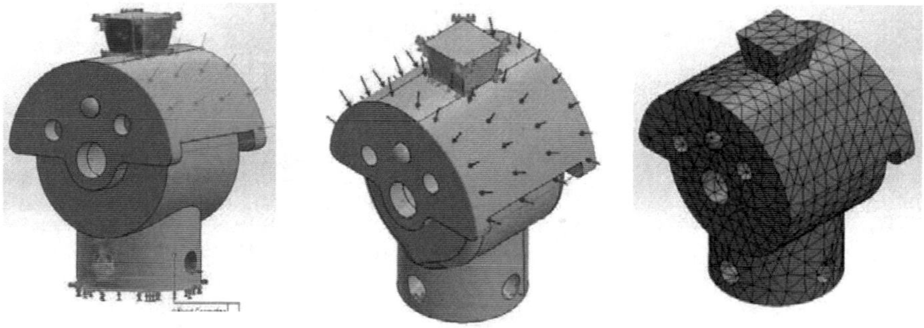

a) Fixtures b) Loads c) Mesh

Figure 2. Ankle joint rotator model.

A force of *3000N* equivalent the user of weight *100kg* was applied on the external curved surface of female component of the ankle joint rotator as the External Loads. The external curved surface of the female component was divided into two parts to enable force to be applied on the entire curved surface. Two different cases of no rotation (standing upright) and ankle rotating (squatting and kneeling) were considered for setting up the boundary condition and running the simulation. Then mesh has been created before running the simulation. Four Points Jacobian solid mesh, with total nodes of 15280, total element of 8944 with size of 4.42311 mm, and a maximum aspect ratio of 7.0568 were chosen for creating the mesh.

The simulation results for stress, strain, displacement analysis, and factor of safety were generated for evaluation of the ankle joint rotator design.

a) Stress distribution without ankle rotation

b) Stress distribution with ankle rotation

c) Strain development without ankle rotation

d) Strain development with ankle rotation

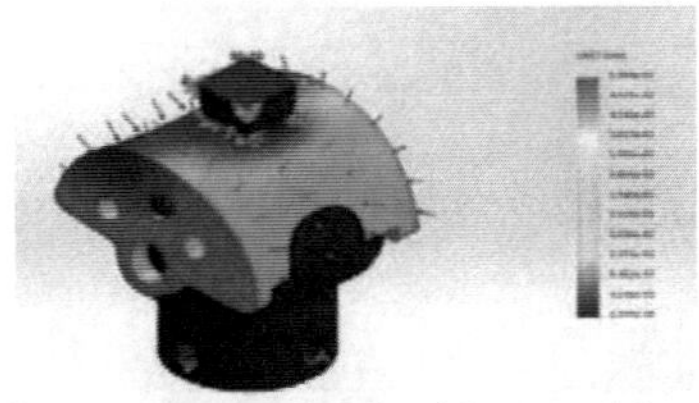

e) Displacement occurred without ankle rotation

f) Displacement occurred with ankle rotation

g) Design FOS without ankle rotation

h) Design FOS with ankle rotation

Figure 3. Stress, strain, displacement and FOS analysis results obtained from simulation of ankle joint rotator.

At the first place, without any rotation at the ankle joint rotator case has been chosen to simulate the standing posture of the prosthesis. Then, with rotation case has been selected to simulate the squatting and keeling conditions. The stress, strain, displacement, and factor of safety analyses have been carried out to see the stress distribution, strain development, displacement occurred and safety of the design respectively.

From the Figure 3, the maximum equivalent stress developed for standing posture is $8.575x10^7\ N/m^2$, which is considerably low comparing to the tensile strength of the titanium alloy Ti-6Al-4V material of $8.27371x10^8\ N/m^2$. The maximum equivalent strain developed in the ankle joint rotator is $6.627x10^{-04}$, which is trivial to make any significant distortion or any quantifiable change in the shape of the components. This indicates that the material used for the ankle joint rotator elements can withstand the load applied by the prosthesis user while standing. The highest displace occurred in the structural member under load is $5.089x10^{-2}\ mm$, which is insignificant and negligible for the ankle joint rotator assembly. The least displacement value indicates that the elements of the ankle joint rotator do not dislocate unnecessarily when in use. If the displacement value was high, it could have caused instability

in the prosthesis arrangement. The minimal factor of safety has obtained for the ankle joint rotator component is *9.649* for the standing position. This indicates that the design is capable of withstanding the weight of the user while standing.
For squatting and keeling postures, the maximum equivalent stress development found to be $1.452x10^8$ N/m^2, which is few times less than the tensile strength of the titanium alloy Ti-6Al-4V material of $8.27371x10^8\ N/m^2$. The maximum equivalent strain developed in the ankle joint rotator is $1.101x10^{-3}$ which is considerably low to make any significant change to the shape of the ankle joint rotator components. Therefore, the material used for the ankle joint rotator elements can sustain the load applied by the user while producing rotation at the ankle joint. The biggest displacement occurred in the components of the rotator under load is $1.198x10^{-1}$ *mm*, which is too small to affect the rotator assembly. The small displacement value directs to the stability of the rotator assembly and the prosthesis arrangement also. For rotation at the ankle joint, the lowest factor of safety of the rotator components has been found to be 5.698. Although the stress during rotation is greater than that at the standing position, the safety factor is still good enough to serve the purpose.
Comparing to the minimum acceptable safety factor of *1.2*, the design is proven to be capable of withstanding an external force up to *3000N* without any occurrence of failure.

4.1. Ankle joint rotator prototype fabrication

After finalizing the design, a prototype has been fabricated using an Alpha T2liFB 3-axes CNC high speed milling machine. Aluminium Grade 6061 material has been used to fabricate the components of the ankle joint rotator. The different components of the ankle joint rotator and their assembly are shown in Figure 4.

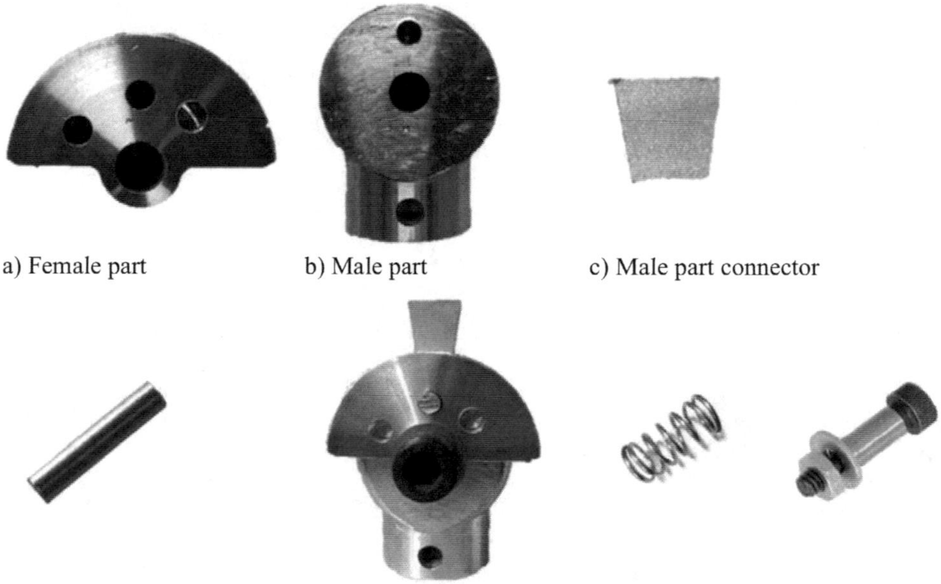

a) Female part b) Male part c) Male part connector

d) Looking pin e) Rotator assembly f) Spring g) Nut bolt

Figure 4. Ankle joint rotator prototype and its components.

Due to the limitations of the CNC machine, the four holes on the cross body extension of the cylindrical male component was machined linearly instead of at an angle of 10° inclination. The rotator pin was fabricated using the lathe machine to a diameter of *5.8mm*. The inverted pyramid part of the female component was prepared using the Precision Wire-Cut EDM machine with a dimension of *15mm x 15mm x 10mm.* The male and female components were assembled using a M10 x 60mm screw. The

Aluminium 6061 made ankle joint rotator prototype has been installed at the ankle joint of the transtibial lower limb prosthesis.
The preliminary investigation has shown that the ankle joint rotator was capable of producing a relative rotational movement between the male and female components to adjust the angle of ankle rotation and lock it at two desired angles of *60°* and *45°*. It has been seen that the locking mechanism was performing well to lock and unlock the male and female components in place when necessary.
However, there were still certain drawbacks in the design. Due to inadequate dimensional data availability, the joint interfacings were not perfectly matched. There was some gap observed between the external curved face of the female component and the curved surface of the pylon end, which caused some disparity in fitting the ankle joint rotator to the pylon of the lower-limb prosthesis. Similar discrepancy has been found at the connection interface between the cylindrical cross-body extension of male component and the prosthetic foot adapter. The minor interface matching disparity between components can be overcome by modifying the interfacing surface of the components based on the precise dimensional data to fit the curved surfaces on both pylon and prosthetic foot adapter. Figure 5 shows the ankle joint rotator prototype installed in a transtibial lower limb prosthesis.

Figure 5. Transtibial lower limb prosthesis equipped with the ankle joint rotator.

The performance of the ankle joint rotator installed in a lower-limb prosthesis is yet to be tested. The stability index, fall risk index, gait cycle analysis, and user feed-backs collections are in the process. The performance of the ankle joint rotator can be stated after obtaining and proper analysis of all these tests results data.

5. Conclusions

The ankle joint rotator designed for providing two angles of rotation for squatting and kneeling postures was successfully developed. The rotator mechanism was seen to be functional; despite some difficulties, the design was found adaptable to the lower-limb prosthesis available in the market. The simulation results have approved the feasibility of the design; the stress, strain and displacement, and factor of safety analysis results confirmed the safety of the ankle joint rotator design. The minimum safety factor of the ankle joint rotator components during standing and rotating was *9.649* and *5.698* respectively, which are considerably higher than the minimum factor of safety requirement. Though, performance test is yet to be done, the simulation results and preliminary investigation endorse the safety and usability of the ankle joint rotator design.

6. References

[1] Gao K, Chen S, Wang L, Zhang W, Kang Y, Dong Q, Zhou H and Li L 2010 Anterior cruciate ligament reconstruction with LARS artificial ligament: a multicenter study with 3- to 5-year follow-up. *Arthrosc* 26(4) p 515-523.

[2] Al Shuaili N, Aslani N, Duff L and McGarry A 2019 Transtibial Prosthetic Socket Design and Suspension Mechanism: A Literature Review. *J. Prosthet. Orthot* 31(4) p 224-245.

[3] Hsu R W, Sim F H and Chao E Y 1999 Reoperation results after segmental prosthetic replacement of bone and joint for limb salvage. *J. Arthroplast* 14(5) p 519-526.

[4] Rodgers M M 1988 Dynamic Biomechanics of the Normal Foot and Ankle During Walking and Running. *Phy.. Ther* 68(12) p 1822-1830.

[5] Bogey R A, Gitter A J and Barnes L A 2010 Determination of ankle muscle power in normal gait using an EMG-to-force processing approach. *J Electromyogr. Kinesiol* 20(1) p 46-54.

[6] Masum H, Bhaumik S and Ray R 2014 Conceptual Design of a Powered Ankle-foot Prosthesis for Walking with Inversion and Eversion. *Procedia. Technol* 14 p 228-235.

[7] Au S K, Weber J and Herr H 2007 Biomechanical Design of a Powered Ankle-Foot Prosthesis. in 2007 *IEEE 10th International Conference on Rehabilitation Robotics*.

[8] Popovic M 2013 Biomechanics and Robotics.

[9] Argunsah Bayram H and Bayram M B 2018 Dynamic Functional Stiffness Index of the Ankle Joint During Daily Living. *J. Foot. Ankle. Surg* 57(4) p 668-674.

[10] Perry J and Schoneberger B 1992 Gait Analysis: Normal and Pathological Function. SLACK.

[11] Major M J, Twiste M, Kenney L P and Howard D 2014 The effects of prosthetic ankle stiffness on ankle and knee kinematics, prosthetic limb loading, and net metabolic cost of trans-tibial amputee gait. *Clin. Biomech* 29(1) p. 98-104.

Acknowledgments
The authors would like to thank the School of Mechanical Engineering, University of Science Malaysia for providing the facilities to carry out this work.

Design and Development of a Service Robot for Wi-Fi RSSI Fingerprint Data Collection

M. Q. Bakri[1*], A. H. Ismail[1,2], M. S. M. Hashim[1], M. S. Muhamad Azmi[1], M. J. A. Safar[1] and M. H. Marhaban[2]

[1]School of Mechatronic Engineering, Universiti Malaysia Perlis (UniMAP), Perlis, Malaysia.
[2]Dept. of Electrical and Electronics Engineering, Faculty of Engineering, Universiti Putra Malaysia (UPM), Selangor, Malaysia.
[*]mohdkhai95@ymail.com

Abstract. We have designed a service robot that can be used for Wi-Fi RSSI Fingerprint database construction for indoor positioning system. This work aims to aid and ease the signal fingerprint database construction process which currently conducted manually by carrying the data acquisition tools around the experimental field. The robot architecture design considered the values and constraint in performance, aesthetic, cost, and expandability. Analysis of the robot's mobile specification was made in order to choose the optimum hardware components. The robot has three main sections which are mobile platform, storage compartment, and user interactive screen that is capable to display facial expression and other useful information.

Keywords- service robot; motor sizing; Wi-Fi fingerprint.

1. Introduction

Positioning system is one of the important aspects for a service robot, especially in a complex indoor environment. This system gives information about the robot's exact position in a closed space that is useful for further action such as path planning and navigation [1]. For outdoor application, positioning can be done by using the information from GPS, which has become the standard in outdoor environment [2]. However, GPS is unreliable when it comes to indoor positioning because of signal attenuation and severe propagation by the building [3]. Therefore it is necessary to consider other approaches such as Wi-Fi positioning system [4]. This approach provide easy and low-cost implementation since the existing access points can be used without the need to make a major work on the infrastructure modification [5].

There are several methods available that can be used to implement wireless positioning system by using the Wi-Fi signal. Such method include triangulation, trilateration, and RSSI (Received Signal Strength Indicator) fingerprint [3][6]. In this study, the focus is on Wi-Fi RSSI fingerprint method. This method works by comparing the average RSSI fingerprint values from a minimum of three access points in the room with the database to predict the exact position of the client such as service robot [7][8]. Database construction is required at the early phase, usually done by placing the data acquisition tools at the exact reference point (RP) to record the RSSI values for each access point, for instance as shown in Figure 1. Up to this point the challenges arise, it is time consuming and huge effort needed for someone to construct the database manually, especially when the number of granularity increases [8].

IOP Conf. Series: Materials Science and Engineering **920** (2020) 012003 doi:10.1088/1757-899X/920/1/012003

Ones need to bring the data acquisition tool at the exact RP in order to record the Wi-Fi RSSI fingerprint data. There are commonly a total of more than 20 RPs in an experimental field and a few times sampling needed to get an average for less data fluctuation [9].

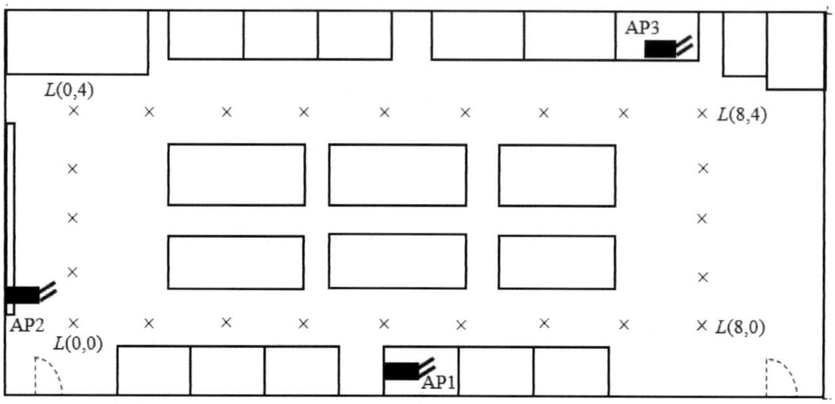

Figure 1. Example layout of a typical experimental field for Wi-Fi RSSI Fingerprint database construction, with the notation × representing the RP and the three Wi-Fi Access Point (AP)

Therefore we have designed a service robot that can do several indoor tasks including construction of Wi-Fi RSSI fingerprint database automatically. The design is a result from our vast review of numerous service robot review around the globe [10]. One key advantage of our robot is that its' generic platform is multipurpose with expendability capability. This robot seeks to aid and ease user effort to place the data acquisition tools around the experimental field. The robot is currently undergoing industrial fabrication process and this paper explains the design process of the robot in terms of design concept, hardware architecture, component selection, and control unit such as electronics and algorithm. The robot is planned to move from one RP to another preferably using line following method, and then record the Wi-Fi RSSI fingerprint on that specific RP from different orientations. The next section of the paper will explain about mechanical design of the robot which include design concept and mechanical specifications. Section 3 discuss the detail about controller hardware design and algorithm in order for the robot to collect data autonomously. In section 4 we discuss the results and findings d of this works. Lastly, section 5 draws the conclusions of this works and briefly suggest the possible action for future works.

2. Robot Mechanical Design

Service robot design is not limited to any specific principle and architecture [11]. Perhaps, the design should consider the scope of application in the first place in order to optimize the budget and lead time especially for academic and research purpose [12]. However, most service robots have two characteristics in common: mobility and user interactive. As service robots need to manoeuvre around, component selection for mobile platform is important to ensure smooth movement from one point to another. One of the process involved in designing the mobile platform is choosing the type of wheel and its configuration. The most common wheel configuration is differential drive due to its simplicity, low-cost, and easy to implement [13]. Apart from that, omni-directional wheel provide greater degree-of-freedom to the mobile platform but usually costs more than differential drive system [14]. User interactive system enable the robot to interact and communicate with the user. Some service robot use LCD screen as the user interactive component to display useful information as well as showing facial expression [15]. There are also service robot that features speech synthesis to communicate with user verbally [16]. In some application, storage compartment is needed for the robot to carry things such as toiletries for hotel guests or medical items in hospital [17][18].

Our design mainly consists of three parts: mobile platform, storage compartment, and user interactive screen as shown in Figure 2. The robot 3D design has been done in Fusion 360 software. For mobile platform, we have decided to implement differential drive with one castor system due to its low-cost and easy implementation. Evaluations has been made for our robot's mobile platform specification in terms of wheel size selection and motor rated power. The evaluation method is further discussed in section 3. Our robot also equipped with a storage compartment to place the on-board computer and data acquisition devices. The storage compartment also can be divided into several partitions to store many other things for future applications. For database construction application, the user interactive screen is used to display information of the data collection process and the interface of the robot operation. There is also a platform for input devices (keyboard and mouse) on the robot to ease troubleshooting process when the robot is in operation.

Several service robot design from the other works has been studied prior to our design. It is to ensure the thorough empathetic on the general requirements for systematic planning process as the designed robot should be flexible for as many applications as possible. As a matter of fact, this mobile robot is targeted in service and hospitality industries, with the competitive development cost in the positioning system utilizing the wireless positioning system via signal fingerprinting method. The robot will be tested in the online phase of indoor positioning using Wi-Fi fingerprinting technique that can be extended to navigation and path planning. This works will eventually need some hardware modification and upgrades in terms of mobile platform, computers, sensors, manipulators, and even the robot appearance for greater aesthetical values. Therefore in designing this robot we consider the expandability in mind.

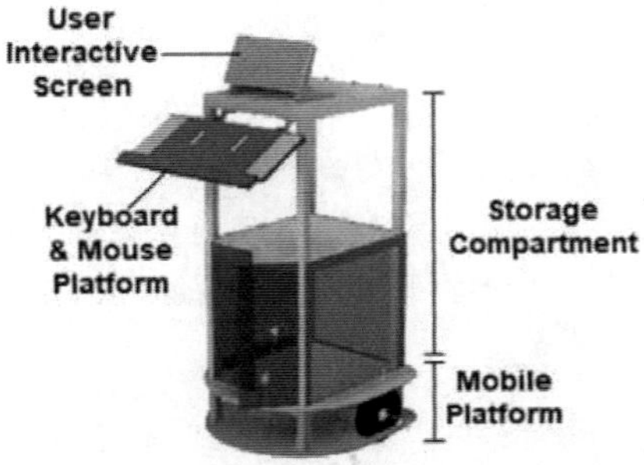

Figure 2. Service robot design for Wi-Fi RSSI fingerprint data collection

3. Design of Control Hardware and Algorithm

The core controller for this robot is a CPU unit running Robot Operating System (ROS). It is connected to several clients to control other mechanisms such as actuator and data acquisition unit. ROS has been widely used in many robotic projects. It provides a development environment that integrates necessary modules and sensors for a robot. There are five main components in our robot control hardware: power supply, main CPU, microcontroller, data acquisition device, and actuator. Each component is integrated over ROS to drive the robot for data collection tasks. Besides electronics, there is also an algorithm for the robot to automatically move from one RP to another and taking the data autonomously.

The power supply of our robot is the rechargeable 24V lead acid battery with a capacity of 7.2Ah. This battery is chosen because of its relatively low price compared to other type of cell with the same capacity. Series of regulators will regulate the voltage from the power supply before connecting it to the other modules or devices on the robot. Main CPU is responsible to control the overall process that occur on the robot systems such as giving command to the microcontroller to move the actuators, reading sensors data, and recording the data from data acquisition device. For Wi-Fi RSSI fingerprint data collection, the data acquisition device that we will use is the ESP8266 Wi-Fi module and general-purpose AP. Both devices were still in evaluation before we can decide to fully implement any of it. The actuator used to drive the robot is 100W VEXTA brushless DC motor.

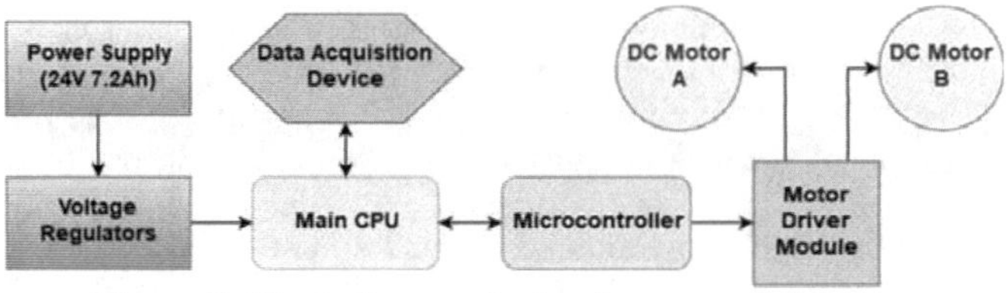

Figure 3. Block diagram for hardware control system

Selection for the DC motor undergo evaluation process by using a certain set of formula to find the suitable parameter of the DC motor. The main objective of this evaluation is to find the minimum required torque $\tau_{\min}$ of a motor that is sufficient to drive the robot in a given wheel radius $\acute{r}$, maximum acceleration $a_{\max}$, and maximum payload $m_{\max}$.

$$\tau_{\min} = F_{\max} \times \acute{r} \tag{1}$$

, where $F_{\max}$ is

$$F_{\max} = m_{\max} \times a_{\max} \tag{2}$$

Based on these equations, we developed a Windows application specifically to perform these calculations with some added features and parameters such as gear ratio and gear power transfer efficiency. We also added a parameter for the number of driving motor so that the application can calculate the torque and power for every independent motor. The snippet of the developed Windows application is shown in Figure 4.

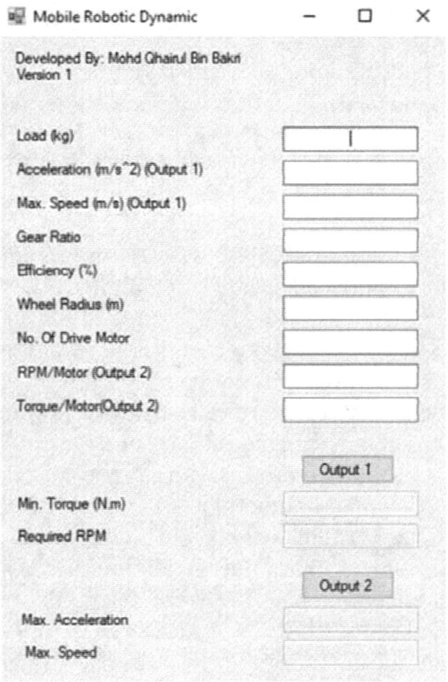

Figure 4. Windows application for motor rated torque evaluation

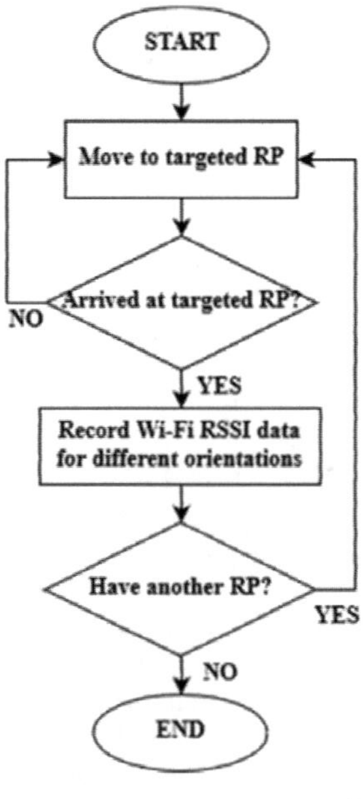

Figure 5. Flowchart of the planned robot algorithm for Wi-Fi fingerprint database construction

The control algorithm defines how the robot behave during operation. For Wi-Fi fingerprint database construction, we planned to program the robot to move from one RP to another in the experimental field, where the RP are connected by lines. Thus, the robot at this time perform similarly to a line following robot. At each RP, the robot will record the Wi-Fi RSSI fingerprint data from a few different orientations. This process is then repeated for the rest of the RPs. The number of RP will be assigned and marked. The greater number of RP, the better the database. Since we are using a robot for this process, increasing the number if RP is no longer a major problem. For a greater coverage, we can improve our database by using an interpolation algorithm [7]. Flowchart of the planned robot algorithm for Wi-Fi fingerprint database construction is shown in Figure 5.

4. Result and Discussion

By using the Fusion 360 software, several design versions have been made evaluate each of the design before we come up with the final version. Each design has its own unique architecture and aesthetic. Our first version of the design as shown in Figure 6(a) has a cantered driving wheel position shown in Figure 6(b) to allow zero radius turning. Later we found out that this design has the risk of castor wheel slip over the uneven floor level. Then we have decided to change the mobile platform in terms of wheel configuration to front wheel differential drive as depicted in Figure 6(c). This design is further improved in terms of aesthetical values, manufacturability, and cost effective aspect. In the final design presented in Figure 6(d), the wheel configuration was changed to rear wheel differential drive in order to provide smooth heading trajectory. The design specification of the final version is shown in Table 1.

Figure 6: The mobile robot design in this works, (a) the first version design, (b) mobile platform of the first design version, (c) second version design, and (d) final version design sent for fabrication.

Table 1. Specification of final version design

Parameters	Value
Dimension (H x L x W)	86 x 40 x 40 cm
Weight	20 kg
Max. Payload	20 kg
Mobile Configuration	Rear differential drive
Battery Capacity	24V 7.2Ah
Type of cell	Lead acid
DC motor	100W VEXTA DC brushless
Max. Translation Acceleration	0.75 ms^{-2}
Max. Translation Velocity	3.142 ms^{-2}
Motor Rated Torque	3.6 Nm
Motor RPM	250

5. Conclusion

This paper described our state-of-the-art service robot for Wi-Fi RSSI fingerprint database construction. The development of this robot seeks to aid the data collection process which is at present is done manually. In designing the robot, we have made a deep study on different aspects of a service robot to make sure that we are able to come out with an optimal design in terms of performance, aesthetic, cost, and expandability. The optimal mechanical design for our research scope is the three layers modular design which are mobile platform, storage compartment, and user interactive display. Component selection such as DC motor was made based on the specific evaluation process. This enable us to know better about the specification and performance of our robot from the design stage. Currently the robot is in fabrication process, while the RP and the line following system are under proof test. It is also worth to note that the facial expression via interactive 'robot face' is also undergoing swift development from the same research group. In the future in development context, we are looking to experiment a few method such as sensor fusion technique [19], RFID or QR code to help our robot navigate more precisely between the RPs. It is possible that the meta heuristic combination of these techniques should enhance the robot capacity especially in the wireless positioning system.

Acknowledgement

The authors would like to acknowledge the support from the Fundamental Research Grant Scheme (FRGS) under a grant number of FRGS/1/2018/ICT05/UNIMAP/02/4 from the Ministry of Education Malaysia.

References

[1] Mingzhe X, Jiabin C, Chunlei S, Nan L and Kong C 2015 The Indoor Positioning Algorithm Research Based On Improved Location Fingerprinting *The 27th Chinese Control and Decision Conference (2015 CCDC)* (Qingdao, China: IEEE) pp 5736–9

[2] Zhao Y, Zhou H, Li M and Kong R 2008 Implementation of indoor positioning system based on location fingerprinting in wireless networks *Wirel. Commun. ...* 8–11

[3] Moghtadaiee V, Dempster A G and Lim S 2011 Indoor localization using FM radio signals: A fingerprinting approach *2011 International Conference on Indoor Positioning and Indoor Navigation, IPIN 2011*

[4] Li B, Salter J, Dempster A G A, Rizos C, Beomju Shin, Jung Ho Lee, Taikjin Lee and Hyung Seok Kim 2006 Indoor Positioning Techniques Based on Wireless LAN *LAN, First IEEE Int. ...* **2** 574–7

[5] Zhao F, Huang T and Wang D 2019 A Probabilistic Approach for WiFi Fingerprint Localization in Severely Dynamic Indoor Environments *IEEE Access* **7** 116348–57

[6] Houria C and Mouhcine C 2017 Indoor Localization Using Wi-Fi Method Based on Fingerprinting Technique *2017 International Conference on Wireless Technologies, Embedded and Intelligent Systems (WITS)* (Fez, Morocco: IEEE)

[7] Ismail A H, Kitagawa H, Tasaki R and Terashima K 2016 WiFi RSS Fingerprint Database Construction for Mobile Robot Indoor Positioning System 1561–6

[8] Thuong N T, Phong H T, Do D T, Van Hieu P and Loc D T 2016 Android application for WiFi based indoor position: System design and performance analysis *Int. Conf. Inf. Netw.* **2016-March** 416–9

[9] Sun W, Xue M, Yu H, Tang H and Lin A 2018 Augmentation of Fingerprints for Indoor WiFi Localization Based on Gaussian Process Regression *IEEE Trans. Veh. Technol.* **67** 10896–905

[10] Bakri M Q, Ismail A H, Hashim M S M and Safar M J A 2019 A Review on Service Robots: Mechanical Design and Localization System *IOP Conference Series: Materials Science and Engineering*

[11] Reiser U, Connette C, Fischer J, Kubacki J, Bubeck A, Weisshardt F, Jacobs T, Parlitz C, Hagele M and Verl A 2009 Care-O-bot® 3 - Creating a product vision for service robot applications by integrating design and technology *2009 IEEE/RSJ Int. Conf. Intell. Robot. Syst. IROS 2009* 1992–8

[12] Van Osch M, Bera D, Van Hee K, Koks Y and Zeegers H 2014 Tele-operated service robots: ROSE *Autom. Constr.* **39** 152–60

[13] Yazdjerdi P and Meskin N 2018 Fault tolerant control of differential drive mobile robots using sliding mode controller *2017 5th Int. Conf. Control. Instrumentation, Autom. ICCIA 2017* **2018-Janua** 270–4

[14] Padgett S T and Browne A F 2017 Vector-based robot obstacle avoidance using LIDAR and mecanum drive *Conf. Proc. - IEEE SOUTHEASTCON* 1–5

[15] Hegel F 2012 Effects of a Robot's Aesthetic Design on the attribution of social capabilities *Proc. - IEEE*

Int. Work. Robot Hum. Interact. Commun. 469–75

[16] Kantharak K, Somboончa C, Trong N and Thinh N T 2017 Design and Development of Service Robot based Human - Robot Interaction (HRI) 293–6

[17] Tasaki R, Kitazaki M, Miura J and Terashima K 2015 Prototype design of medical round supporting robot "Terapio" *2015 IEEE International Conference on Robotics and Automation (ICRA)* (IEEE) pp 829–34

[18] Zhaohui Z, Mei X, Bian X, Cai H and Ti J 2017 Development of an intelligent interaction service robot using ROS *Proc. 2016 IEEE Adv. Inf. Manag. Commun. Electron. Autom. Control Conf. IMCEC 2016* 1738–42

[19] Dobrev Y, Gulden P and Vossiek M 2018 An Indoor Positioning System Based on Wireless Range and Angle Measurements Assisted by Multi-Modal Sensor Fusion for Service Robot Applications *IEEE Access* **6** 69036–52

Characterization of Environmental Noise Pollution Based on Noise Measurement and Mapping at USM Engineering Campus

O F S Khasawneh, H Halim*, S N Abdullah, S A Razali, H R F Algburi, and A H Salleh

School of Civil Engineering, Engineering Campus, Universiti Sains Malaysia 14300 Nibong Tebal, Penang, Malaysia

*Corresponding author: ceherni@usm.my

Abstract. This paper describes an analysis of visualizing noise impact on the environment by producing the noise mapping, determining the noise sources and characterization of sound levels measurements at USM Engineering Campus, Nibong Tebal, Penang. Measurements were taken at a total of 30 points with the duration of 3 minutes during working hours using the sound level meter. The ArcMap 10.3.1 software and Google Maps were used to create the noise mapping based on the coordinates and each sampling point. The results revealed that 23 points exceeded the maximum permissible limit, 55 dB(A). The maximum measured sound level was obtained at point 4 (68.3 dB(A)), while the lowest sound level was 43.8 dB(A) at point 26. Characterization of sound levels by measuring the noise levels and performing noise mapping is substantial to analyse and determine the level of noise pollution in institutional and educational areas. In addition, such practices would help in understanding the activities that contribute to the increase in sound level.

1. Introduction

Noise can be defined as the excessive or annoying unwanted sound in a specific area. However, noise pollution occurs when the noise level is above the allowable limit for a given environment [1]. Noise acts as an environmental stressor, which can lead to physiological health issues that can influence human's daily activities and interfere with communications. Excessive exposure to noise has also been found to cause sleeplessness and reduce the sleeping efficiency [2]. According to the world health organization (WHO), noise can be observed at different areas such as schools, preschools, educational institutions, industrial and commercial buildings, and traffic areas. Countries have been encouraged to assess and determine the levels of noise pollution in their territories by conducting noise mapping studies [3].

In educational institutions, noise-induced leads to complications hinder in communication between the teacher and the student. Moreover, noisy learning environments may disrupt the performance of students, as it can affect their focus in the class, which can lead the students to achieving lower grades and lower evaluation. Thus, the teacher's sound is important to be heard by receptors without any outside interferences; as the great task of knowledge transmissions must be clear and harmonious than any competitive noise [4]. Several authors have performed noise measurements studies in educational institutions. Ibrahim [5] assessed the noise pollution at the University of Technology in Baghdad, Iraq using noise mapping. His findings stated that the main sources of noise at the campus were caused by the electrical generators, traffic movements, parking lots, and engineering workshops with open doors. The maximum noise levels observed in his study ranged from 85 dB(A) to 95 dB(A). In a similar study, Ibrahim [3] focused on noise mapping of the campus of the School of Engineering at the University of Al-Mustansiriyah. Based on his noise mapping studies, it was observed that noise levels greater than 90 dB(A) were recorded when the generators are turned on, and the lowest levels overlapped with noise levels generated by the industrial buildings, Bab Al-Moatham road industrial buildings and parking lots. The high sound levels were recorded in and around the areas where students would gather, near the generators and other noise sources, without awareness of the high noise levels to which the students are being exposed to.

Zannin and Ferraz [6] conducted a study in Hospital de Clínicas of the Federal University of Parana and found that measured noise levels were mostly in the range of 55 to 60 dB(A) and reached up to 60 and 65 dB(A) in the central part of the frontal facade of the university hospital. The noise maps presented showed that the source of noise pollution was mainly due to the traffic congestions and heavy vehicles. Zannin et al. [7] stated on a case study performed at the campus of the Polytech-nic Center of the Federal University of Paraná, Brazil, that the road that connects the two entrances to the campus displayed dominant noise levels above 70 dB(A), thus being the main noise source inside the campus. From these observations, it can be indicated that noise can affect the communication and concentration, particularly during learning and teaching sessions. In addition, it was revealed that a variety noise sources can be detected in campuses, such as noise generated due to events and different activities at the campus.

The purpose of this study is to visualizing noise impact on the environment by producing noise maps that can easily illustrate the measured sound levels at the studied area. Furthermore, this study also aims on determining the sources that affect the noise levels and to characterize sound levels measurements to find the noise levels at which people are exposed during the working day at USM Engineering Campus.

2. Materials and method

The study was carried out at Universiti Sains Malaysia, USM Engineering campus at Nibong Tebal. USM is located at the south of Seberang Perai. The campus covers an area of 320 acres and currently has approximately 4000 registered students. The average temperature is 32.2°C, with an approximate annual rainfall of 267cm^3. The locations of measurements included the internal streets with all main roads, such as the main gate, and some of the branch roads. Type of buildings included: (1) educational (such as college buildings), (2) administrative (such as the main administrative building of the campus) and (3) services buildings (such as library and restaurants).

2.1. Sampling method

Thirty measurement points were selected as in Figure 1. For each point, the measurement duration was 3 minutes, and the measurement period for all point were between 12:30 pm and 6:00 pm [7]. Meanwhile, the human activities, vehicles, and any noise source that can affect the readings in each location were recorded. The exact locations coordinates were taken using Google Maps (Table 1).

Figure 1. Map pinpoint the location of 30 noise measurement at USM Engineering Campus.

Levels of noise intensity were measured using Cirrus Research plc type CR:1710 Sound Level Meter (SLM) as shown in Figure 2.

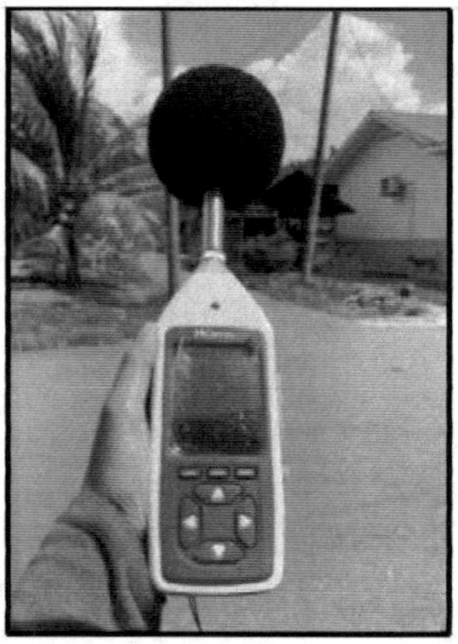

Figure 2. Sound Level Meter

This SLM has the ability to measure the following values instantaneously (which represents the now live noise level, in which, a reading every 5 seconds is automatically recorded), L_{AFmax} and L_{AFmin} (which represent the maximum and minimum noise levels during the overall period of measurement), and the equivalent continuous level L_{Aeq} (which represents the measured value which needs to be averaged over a period of time for all L_{AF} readings [8]. The instrument was calibrated on 39.7 dB (A) before taking noise measurements and was set to record noise samples at 5-sec intervals during the 3 minutes exposure time. The sound level meter was placed at a height of 1.5 meter above the ground level and at a minimum distance of 3 meter from the building facade [9]. Acoustic descriptors such as A-weighted equivalent sound pressure levels (L_{eq}) were used and were expressed in decibel (dB). The start-up time of this SLM depends on the state the instrument was in when last switched off, and it may take up to 2 min from a cold start, or up to 10 s if the instrument is already in standby mode, (i.e., from a warm start). The collected data was analysed by using ArcMap 10.3.1 software to create a noise map.

2.2. Noise mapping

Noise mapping is a graphic representation of the sound level distribution existing in a given region, for a defined period which is shown on map of that region which is coloured according to the noise levels in the area [10]. Measured noise level was subjected to noise mapping based on the data in Table 1. The ArcMap 10.3.1 software was used in analyses of the data. The values were mapped as the coordinate X and coordinate Y including one parameter of sound level measurement (L_{10}, L_{50}, L_{90}, L_{eq}, L_{max} and L_{min}). The contouring method by using Inverse Distance Weighted (IDW) spatial interpolation technique were used to capture the adjoining areas and to cover the entirety of the study area. Generally, interpolation predicts cell values in a raster format using a given albeit limited number of sample data. It is a veritable tool for prediction of unknown values for a given geographic point data which in this study is noise [10]. The Google Maps were used to provide the real photo of the location by taken 4 points of each location at the boundary on the study area.

3. Results and discussion

The noise level measurements at USM Engineering Campus were determined and compared with the noise limits recommended by the Malaysian Standards of Planning Guidelines for Environmental Noise Limits and Control (2004) [11] and also WHO [1] for institutional areas. Figure 2 shows the equivalent sound pressure level (L_{eq}) measured on the campus with the noise limit recommended by Malaysian Standard 50 dB(A) and WHO limit 55 dB(A). The measurement indicated 23 of the 30 measured points exceeded the maximum permissible sound level, 55 dB (A) of the L_{eq}. The remaining 7 points are up to 55 dB(A) in the range of sound levels. The maximum sound level recorded was 68.3 dB(A) at point 4 and followed by 65.9 dB(A) at point 12 (Table 1). These measured high noise levels are attributed to the traffic movements and other activities nearby the points, such as the laboratory works, where noisy equipments were in operation. Moreover, the mosque located near to point 4 contributed to increased sound levels, due to the elevated number of pedestrians passing by especially during the prayers. The noise limit above 55 dB (A) is generally considered to be an anxiety level and above 65 dB(A) is a noise level that affects the quality of life if the noise is continuous in the region [12].

Figure 4 until Figure 6 show the noise mapping of the USM engineering campus, identifying the six parameters for each of the sampling points throughout the campus. The measured parameters are L_{10}, L_{50}, L_{90}, L_{eq}, L_{max}, and L_{min}. Figure 4 shows the noise mapping of the USM campus for the L_{eq} measurement, which the highest equivalent continuous level is 68.3 dB(A) with the constant noise level over a given time period. The mechanical school generated noise from the laboratory machineries and compressors. This noise level would contribute greatly to diminishing people's quality of life inside the campus. Unwanted sound of sufficient intensity and duration can cause temporary and permanent hearing loss [13]. It can also interfere with speech communication, the transmission of other auditory signals, disturb sleep, act as a general source of annoyance or disturbance, and interfere with the performance of complicated tasks and the opportunity for privacy [12].

Exposure of people to daytime noise levels above 65 dB(A) can cause severe health problems [13]. Thus, the noise level should be well controlled in order to ensure that people's lives can be safe without causing any health issues or damaging their hearings. The collected data indicated that the mechanical school building is subjected to sound pressure levels exceeding the limit established by guidelines for the educational areas above the limit of 55 dB(A). Furthermore, an analysis of noise maps in Figure 6 indicates that the schools of civil, electrical, aerospace, chemicals, and materials have generated noise within 60 to 64 dB(A). These areas also impacted by the highest sound pressure level which exceeded the limits for noise level in Malaysia.

IOP Conf. Series: Materials Science and Engineering **920** (2020) 012004 doi:10.1088/1757-899X/920/1/012004

On the other hand, the hostel and sports center are located far from the school center. Therefore, those areas are not being affected by the noise generated from the school center which has more students, lecturers, and staff during the working hour. Most of the student spent most of their time at the school center instead of staying in the hostel during the class sessions. Thus, the school center and cafeteria contributed to being the main noise source inside the campus as shown in Figure 6 with the noise level measured around 57.5 and 60.6 dB(A). However, the noise is also caused by airplanes, air-conditioning, and laboratory machinery in the school's buildings. The generated noise is about 53.5 dB(A). Additionally, as the main roads on the campus allow the distribution of pedestrians, bicycles, light vehicles, utility vehicles and buses, the noises are not too affected in comparison with residential areas located near the campus. The main roads are still under control and not busy with vehicles around the campus [7].

Table 1 shows the raw data and converted to passenger car unit (PCU) for the number of vehicles and pedestrians for each sampling point at the USM engineering campus during the measurement period. The table reveals that traffic volume is considered as a major source of noise at USM engineering campus. Generally, traffic contributed to the increase in noise levels, as shown in Figure 3 which contributed to higher noise levels at the sampling points. It is worth mentioning that the number of pedestrians also contributed to higher recorded values of noise levels. For instance, at sampling point 4, the number of people passed that point reached 12 during that period, and the measured L_{eq} was 68.3 dB(A), compared to only 43.8 dB(A) at sampling point 26, where no pedestrians existed at that point.

The influence of traffic on the measurement of noise levels was in accordance with several studies [7][9] conducted in different cities around the world. Zannin et al. [7] studied the characterization of environmental noise at a university campus in Brazil and showed that elevated noise levels were measured above 70 dB(A) in the areas near the main roads inside the campus, specifically near the main gate, where more circulation of vehicles was allowed in that area. In addition, these findings were in accordance with a study conducted in Kolhapur, India, where the observed results indicated that high Leq measurements that reached up to 63.71 dB (A) were recorded in the educational and institutional zones [14].

Table 1. GPS coordinate and noise descriptors of each point in this study.

No.	Cars	Motorcycle	Medium Lorries	Busses	Pedestrian	PCU	No.	Cars	Motorcycle	Medium Lorries	Busses	Pedestrian	PCU
1	3	9	0	0	13	10	**16**	4	1	0	0	16	5
2	8	5	0	0	0	12	**17**	6	3	0	0	6	9
3	3	5	0	1	11	10	**18**	0	0	0	0	0	0
4	3	1	0	0	12	4	**19**	4	9	0	0	0	11
5	2	1	0	0	2	3	**20**	5	2	0	0	0	7
6	3	3	0	0	0	6	**21**	8	10	0	0	0	16
7	6	1	1	0	2	10	**22**	4	1	0	0	0	5
8	6	1	0	0	5	7	**23**	0	1	0	0	0	1
9	0	0	0	0	0	0	**24**	4	5	0	0	1	8
10	4	2	0	0	0	6	**25**	0	1	0	0	0	1
11	0	1	0	0	4	1	**26**	0	0	0	0	0	0
12	0	1	0	0	1	1	**27**	4	3	0	0	0	7
13	1	0	0	0	2	1	**28**	9	8	0	0	0	15
14	0	1	0	1	2	4	**29**	3	3	0	0	0	6
15	1	1	0	0	0	2	**30**	0	2	0	0	0	2

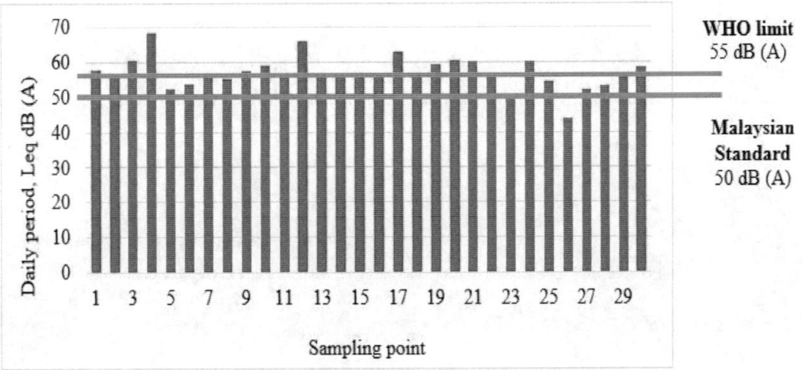

Figure 3. The equivalent sound pressure noise level measured at USM Engineering Campus.

Figure 4. Noise mapping of the USM Engineering Campus for L_{eq}.

Figure 5. Noise mapping of the USM Engineering Campus for L_{max}.

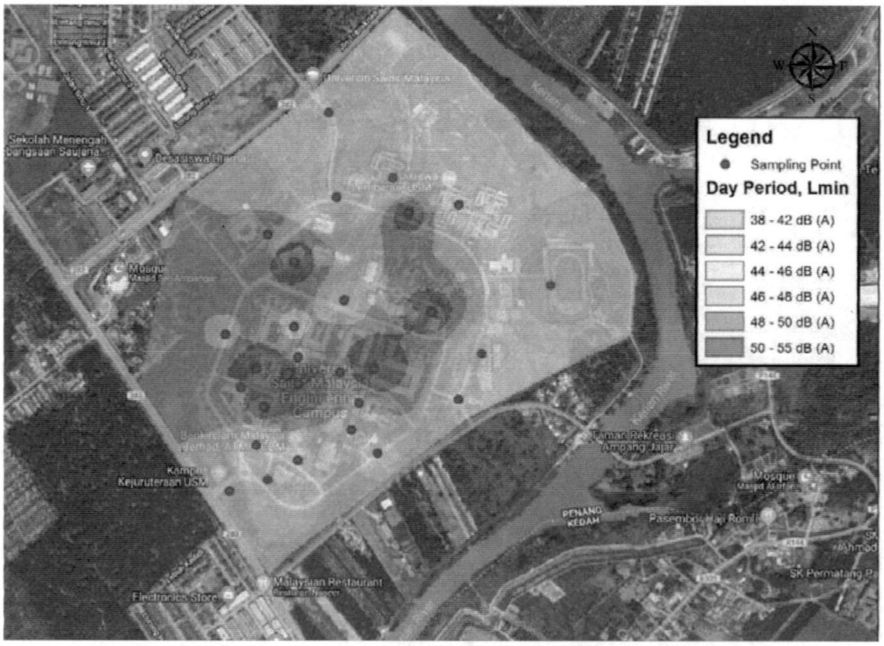

Figure 6. Noise mapping of the USM Engineering Campus for L_{min}

4. Conclusions

In conclusion, this case study characterized the noise levels at USM engineering campus for six parameters: L_{10}, L_{50}, L_{90}, L_{eq}, L_{max}, and L_{min}, at 30 different points distributed around the campus, as shown in the noise mapping Figures. Such work is important to be conducted to be able to analyse the level of noise pollution in educational areas. Throughout this study, we have observed that most of the measured sampling points experienced high noise levels that exceeded the maximum permissible sound level, 55 dB(A) of L_{eq} for institutional areas. However, only seven sites were within the allowable standards of the noise level. The high sound levels measured at USM campus was mainly affected by the traffic volume inside the premises of the campus. Additionally, other activities like the use of noisy equipment and machineries at the engineering laboratories also contributed to high levels of noise. Moreover, the gathering of students in certain areas like the cafeteria, or assembling points, increased the values of sound levels.

Acknowledgement
The authors would like to thank School of Civil Engineering, Universiti Sains Malaysia for the support of this project. Thanks also to Universiti Sains Malaysia for providing financial support under the short term grant (304/PAWAM/6315066) to carry out this study.

5. References

[1] World Health Organization 2018. Environmental Noise Guidelines for the European Region (Copenhagen-WHO).
[2] Kusuma S 2011 *Int. J. Earth Sci. Eng.* **4** 459.
[3] Ibrahim, A S 2015 *J. Env. Earth Sci.* **5** 108.
[4] Josta, L N 2018 *Afr. Res. J. Educ. Soc. Sci.* **5**.
[5] Ibrahim, A S 2016 *J. Eng. Sustain. Dev.* **20** 115.
[6] Zannin, P H T and Ferraz F 2016 *Open J. Acoust.* **6** 71.
[7] Zannin, P H T, Margret, S E Fiedler P E K and Bunn F 2013 *Cities* **31** 317.
[8] El-Sharkawy M and Alsubaie A 2014 *Saudi J. Med. Med. Sci.* **2**.
[9] Phukan B and Kalita K 2013 *Int. J. Env. Sci.* **3**.
[10] Akintuyi A, Raji A Adewuni D and Wunude E 2014 *J. Soc. Sci.* **4** 154.
[11] Ministry of Natural Resources and Environment, Malaysia. The Planning Guidelines for Environmental Noise Limits and Control.
[12] Jan C J 2008. *Malays. Music J.* **3** 32.
[13] Toh, Phyllis, CW 2014. *Profiles and residents' perception of environmental sound and music activities in Bandar Dataran Segar, Port Dickson, Malaysia*, Master thesis. (Serdang: Universiti Putra Malaysia).
[14] Hunashal R B and Patil Y B 2012. *Proc. Soc. Behav. Sci.* **37**, 448.

IOP Conf. Series: Materials Science and Engineering **920** (2020) 012005 doi:10.1088/1757-899X/920/1/012005

Performance Evaluation of Composite Cross-Arm Structure Under Different Magnitude of Loading

Daud Mohamad[1], Salmia Beddu[1], Agusril Syamsir[2], Nazirul Mubin Zahari[1], Sareh Aiman Hilmi Abu Seman[3,*], Muhammad Fauzinizam Razali[3], Aizat Abas[3] and Fei Chong Ng[3]

[1] Department of Civil Engineering, Universiti Tenaga Nasional, 43000 Kajang, Malaysia
[2]Institute of Energy Infrastructures (IEI), Universiti Tenaga Nasional, 43000 Kajang, Malaysia
[3]School of Mechanical Engineering, Universiti Sains Malaysia, Engineering Campus, 14300, Nibong Tebal, Penang, Malaysia.

*E-mail: sarehaiman@usm.my

Abstract. In this paper, the effects of load with increasing factor of safety on the mechanical performance and intra-laminar damage formation of cross-arm structure made up of two different schemes are studied. The composite structure has been numerically modelled by a finite element (FE) model. Research was carried out using a VUMAT subroutine to model intra-laminary damage formation in the analysed composite structure, and Hashin's failure criteria have been used. This investigation revealed that with a good combination of stacking sequence and ply properties as shown by scheme A which has combination of 90 and 45-degree plies at the both ends and much higher value of ply properties, a safety reserve can be obtained that provides a margin between actual load conditions and those leading to cross-arms structure failure.

1. Introduction

Different kinds of loads during service life such as a windstorm, rain and other serious climate can endanger composite structures commonly used in electricity grid structure building. The extreme and very rare weather and climate induces significantly higher load than the actual grid load. Once the grid fails, the transmission line overload will lead to a cascade failure that will weaken the power system successively and could lead to electric instability and large-scale blackouts [1, 2]. A lot of studies into the susceptibility of power grids in severe conditions has been driven by massive financial and social impacts of these occurrences [3-5]. Therefore, it is important to understand how such loads affect distinct types of grids failure if the grid is designed to become solid and more resilient to failure.

As protective systems, laminate composite structures are usually utilized for resisting severe loads. During the past two decades, a number of studies have focused on the failure response of laminate structures under different load types [3, 6, 7]. Tuo, Lu [8] found that the compressed effects damage to the matrix and fibers of composite laminate depends on impact energy levels as damage is severe when it comes to higher impact energy.. The effect of different loading magnitudes based on three different safety factors on the structural deflection and the development of matrix and fibers damage of the cross-arm structure with distinct designed laminate scheme was evaluated in this analysis. In the ABAQUS / EXPLICIT environment, a detailed FE model for the laminate composite was created. A VUMAT subroutine provided by ABAQUS for the prediction of intra-lamina damage is used to execute the Hashin failure criteria [9]. Two different schemes have been considered to develop the cross-arm structures, in order to better understand which scheme is more solid and resilient to structural deflection and intra-laminar damages formation under much higher load that a structure is required to withstand.

2. Damage modelling

In order to model intra-laminar damages including breakage of fibers and progressive matrix damages, the progressive failure strategy implemented in ABAQUS/EXPLICIT solver algorithm was considered. This analysis employs the Hashin criteria [9] to detect the initiation of damage. A bilinear constitutive law as shown in Figure 1 can be used to represent the intra-laminar failure mode. In Figure 1, point A indicates the activating value of damage and the material before this point is undamaged. Then, damage propagates until point B, where the elements stress has exceeded the limit values and degraded completely. The criteria for damage initiation is based on the theory of Hashin which takes into account different intra-laminar modes of failure. Tension and compression fiber fracturing and matrix cracking in transverse tension and compression are four different types of failure considered. For each failure mode, the damage variable is specified as set out in Table 1.

Table 1. Initiation criteria for Hashin failure in each mode of failure

Failure criteria		Failure modes
$(\frac{\sigma_{11}}{X_{1T}})^2 + (\frac{\sigma_{12}}{S_{12}})^2 + (\frac{\sigma_{13}}{F_{13}})^2 = 1$	(1)	Fiber tensile failure
$(\frac{\sigma_{11}}{X_{1C}})^2 = 1$	(2)	Fiber compressive failure
$\frac{(\sigma_{22}+\sigma_{33})^2}{X_{2t}^2} + \frac{\sigma_{23}-\sigma_{22}\sigma_{33}}{S_{23}^2} + \frac{\sigma_{12}^2+\sigma_{13}^2}{S_{12}^2} = 1$	(3)	Matrix tensile failure
$\left[\left(\frac{X_{2c}}{2S_{23}}\right) - 1\right]\frac{(\sigma_{22}+\sigma_{33})}{X_{2c}} + \frac{(\sigma_{22}+\sigma_{33})^2}{4S_{23}^2} + \frac{(\sigma_{23}^2-\sigma_{22}\sigma_{33})}{S_{23}^2}] + \frac{\sigma_{12}^2+\sigma_{13}^2}{S_{12}^2} = 1$	(4)	Matrix compressive failure

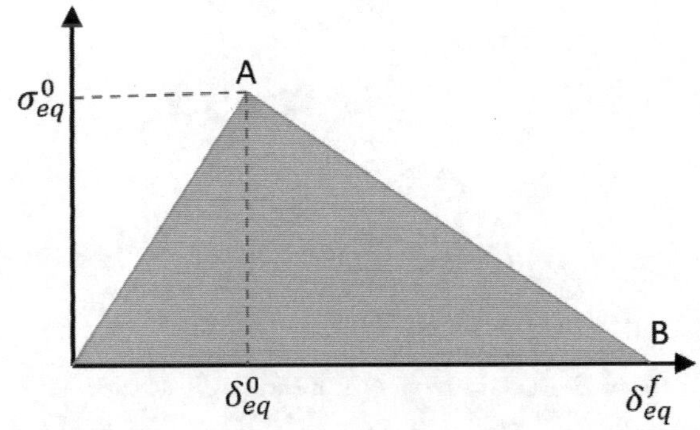

Figure 1. Damage initiation and propagation points.

3. Finite element model

The ABAQUS/Explicit FE code has been employed to build the finite element model. Due to the high calculation time needed for full model evaluation, this alternative method was chosen for quasi-static modeling. A successful balance of loading rates and mass scaling methods was therefore implemented to reduce the time for the explicit time integration process [10]. The composite laminate was modelled in this study as a cross-arm structure. Two cross arm structures have been modelled based on scheme A and B as shown in Figure 2 which each arm was modelled with 9 and 7 equal thickness plies respectively. Nevertheless, the collective laminate thickness of both schemes is equal. Composite laminate was developed using Abaqus database's solid element formulation. Every element has eight nodes and at each node with three degrees of freedom and an integration point. Each ply is modelled along the thickness with two solid elements to better predict the onset of intra-laminar damage. The deletion of the element during the analysis has assured localized stiffness reduction. A perfect bonding between the plies was assumed for the interface conditions. Figure 2 displays the two different stacking sequence schemes used in this analysis. The arm framework has been fixed at four ends and increased multiaxial loads have been applied to the front end with the safety factor of 1, 2 and 4 (load 1, load 2 and load 3) as shown in Figure 3. The magnitude of load 1 (safety factor of 1) which consisting of longitudinal force (Fl), a vertical downward force (Fv) and a transverse force (Ft) are 6524 N, 84992 N and 46416 N respectively. Friction between all contact surfaces was implemented with an average friction coefficient of 0.25 as it gave good result convergence and shorter computer run time. The material properties given in Table 2 and 3 have been defined to each ply in scheme A and B.

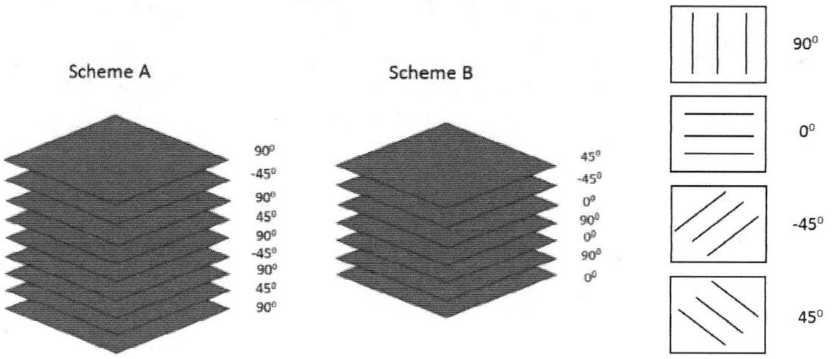

Figure 2. Laminate structure of scheme A and B.

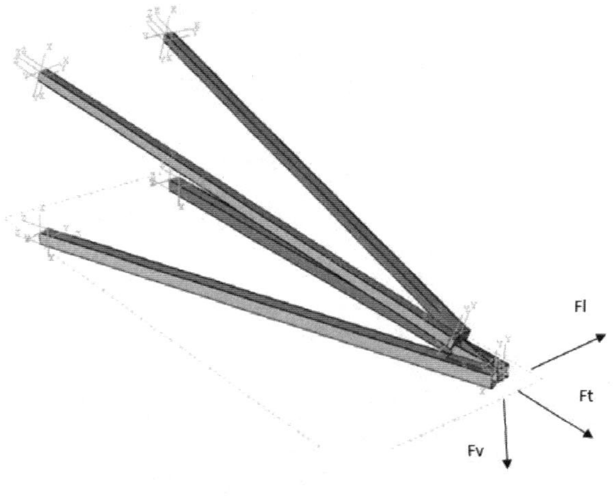

Figure 3. FE model of cross-arm structure.

Table 2. Material properties of each ply in scheme A.

Intra-laminar properties of scheme A	
Density	1900 kg/m^3
Orthotropic properties	E_1= 18.3Gpa E_2 = 1.089Gpa E_3= 0.327GPa $G_{12} = G_{13} = G_{23}$= 4Gpa $v_{12} = v_{13} = v_{23}$= 0.28
Ultimate stress	X^T= 429Mpa Y^T = 100Mpa X^C= 320Mpa Y^C = 76Mpa $Z^T = Z^C$= 80Mpa S_{12}= 89Mpa $S_{13} = S_{23}$= 50Mpa

Table 3. Material properties of each ply in scheme B.

Intra-laminar properties of scheme B	
Density	1900 kg/m^3
Orthotropic properties	E_1= 17Gpa $E_2 = E_3$ = 5.1Gpa
	$G_{12} = G_{13} = G_{23}$= 4Gpa
	$v_{12} = v_{13} = v_{23}$= 0.26
Ultimate stress	X^T= 321Mpa Y^T = 80Mpa
	X^C= 151Mpa Y^C = 65Mpa
	S_{12}= 89Mpa $S_{13} = S_{23}$= 50Mpa

4. Results and discussions

Simulations of the two separate scheme structures studied under the multiaxial loading are described and discussed. The numerical models' mechanical and damage behavior is expressed by the static displacement contour and the position of intra-laminar damage formation. In Figure 4(a), 5(a) and Table 4, total structure displacement for different composite laminate schemes are shown. As shown in Figure 4(a) and 5(a), at the front end close to the applied load area, the peak deflection of the two structures was predicted and it was more obvious as much higher load being applied. Based on Table 4, the maximum deflection of both schemes increases with different value as load increases. The value of maximum deformation is bigger in scheme B compared to scheme A. The less deformation suggests stronger characteristics of quasi-static loads. Scheme A is the ideal lay-up scheme with a minimum deformation as load increases.

Referring to Figure 4(b), 4(c), 5(b), 5(c) and Table 5, both schemes structure able to withstand the load 1 and load 2 without any fiber tensile and compressive damages. However, as much higher load being applied (load 3), the compressive fiber damage has been formed as indicated by red elements on both lower arms of Scheme B close to the area in which the highest displacement is occurring and clustered around the area of pinned arms. Meanwhile, on both upper arms of scheme B, the fiber tensile damage was formed as indicated by red elements near the area where the arms were attached to a connector. The fiber damage results have been tabulated in Table 5. The value 1 indicates that the criterion of damage initiation has been met. Based on the results shown in Figure 4 and 5, stacking sequence and properties of plies have obvious effect on maximum deflection value and damage formation of both schemes as the load increased. Comparison of results between cross-arm structure of scheme A and B show that the combination of 90 and 45-degree plies at the both end has encourage the ability of composite laminate structure to reduce deflection and resist both the fiber tensile and compressive failure formation under much higher loading (load 3) compared to scheme B which either 90 or 45-degree ply is presented at the extreme end. A huge leap in performance of the composite laminate structure has also been reported by Mohamad, Syamsir [11] when a ±45-degree ply is presented at both ends and Riccio, Di Felice [12] also confirmed that plate composed of 45 and -45 oriented plies would withstand significantly higher impact intensity than the 0/90 oriented ply configuration.

IOP Conf. Series: Materials Science and Engineering **920** (2020) 012005 doi:10.1088/1757-899X/920/1/012005

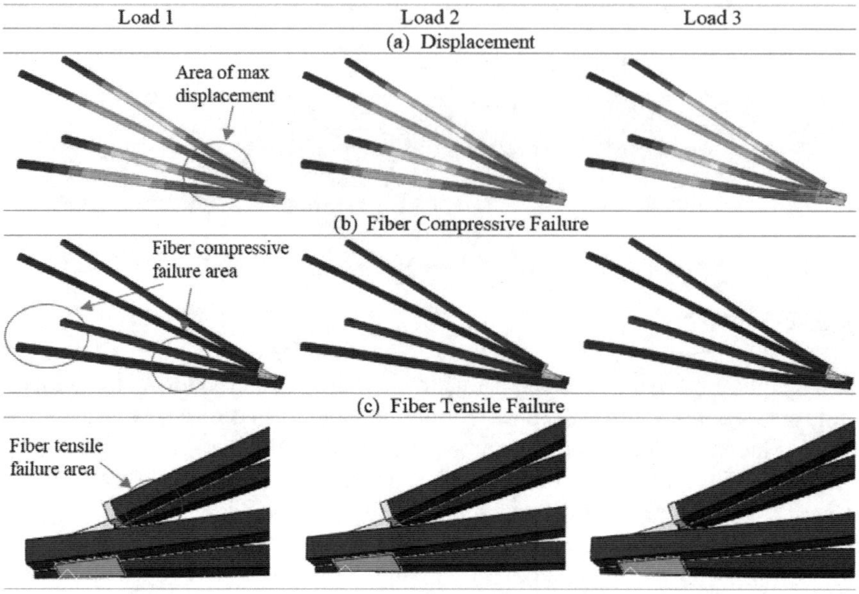

Figure 4. Numerical results of scheme A under different loading a) Displacement contour b) Fiber compressive failure location c) Fiber tensile failure location

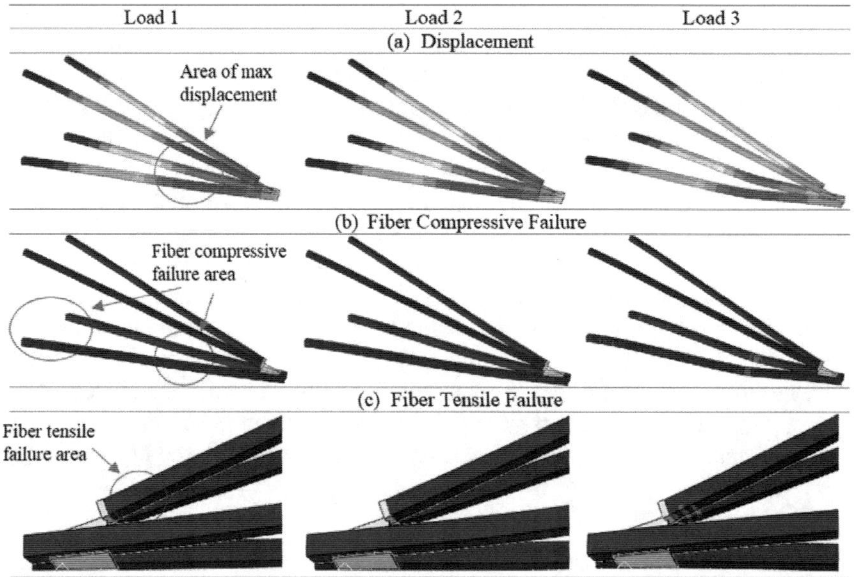

Figure 5. Numerical results of scheme B under different loading a) Displacement contour b) Fiber compressive failure location c) Fiber tensile failure location

Table 4. Maximum deflection of scheme A and B structures.

	Maximum deflection of structure (mm)	
Load	Scheme A	Scheme B
1	33	46
2	68	97
3	159	259

Table 5: Failure modes and indicators of Scheme A and B structures.

	Load 1		Load 2		Load 3	
	Scheme A	Scheme B	Scheme A	Scheme B	Scheme A	Scheme B
Fiber tensile failure	0	0	0	0	0	1
Fiber compressive failure	0	0	0	0	0	1

Properties of ply also plays a major part in deciding the performance of composite laminate structures. Scheme A which defined with higher value of Young's Modulus and ultimate stresses compared to scheme B showed smaller deflection and none of fiber tensile and compressive were formed when it was being subjected beyond the intended load. The effect of ply properties on composite laminate structure performance has been studied by [Mohamad, Syamsir [13]]. They found that composite laminate structures which defined with much higher properties gave a good overall performance improvement when subjected to multi-axial loading. Ozsoy, Mimaroğlu [14] have also found that composite laminate with higher value of properties able to withstand much higher tensile and bending loads. Based on the above discussions, through a good combination of stacking sequence and ply properties, a much stronger composite laminate structure performance can be achieved which it can bear extra load beyond what is intended a structure will actually take which is shown by scheme A.

5. Conclusions

FE analyses have been carried out to identify the capability of the cross-arm structure which made up of two different schemes when subjected to multi-axial static loading with increasing factor of safety based on the displacement value and formation of damage. Based on the above results, both structures with different scheme were able to bear the load 1 and load 2 successfully. However, when load 3 being applied, stacking sequence and ply properties have huge effect on the overall structural deflection as well as damage formation. Combination of 90 and 45-degree plies at the both ends, and selection of much higher ply properties have improve the overall performance of composite laminate structure to reduce deflection and resist the damage formation even it is being applied with much higher load than it needs to be for an intended load. In this study, scheme A has proved to be the ideal solution to the safe design of cross-arm structure which show smaller and no permanent deformation under three different load magnitude.

6. Acknowledgement

The authors acknowledge Tenaga Nasional Berhad (TNB), UNITEN R&D, and Institute of Energy Infrastructure (IEI) for the lab facilities and financial support (TNB Seeding Fund: U-TS-RD-19-03). Special thanks to those who contributed to this project directly or indirectly.

7. References

[1] Negnevitsky M, Tomin N, Panasetsky DA, Voropai N, Kurbatsky V, Hager U, et al., editors. *Proc-Power Systems Computation Conf, PSCC 2014*; 2015 02/10; Wroclaw, Poland: Institute of Electrical and Electronics Engineers (IEEE);
[2] Haes Alhelou H, Hamedani Golshan ME, Njenda T, Siano P. 2019 *Energies*.**12**.
[3] Greif R, Chapon E. 1993 *Journal of Reinforced Plastics and Composites*.**12**(5):602-21.
[4] Cadini F, Agliardi GL, Zio E. 2017 *Applied Energy*.**185**:267-79.
[5] Xue Y, Xiao S. 2013 *Journal of Modern Power Systems and Clean Energy*.**1**(2):91-100.
[6] Nyambeni N, editor. *18th Int Conf on Experimental Mechanics (ICEM18)*; 2018; Brussels, Belgium MDPI;
[7] Wu LC, Lo CY, Nakamura T, Kushner A. 1998 *International Journal of Solids and Structures*.**35**(12):1137-61.
[8] Tuo H, Lu Z, Ma X, Zhang C, Chen S. 2019 *Composites Part B: Engineering*.**167**:329-41.
[9] Hashin Z. 1980 *Journal of Applied Mechanics*.**47**(2):329-34.

[10] Wang J, Chen T, Sleight D, Tessler A, editors. *45th AIAA/ASME/ASCE/AHS/ASC Structures, Structural Dynamics & Materials Conf*; 2004; Palm Springs, California: Aerospace Research Central (ARC);
[11] Mohamad D, Syamsir A, Sa'don SN, Zahari NM, Seman SAHA, Razali MF, et al. 2019 *IOP Conf Series: Materials Science and Engineering*.**530**:012030.
[12] Riccio A, Di Felice G, Saputo S, Scaramuzzino F. 2014 *Procedia Engineering*.**88**:222-9.
[13] Mohamad D, Syamsir A, Beddu S, Kamal NLM, Zainoodin MM, Razali MF, et al. 2019 *IOP Conf Series: Materials Science and Engineering*.**530**:012029.
[14] Ozsoy N, Mimaroğlu A, Ozsoy M. 2015 *Acta Physica Polonica A*.**127**:1032-4.

Influence of grain size distribution towards improvements of turbidity, colour and suspended particles in a riverbank filtration process - a column study

M F Murshed[1, 2, 3], A Kamaruzaman[1], N A Ab Aziz[1] and N H Mokhtar Kamal[1,2,3 *]

[1]School of Civil Engineering, Universiti Sains Malaysia, Engineering Campus, Nibong Tebal, Pulau Pinang, Malaysia
[2]Solid Waste Management Cluster, Engineering Campus, Universiti Sains Malaysia, 14300 Nibong Tebal, Pulau Pinang, Malaysia
[3]Water Security Cluster, Engineering Campus, Universiti Sains Malaysia, 14300 Nibong Tebal, Pulau Pinang, Malaysia
[*] cehana@usm.my

Abstract. Riverbank filtration (RBF) system is an efficient and natural treatment technology which, consists of high capacity pumping wells, located adjacent to river. It takes advantage of natural attenuation processes that occur as surface water infiltrate through the riverbed to produce potable water. This research determined the influence of media size distribution, and its hydraulic conductivity and porosity as filter media, of sand and gravel. The efficiency of column model for riverbank filtration system to remove turbidity, colour and suspended solids was also determined. The best condition of column model to remove contaminants with different flowrates and different types filter media used was identified through this research. The column study was conducted with three different types of filter media; C1 (sand), C2 (gravel) and C3 (sand + gravel) and with three different flowrates; 20 mL/min, 25 mL/min and 40 mL/min. Based on the results, C3 (sand + gravel) shows the highest percentage removal for turbidity, colour and suspended solids, which were 28.5%, 59.1%, 71.4% and 29.4%, respectively, although at difference flowrates. The highest percentage of removal of suspended solids (71.4%) occurs at 25 mL/min. The highest percentage of removal of colour (59.1%) occurs at 40 mL/min. Longer contact time might improve the removal of these contaminants.

Keywords: Riverbank filtration, column study, grain size distribution, river water abstraction

1. Introduction

Riverbank filtration (RBF) system is an efficient and natural treatment technology, which consists of high capacity pumping wells that located adjacent to rivers [1, 2] During riverbank filtration, a reduction in the concentration of pollutants is achieved by physical, chemical and biological processes that take place, between the surface water and underground, and with the substrate [3, 4] found that the RBF system is mostly constructed at alluvial aquifers that are located along the riverbanks. There are a variety of aquifers used in RBF system ranging from sand, sand and gravel, large cobbles and boulders. The ideal condition for this system typically includes coarse-grained and low permeability zones (such as clay and silt layers). A riverbank filtration can also occur under natural conditions or be induced by lowering the groundwater table below the surface water level by abstraction from adjacent boreholes [5]. There are three types of RBF wells that can be designed, which are vertically (as the most common practice especially for the extraction of low water quantities), horizontally (for higher extraction rates) or pit well (for individual residential water supply) [3, 4]. A study in 2013 indicated that RBF system is the traditional drinking water treatment technology with low cost and sustainable alternatives that have the potential to supply water to 120 million people in the United States [6]. A study in 2012 [3] shows that the pollutants (particles, microorganisms, organic, and inorganic compounds) can be removed as the river water flow through the aquifer, which acts as a barrier against concentration peaks. The concept of RBF system is similar to some extent, to slow-sand filtration (SSF) [7]. A study in 2012 mentioned that the RBF systems have been used in Europe to supply drinking water to communities along the Rhine, Elbe, Danube, and Seine Rivers for more than 100 years. 75 % of the city of Berlin depends on RBF, as the main drinking water supply [3]. About 50% of potable water was supplied by RBF in Slovak republic, 45% in Hungary, 25% in Switzerland, 16% in Germany and 5% in Netherlands [8, 9]. Nowadays, there are many countries around the world that have started to use RBF system for water treatment including India, South Korea, China and Jordan [7]. Regarding the suitability of grain size for an RBF system, it was found that the quality of water produced was satisfied by filtering with sand sizes larger than 0.2mm up to 0.45mm [10]. However, in other studies, the grain size of sand used by in their column study was between the range of 0.125 mm until 0.250 mm [9, 11].

Malaysia is blessed with rainfall throughout the year and surface water has been the major option of drinking water source. However, water shortages still occur at certain time in the year during the drought season which might be due to uneven rainfall distribution and poor water management due to extensive development of certain areas. RBF can be an option to increase the water abstraction quantity from an existing drinking water treatment plant, however the suitability of its soil as filter media and its abstraction capability must be tested. RBF has been practiced since 2013 in Malaysia [12, 13] however this technology must be tested locally for the suitability of the existing soil for RBF application. In Kota Lama Kiri, Kuala Kangsar the sieve analysis for the soil test found that the soil consisted mostly of medium sand, with fine sand, coarse sand and gravel present. The hydraulic conductivity for Kota Lama Kiri soil sample showed to be in the similar range to other alluvial aquifers in Ohio river (Louisville) and the Great Miami river (Cincinnati) in the USA [13]. This indicate the potential application of Kota Lama Kiri for RBF application. Thus, in this study, the suitability of sand and gravel from a river in Kelantan was studied to determine its suitability and capability for a possible RBF application.

2. Methodology

2.1. Site location

In this column study, river sand from Kampung Semuba, Pasir Mas, Kelantan and gravel was used as a filter media. The water sample was taken from the site on Kerian River at coordinates 5°7'38.79" N, 100°35'44" E, Lubuk Buntar, Kedah, located near to a palm oil and pineapple plantation.

2.2. Particle size distribution (PSD)

Sieve analysis was conducted for particle size distribution for particle sizes larger than 63 μm in diameter. The opening sieve sizes used was according to British Standard (BS), and were set in the order of 14 mm, 10 mm, 6.30 mm, 5.00 mm, 3.35 mm, 2.00 mm, 1.18 mm, 0.600 mm, 0.425 mm, 0.300 mm, 0.212 mm, 0.150 mm, 0.063 mm and pan. The percentage of passing for each sieve was calculated and particle size curve was plotted. By examining particle size curve, the soil type can be determined by referring soil mass fraction percentages that have been developed by the U.S. Department of Agriculture (USDA). This experiment was conducted to classify the soil sample, either it is well or poorly graded soil.

2.3. Hydraulic conductivity

In order to determine the hydraulic conductivity of the filter media used, constant head permeability test was conducted. Permeability is a measure of the capacity of soil to allow the flow of water through the pore spaces between solid particles [14]. The degree of permeability is determined by applying a hydraulic pressure gradient in a sample of saturated soil and measuring rate of flow. The constant head test method is used when the soils are said to be permeable soils (k > 10^{-4} cm/s). The sample must be dried before placing in the permeameter cell and the test was run according to the standard procedure. The principle behind this test is Darcy's law for laminar flow. Thus, to calculate permeability, the Darcy's law was used. The general equation of Darcy's law is in Equation (1) is used to find the flow, Q. The calculation of coefficient of permeability, k value using constant head test was measured using Equation (2).

$$Q = k.A.i \tag{1}$$

$$k_T = \frac{VL}{Aht} \tag{2}$$

Where,

k = Darcy's coefficient of permeability
i = Hydraulic gradient = h/L
k_T = Coefficient of permeability at temperature T (cm/s)
V = Volume of collected water (cm^3)
L = Length of specimen (cm)
t = Time of discharge (s)
A = Cross-sectional area of permeameter (cm^2)
h = Hydraulic head difference, h_1-h_2 (cm)

2.4. Porosity

Two 1000 mL beakers were prepared (beaker 1 for soil sample and beaker 2 for water). Soil sample and water were added until 500 ml into each beaker. Water from beaker 2 was added into beaker 1 until the level is the same as the soil sample. The remaining volume of water was recorded. The total pore volume in the soil sample was calculated by subtracting the initial volume of water with the remaining volume of water. Lastly, the porosity of the soil sample was calculated by dividing the total pore volume by the total volume (Equation 3).

$$Porosity\ (\%) = \frac{Total\ pore\ volume}{Total\ volume} \times 100 \tag{3}$$

2.5. Column experiment

The column properties in this study are presented in Table 1.

Table 1. Column properties.

Parameter	Column 1 (sand)	Column 2 (gravel)	Column 3 (sand + gravel)
Diameter of the column (mm)	76	76	76
Length of the column (mm)	180	300	300
Depth of sand in the column (mm)	160	0	140
Depth of gravel in the column (mm)	0	210	70
Porosity (%)	32	41	32 & 41
Grading of soil	Poorly graded	Poorly graded	Poorly graded
Flowrate, mL/min	20, 25, 40	20, 25, 40	20, 25, 40

A schematic representation of the setup of the column experiment is as shown in Figure 1. The column was completed with two sampling ports (SP), which are SP 0 for the inlet and SP 1 for the outlet. The grain size of the river sand used in this experiment was 0.15 mm to 0.6 mm, while the grain size of the gravel used was between 3.35 mm to 6.3 mm. Three sets of column experiment were conducted; Column 1 (sand), Column 2 (gravel) and Column 3 (sand + gravel). These experiments were conducted to evaluate the performance of the three types of filter media to remove contaminants. First, media were washed with tap water and distilled water to remove the dirt off the media. Then, the washed media were dried in the oven for ±24 hours. Each set of columns were packed with the media. River water from Kerian River, Lubuk Buntar was used as feed water as this is the nearest source of river water to the lab. It was pumped from the bottom to the top of the column by using the peristaltic pump at 20, 25 and 40 mL/min flowrate. The experiment was conducted continuously for 3 hours and filtered water was taken at 15, 30, 60, 120 and 180 minutes and the filtered water were tested for the selected water quality parameters.

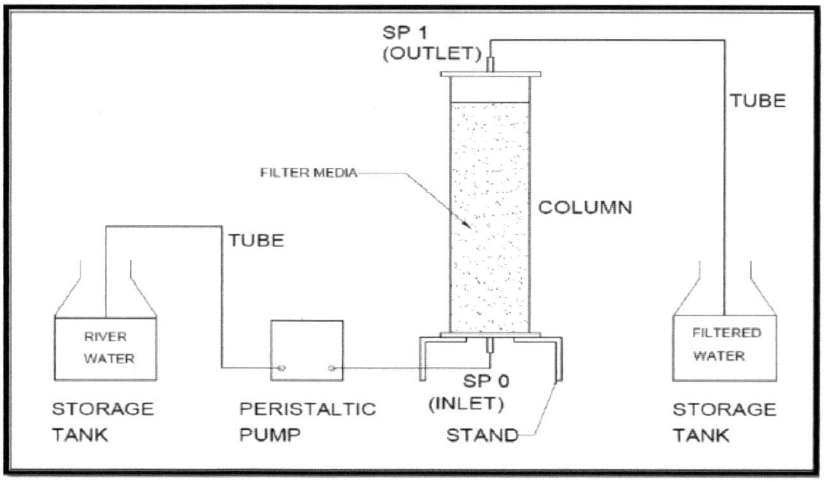

Figure 1. Schematic representation of the setup of the column experiment.

2.6. Analytical experiments

In the analytical experiment, the parameters of water quality measured were turbidity, colour and suspended solids. Turbidity was measured by using a HACH 2100N portable turbidimeter. For true colour (455nm), the filtered water was filtered through 0.45 μm cellulose nitrate filter by using a vacuum pressure pump an then was measured by using HACH DR 2800 Spectrophotometer. For suspended solids, HACH DR 2800 Spectrophotometer was used to measure the concentration of suspended solids in raw and treated river water.

3. Results and discussions

3.1. Particle size distribution (PSD)

Figure 2 shows the particle size distribution curve of soil sample from Kampung Semuba, Pasir Mas, Kelantan. D_{10} represents 10% of the particles finer and 90% of the particles coarser than that particular size of D_{10}. D_{30} represents 30% of particles are finer and 70% of the particles are coarser than that particular size of D_{30}. D_{60} represents 60% of particles are finer and 40% of the particles are coarser than that particular size of D_{60}.

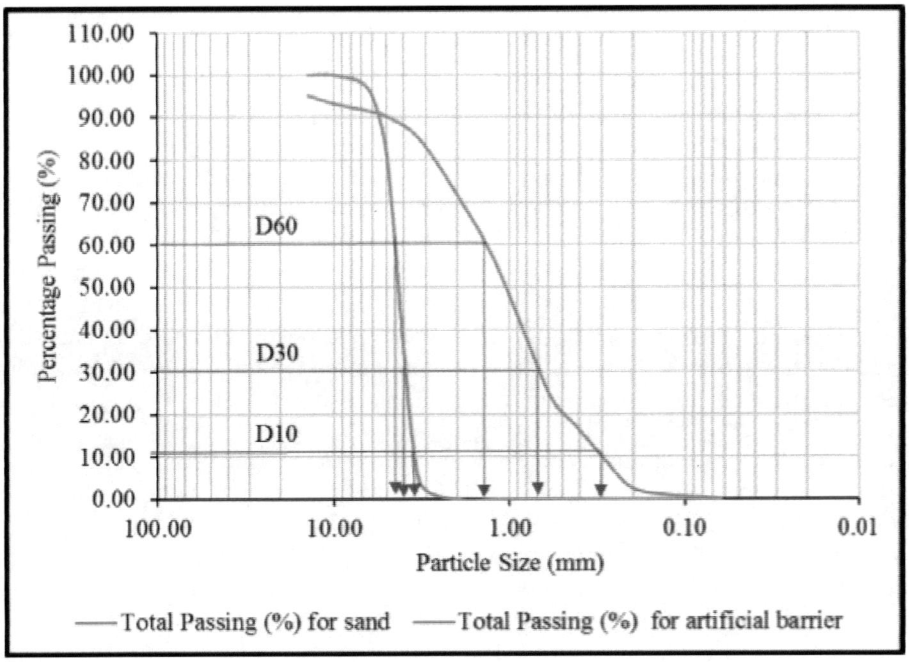

Figure 2. Graph of log sieve size versus percentage of passing for the soil sample.

The results of C_u for sand and gravel are 5.00 and 1.31 respectively. For a gravel to be classified as well graded, C_u and C_c should range between $C_u \geq 4$ and $1 \leq C_c \leq 3$. For a sand to be classified as well graded, C_u and C_c should range between $C_u \geq 6$ and $1 \leq C_c \leq 3$. In cases where both characteristics are not met, the soil is classified as poorly graded soil. Therefore, both sand and gravel media in this study were classified as poorly graded as sand has the value of C_u less than 6 and gravel has the value of C_u less than 3. The range of size of the particles that were obtained from the sieve analysis of the sand was 0.063 mm to 14.00 mm while the range of size of particles for the gravel was 3.35 mm to 6.3 mm.

3.2. Hydraulic conductivity

According to Das (2010), permeability or hydraulic conductivity, k, is a measure of the capacity of soil to allow the flow of water through the pore spaces between solid particles. The hydraulic conductivity, k value, for sand and gravel were 0.06 cm/s and 3.07 cm/s, respectively. The hydraulic conductivity of the gravel was greater than the hydraulic conductivity of the sand, with a difference of 3.01 cm/s. Hydraulic conductivity for sand at Pasir Mas showed to be similar to the sand in Kota Lama Kiri [13] that ranges between 0.10 and 0.91 cm/s. A large hydraulic conductivity value indicates high permeability and filtration [6]. Finer sediments exhibit low conductivity because water cannot filter as well through the pore spaces. The range of permeability indicates its high permeability and filtration. Study of filtration media is important to determine the suitability to establish a riverbank filtration system of a location [15] in their research also verify that this site (Pasir Mas, Kelantan) is suitable for riverbank filtration due to the presence of sand and gravels that come from granite and quartz [16].

3.3. Porosity

Porosity is normally calculated from measurements of bulk density. This varies with water content in soils that reflect the performance of filtration process. Based on the results obtained, it was found that the porosity for the river sand and gravel was 32% and 41%, respectively. It influences the velocity at which water moves through an aquifer under a given hydraulic gradient [17]. The porosity of the river sand is lower than gravel because of the size of the river sand was smaller compared to the gravel. As the size of the river sand was small, the void ratio obtained was also small. Thus, river sand which has low porosity compared with gravel will has less void within, to hold the flocs and solids removed from the water.

3.4. Column experiments

Column study was conducted regarding the effect of different filter media and flowrate used during column experiment. Flowrate used in in this experiment was set at 20, 25 and 40 mL/min for all three column conditions. Here, three types of filter media were used, sand (C1), gravel (C2) and combination of sand and gravel (C3). The type of filter media used affected the performance of water quality during column experiment due to porosity, hydraulic conductivity and particle size distribution of the filter media. The lower the porosity, the lower the void space between particles, which allows it to better filter the particles present in the water. However, this might also affect its outflow rate, due to less porosity. Flowrate is also one of the factors that affect the percentage of removal in the continuous flow of column experiment. This flowrate for this research has been set at 20, 25 and 40 mL/min. A study indicated that the lower the flowrate used, the more effective the diffusion process was and the higher the residence time is, thus resulted in high sorption capacity [18].

Table 2 shows the summary of results of the percentage of removal of contaminants for the column experiment. From this study, the percentage of removal was relatively low. Here, three types of filter media were used; C1 (sand), C2 (gravel) and C3 (sand + gravel). In general, C1 has better removal of the pollutants and C3 also shows a comparable result. Results show that at the lowest flowrate (20 mL/min), percentage removal of turbidity was the highest for C1, C2 and C3 (28.5%, 22.2% and 28.5%) followed by 25 mL/min and 40 mL/min. For colour removal, it was the highest when high flowrate (40 mL/min) was applied for C1, C2 and C3 (50.0%, 50.0% and 59.1%) followed by 20 mL/min and 25 mL/min. At medium flowrate (25 mL/min), percentage removal of suspended solids is the highest for C1, C2 and C3 (67.4%, 65.3% and 71.4%) followed by 20 mL/min and 40 mL/min.

Sand media has been frequently used due to their advantages and its efficiency to remove pollutants [19]. Gravel is commonly used to support the filter media. Gravel role is not just to avoid filter media from flowing out of the filter in filtration process but also to distribute clean water flowing into the filter in backwashing process. Based on the results, it is not suitable to only use gravel as a filter media to remove contaminants. For the removal of turbidity, the percentage removal is higher when the flowrate is 20 mL/min. For the removal of colour, the percentage removal is higher when the flowrate is 40 mL/min. For the percentage removal of suspended solids, the percentage removal is higher when the flowrate is 25 mL/min. Thus, different flowrates used in this column experiment does not significantly affect the efficiency to remove these selected water contaminants. This might also be affected with the short column run time, only 3 hours. A longer period of experiment might be needed to reach the stability and better performance of the filtration system. This is supported by the finding in Egypt [20], where it took 2 to 8 months to stabilise the RBF for hydrochemical parameters monitoring, with a continuous operation of the RBF. The time for column study to stabilised might be lesser than on-site study as the scale is smaller as well. However, it still requires continuous flow of water to 'ripen' the column condition. In a column study of sand filtration and schmutzdeckes layer function on E.coli removal by Unger (2006), between 4 hours and 2 weeks of 'ripening' time was applied to understand the effects of the layer on bacteria removal in water [21]. With a longer continuous run time, the development of schmutzdeckes layer will be faster. It was shown that over time, particles clogging, and the layer development slows the water seepage in the media and thus improved the discharged water quality. Overall, different types of filter media used affect the efficiency to remove water contaminants in which C1 (sand) and C3 (sand + gravel) show a better performance compared with C2 (gravel), but the flowrates did not affect the contaminants removal performance.

Table 2: Summary of percentage removal of contaminants for column experiment

	Percentage removal (%)								
Flowrate (mL/min)	20			25			40		
Filter media	C1	C2	C3	C1	C2	C3	C1	C2	C3
Turbidity (NTU)	28.5	22.2	28.5	20.2	19.9	23.7	17.2	16.6	22.2
Colour (TCU)	47.1	11.8	29.4	6.7	13.3	20.0	50.0	50.0	59.1
Suspended solids (mg/L)	50.0	42.0	48.0	67.4	65.3	71.4	35.6	30.1	35.6

4. Conclusions

Sand and gravel media used in this study were classified as poorly graded as sand has the value of C_u less than 6 and gravel has the value of C_u less than 3. The sand sample has a low hydraulic conductivity and porosity compared to gravel. Based in the column study, the percentage removal was relatively low, the maximum removal in all condition was only 71.4% of suspended solids in the gravel only column, that might be due to the duration of time to conduct this experiment for each different flowrate is only 3 hours. Basically, a longer duration of time is required in order to achieve higher percentage removal of contaminants. Types of filter media used is one of the factors that affects the percentage removal of water quality. Here, three types of filter media were used; C1 (sand), C2 (gravel) C3 (sand + gravel). From these three types of filter media, C1 and C3 have a better potential to remove contaminants compared with C2. For the removal of turbidity, the percentage removal is higher when the flowrate is 20 mL/min. For the removal of colour, the percentage removal is higher when the flowrate is 40 mL/min. For the percentage removal of suspended solids, the percentage removal is higher when the flowrate is 25 mL/min. Thus, different flowrates used in this column experiment does not significantly affect the efficiency to remove water contaminants.

Acknowledgements

The authors would like to acknowledge to all persons who, directly or indirectly, have lent their helps and supports in this research. This study was funded by a research grant under Universiti Sains Malaysia (RUI: 1001/PAWAM/814259).

References

[1] Massmann G, Nogeitzig A, Taute T and Pekdeger A 2008 Seasonal and Spatial Distribution of Redox Zones during Lake Bank Filtration in Berlin, Germany *Environmental Geology* **54** p 53-65

[2] Ulrich C, Hubbard S S, Florsheim J, Rosenberry D, Borglin S, Trotta M. and Seymour D 2015. Riverbed Clogging Associated with a California Riverbank Filtration System: An Assessment of Mechanisms and Monitoring Approaches *Journal of Hydrology* **529** p 1740-53

[3] Jaramillo M 2012 Riverbank Filtration: An Efficient and Economical Drinking-Water Treatment Technology *Dyna* **79**(171) p 148-157

[4] Hunt H 2002 American Experience in Installing Horizontal Collector Wells. *Riverbank Filtration* Springer Netherlands. p 29-34

[5] Hiscock K M and Grischek T 2002 Attenuation of Groundwater Pollution by Bank Filtration *Journal of Hydrology* **266** p 139-144

[6] Vozza G M 2013 Riverbed Filtration Clogging at Wohler on the Russian River, Sonoma County, California p 1-25.

[7] Shamrukh M and Abdel-Wahab A 2008 Riverbank Filtration for Sustainable Water Supply: Application to a Large-Scale Facility on the Nile River *Clean Technology Environment Policy* **10** p 351-358.

[8] Dash R R, Prakash E B, Kumar P, Mehrotra I, Sandhu C and Grischek T 2010 River Bank Filtration in Haridwar, India: Removal of Turbidity, Organics and Bacteria. *Hydrogeology Journal* **18** p 973-983
[9] Diem S, Von Rohr M R, Hering J G, Kohler H P E, Schirmer M and Von Gunten U 2013 NOM Degradation during River Infiltration: Effects of the Climate Variables Temperature and Discharge *Water Research* **47** p 6585-95
[10] Muhammad N, Ellis K, Parr J and Smith M D 1996 Optimization of Slow Sand Filtration. WEDC, Loughborough University, Leicestershire Le 11 3 TU(UK) p 283-285
[11] Von Rohr M R, Hering J G, Kohler H P E and Von Gunten U 2014 Column Studies to assess the Effects of Climate Variables on Redox Processes during riverbank filtration *Water Research* **61** p 263-275

[12] Rashid, N. A., Abustan, I., and Adlan, M. N. 2016. River Bank Filtration Artificial Barrier to Remove Escherichia Coli. *International Journal of Scientific Research in Knowledge,(IJSRK)*, 4**,** pp. 45-54.
[13] Adlan M N, Mohd Ariff Zainol M R R, Ghazali M F, Selamat M R and Othman S Z 2016. A study on the soil characteristic and properties of riverbank soil samples from Sungai Perak, Kota Lama Kiri, Kuala Kangsar, Malaysia *IOP Conference Series: Materials Science and Engineering*, ***133(1)*** p 012003
[14] Das B M 2010 Principles of Geotechnical Engineering. 7th Edition. p 1-666.
[15] Sharma B, Uniyal D P, Dobhal R, Kimothi P C and Grischek T 2014. A Sustainable Solution for Safe Drinking Water through Bank Filtration Technology in Uttarakhand, India. *Current Science* **107** p 1118-24
[16] Adlan M N, Nawawi M, Murshed M F, Palaniandy P, Abustan I, Abdul Aziz H and Mohd Ariff Zainol, M R R 2015 Preparedness on Water Abstraction and Treatment on Flood Prone Areas in Kelantan River Basin *Jurnal Teknologi* p 1-8.
[17] Caldwell T G 2006 Presentation of Data for Factors Significant to Yield From Several Riverbank Filtration Systems in the U.S. and Europe *Springer Netherlands*. p 299-344.
[18] Bharathi K S and Ramesh S P T 2013 Fixed-Bed Column Studies on Biosorption of Crystal Violet from Aqueous Solution by Citrullus Lanatus Rind and Cyperus Rotundus *Applied Water Science* **3** p 673-687.
[19] Segismundo E Q, Kim L H, Jeong S M and Lee B S 2017 A Laboratory Study on the Filtration and Clogging of the Sand-Bottom Ash Mixture for Stormwater Infiltration Filter Media. *Water* **9** p 32.
[20] Wahaab R A, Salah A, Grischek T 2019 Water Quality Changes during the Initial Operating Phase of Riverbank Filtration Sites in Upper Egypt. *Water* **11(6)** 1258.
[21] Unger, M C 2006 The Role of the Schmutzdecke in Escherichia Coli Removal in Slow Sand and Riverbank Filtration. Master's Theses and Capstones. 251. https://scholars.unh.edu/thesis/251

Impact of high particulate event on the indoor and outdoor fine particulate matter concentrations during the Southwest monsoon season

A Suroto, S Shith, N F F Md Yusof* and N A Ramli

School of Civil Engineering, Engineering Campus, Universiti Sains Malaysia, 14300 Nibong Tebal, Pulau Pinang, Malaysia.

*Corresponding author: noorfaizah@usm.my

Abstract. Haze event, also known as the high particulate event (HPE) normally occurs during the Southwest monsoon season due to the high concentrations of atmospheric particulate matters including $PM_{2.5}$ which are majorly caused by anthropogenic activities such as open forest and biomass burning. The impact of haze, not only on ambient air but also indoor environment, elevates with the emission of particulate matters from local transportation and industrial sources. . Polluted indoor environment could impose adverse effects towards the occupants, particularly those categorized in the sensitive group. School children are among the sensitive group whom spend the majority of their time indoors, especially in school classroom and home. The present study aims to determine the indoor/outdoor ratios of $PM_{2.5}$ during non-HPE and HPE. The concentrations of indoor and outdoor $PM_{2.5}$ were measured continuously using a direct optical monitor (E-Sampler) and portable environmental beta-attenuation monitor (E-BAM) respectively for 24 hours. The findings revealed that the I/O ratio during HPE (0.35) was higher than the non-HPE (0.26). The Pearson correlation analysis exhibited a significant relationship ($p < 0.01$) between the indoor and outdoor $PM_{2.5}$ concentrations with a strong correlation ($r = 0.75$). Therefore, it is evident that the outdoor concentrations of $PM_{2.5}$ during HPE inflicted a significant impact on the indoor environment of naturally ventilated classroom due to the high I/O ratio in comparison to non-HPE, in addition to the less practical design of the classroom (open system).

Keywords: *children, health effects, indoor/outdoor ratio,* $PM_{2.5}$*, school*

1. Introduction

Haze episodes have been recorded to affect several countries in the Southeast Asia (Singapore, Malaysia, Indonesia, Brunei Darussalam, and Thailand) since 1982 [1]. [2] stated that the major contributor to this problem is the transboundary pollution due to uncontrolled biomass burning. Haze events mostly take place during the dry season, which is also known as the Southwest monsoon (from June to September) [3], and the haze impacts aggravate with the low recorded rainfall during this season [4]. Recently, haze has become an annual event in the Southeast Asia region especially in Indonesia, Malaysia, and Singapore [5, 6, 7, 8]. This phenomenon occurs upon the gathering of high-level air pollutants originated from various anthropogenic activities such as peatland and wood burning, transportation, and industrialization [8, 9, 10].

Particulate matter (PM) has become the main focused parameter in a haze evaluation, where the Malaysia Air Pollution Index (API) is invariably dependent on the $PM_{2.5}$ (particles with aerodynamic diameters of less than 2.5 μm) and PM_{10} (particles with aerodynamic diameters of less than 10 μm) concentrations [11, 12, 13]. A previous study exercised the term 'high particulate event' (HPE) to define this condition [13]. HPE has become a major concern due to its adverse impacts towards the human health, including complications in children respiratory system (i.e. upper respiratory zone infection and asthma) and conjunctivitis [5, 14]. In this matter, teaching and learning sessions are also affected whereby schools are encouraged to close due to the very unhealthy and hazardous air quality during HPE based on the API record [11]. In addition, economic sectors are also influenced by HPE due to the output cutback particularly in manufacturing, construction, and tourism earnings [8].

Outdoor air pollution in school environment should be taken as a serious matter because of its detrimental effects on school children who are still vulnerable and less resilient towards polluted air. [15, 16] reported that relationship between the indoor and outdoor $PM_{2.5}$ concentrations in a school environment significantly existed. Infiltration process is one the transportation mechanism that could transport the outdoor $PM_{2.5}$ into indoor environment, thus contributing to its suspension especially in naturally ventilated classrooms wherein windows and doors are left open throughout the school session in order to encourage ventilation. However, during HPE, such ventilation system is inappropriate since particulate matter, typically the $PM_{2.5}$ could be transported to a long distance and suspended for an extended period in the atmosphere. The Ministry of Education (MOE) Malaysia has enforced the closure of schools when the API reaches very unhealthy (101-200) and hazardous (above 300) levels in order to avoid risks to school children and teachers [17, 18].

Most school classrooms in Malaysia practice a mix-mode ventilation system (natural ventilation with mechanical aids i.e. fans), that is also exposed to outdoor $PM_{2.5}$ infiltration. Consequently, school children may suffer from visibility problem, difficulty in breathing, and other respiratory problems during HPE. [19, 20, 21] found that ventilation system gives significant impacts on the infiltration of outdoor air pollutants including $PM_{2.5}$ into indoor. Thus, the aim of present study is to find out the ratios of indoor/outdoor $PM_{2.5}$ concentrations during non-HPE and HPE.

2. Materials and Methods

2.1. Description on sampling site

The monitoring of indoor and outdoor concentrations of $PM_{2.5}$ was performed at a school located in Johan Setia, Klang at coordinate 2°58'42.54"N, 101°29'24.25"E. The study area is surrounded by agricultural area known as Green Revolution with an area of 4000 acre [22], and several residential areas. The classrooms are ventilated with a mix-mode system (i.e. natural ventilation with a mechanical aid; fans) where each classroom is set-up with louver windows and two doors. The selected classroom for $PM_{2.5}$ monitoring which is located in the second floor was occupied by Standard 5 primary school children in the morning and Standard 3 children in the evening school session. The distance between the agricultural area and the school was in the range of 0.5 – 5 km radius (red circle). Figure 1 illustrates the location of the sampling site in Johan Setia, Klang.

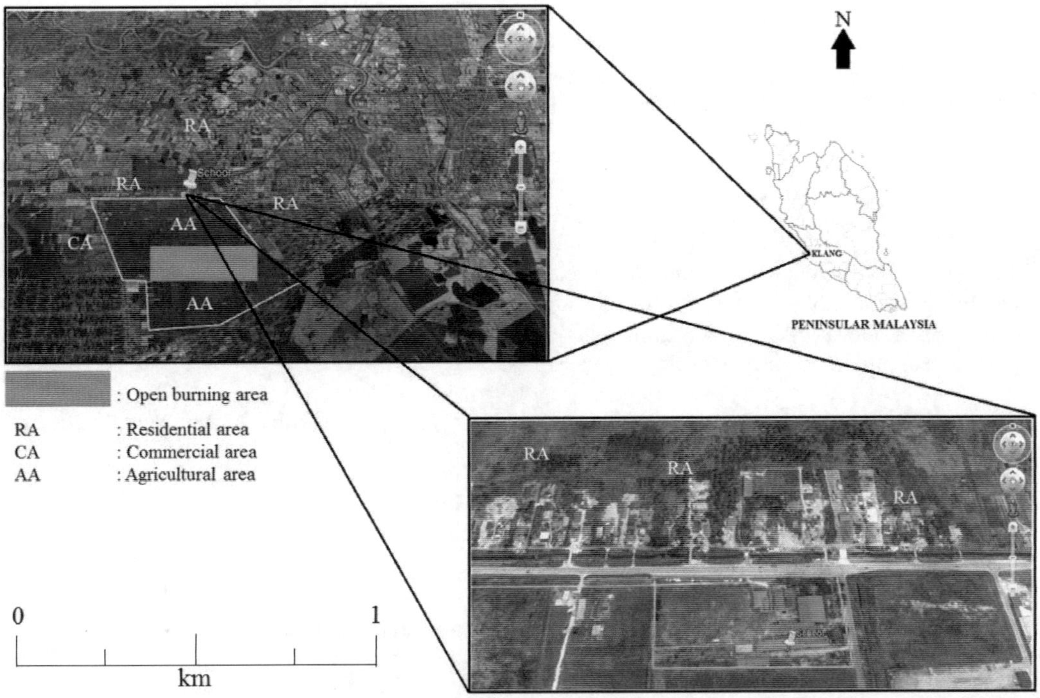

Figure 1. Sampling site in Johan Setia, Klang

2.2. Data monitoring for indoor and outdoor $PM_{2.5}$

Measurement of indoor and outdoor $PM_{2.5}$ concentrations was conducted continuously in non-HPE and HPE for 24 h with 1 min interval during the Southwest monsoon season (July and August 2019). The $PM_{2.5}$ measurement was carried out for durations of 2 days (48 hours) and 6 days (144 hours) for non-HPE and HPE respectively. A direct optical monitor (E-sampler) was allocated in the assigned classroom (indoor). Meanwhile, for outdoor evaluation, samples of $PM_{2.5}$ were collected and measured using an environmental beta attenuation monitor (E-BAM).

2.3. Variations of $PM_{2.5}$ concentration, meteorological parameter, and I/O ratio analysis

Descriptive statistic was performed to determine the minimum, maximum, mean, and standard deviation values for indoor and outdoor concentrations of $PM_{2.5}$ during non-HPE and HPE. In addition, the I/O ratio was implemented to identify the fraction of outdoor $PM_{2.5}$ that penetrates into the indoor classroom area [23, 24, 25, 26]. This analysis is also supported by [27] whom utilized the I/O ratios in the determination of relationship between indoor and outdoor particles.

2.4. Pearson correlation analysis for indoor and outdoor $PM_{2.5}$ relationship

The linear relationship between the two variables was analyzed using the Pearson correlation analysis. The Pearson correlation coefficient (r) could be classified into three types, namely positive correlation, negative correlation, and no correlation. A positive correlation indicates a directly proportional relationship between two variables whereas a negative correlation represents an indirectly proportional relationship wherein if the factor (x) increases; the other variable (y) will decrease. No correlation on the other hand demonstrates a condition of no influence between the variables (x) and (y). Correlation coefficient (r) is used to determine the strength of correlation between variables in the range between -1 and +1 as shown in Table 1 [28, 29]. The Guildford's Rule of Thumb that established by Guildford [29] is referred in order to determine the strength of linear relationship [28].

$$r = \frac{\sum\left(X-\overline{X}\right)\left(Y-\overline{Y}\right)}{\sqrt{\left(X-\overline{X}\right)^2\left(Y-\overline{Y}\right)^2}} \quad (1)$$

Where r = coefficient for Pearson correlation, X = outdoor $PM_{2.5}$ concentration, outdoor meteorological variables; Y = indoor $PM_{2.5}$ concentration, indoor meteorological variables; $\overline{X}$ = average for outdoor $PM_{2.5}$ concentration, outdoor meteorological variables; and $\overline{Y}$ = average for indoor $PM_{2.5}$ concentration, indoor meteorological variables.

Table 1. Range of correlation coefficient by Guildford's Rule of Thumb (Dominick et al., 2012; Guilford, 1973)

Size of Correlation	Interpretation
0.90 to 1.00 (-0.90 to –1.00)	Very strong positive (negative) correlation
0.70 to 0.90 (-0.70 to -0.90)	Strong positive (negative) correlation
0.50 to 0.70 (-0.50 to -0.70)	Moderate positive (negative) correlation
0.30 to 0.50 (-0.30 to -0.50)	Low positive (negative) correlation
0.00 to 0.30 (0.00 to -0.30)	Little if any correlation

3. Results and Discussion

3.1. Variations of indoor and outdoor $PM_{2.5}$ concentrations and meteorological parameters

Table 2 shows the variations of 24 h-averaged indoor and outdoor $PM_{2.5}$ concentrations and meteorological parameters recorded during the non-HPE and HPE. As expected, the mean indoor and outdoor $PM_{2.5}$ concentrations ± standard deviation during HPE (of 60.40 ± 23.83 μgm^{-3} and 21.10 ± 10.01 μgm^{-3} respectively) were higher than those during the non-HPE.

The outdoor mean $PM_{2.5}$ concentration in HPE exceeded the limit set by the United State of Environmental Protection Agency (USEPA) and Malaysian Ambient Air Quality Standard (MAAQS) of 35 μgm^{-3} and 50 μgm^{-3} respectively [11, 30]. The indoor $PM_{2.5}$ concentration is therefore influenced by indoor activities as well as the infiltration of outdoor air [31]. In comparison to other studies conducted during the Southwest monsoon season [32, 33, 34, 35], the outdoor $PM_{2.5}$ concentrations obtained in this study were much higher (28.5 – 51.2%). The indoor temperatures (Temp.) were found higher during HPE (30.60°C) and non-HPE (30.20°C), compared to outdoor temperatures of 27.85°C and 29.06°C respectively. This condition might be due to the low humidity and rainfall events during the Southwest monsoon season [36]. Meanwhile, during HPE, the relative humidity (RH) was revealed higher compared to the non-HPE, of 76.60% and 74.32% for indoor and outdoor humidity accordingly.

Table 2. Variations of 24 h-averaged indoor and outdoor $PM_{2.5}$ concentrations and meteorological parameters recorded during non-HPE and HPE

	Parameter	Mean	Standard Deviation	Min.	Median	Max.
Non-HPE	**Indoor**					
	$PM_{2.5}$ Concentration (μgm^{-3})	9.70	4.68	2.63	9.75	19.00
	Temperature (°C)	30.60	0.89	28.80	30.65	31.90
	Relative Humidity (%)	72.20	6.70	58.75	75.30	81.35
	Wind Speed (ms^{-1})	0.68	0.43	0.30	0.43	1.51
	Outdoor					
	$PM_{2.5}$ Concentration (μgm^{-3})	37.50	13.90	16.50	34.00	59.5
	Temperature (°C)	29.06	2.83	25.70	28.05	34.10
	Relative Humidity (%)	67.30	12.07	44.80	72.68	81.50
	Wind Speed (ms^{-1})	0.75	0.39	0.34	0.61	1.50
HPE	**Indoor**					
	$PM_{2.5}$ Concentration (μgm^{-3})	21.10	10.01	11.10	19.14	46.30
	Temperature (°C)	30.20	1.04	28.75	30.35	31.85
	Relative Humidity (%)	76.60	6.68	64.40	78.10	84.75
	Wind Speed (ms^{-1})	0.69	0.41	0.31	0.42	1.43
	Outdoor					
	$PM_{2.5}$ Concentration (μgm^{-3})	60.40	23.83	27.82	53.50	105.20
	Temperature (°C)	27.85	2.73	24.80	26.90	32.80
	Relative Humidity (%)	74.32	10.86	54.40	78.10	85.63
	Wind Speed (ms^{-1})	0.87	0.54	0.32	0.56	1.76

***Note: Min.: Minimum; Max.: Maximum**

3.2. Comparisons between HPE and non-HPE variations for indoor and outdoor $PM_{2.5}$ concentrations

Figure 2 presents the diurnal 24 h-averaged indoor and outdoor $PM_{2.5}$ concentrations recorded during HPE and non-HPE at the study area. The trend for both non-HPE and HPE was similar but different in magnitude. $PM_{2.5}$ in HPE exhibited a higher level of concentration compared to the non-HPE for both indoor and outdoor $PM_{2.5}$. During HPE, the maximum indoor and outdoor $PM_{2.5}$ concentrations were recorded in the early morning as 46.30 μgm^{-3} (12 a.m.) and 105.20 μgm^{-3} (1 a.m.), respectively. Even though it clearly occurred during HPE, there was an expected increase in biomass burning activities from the existing local agricultural area that induced the $PM_{2.5}$ generation. In fact, biomass burning is one of the main anthropogenic sources for $PM_{2.5}$ origination as reported by [8]. This scenario is also linked to the haze episodes occurring in the Southeast Asia (SEA) region particularly during the Southwest monsoon season (between June and September). [37, 38, 39] reported that during HPE, high densities of biomass fire and wildfire hotspots were identified at several sampling locations in Klang Valley.

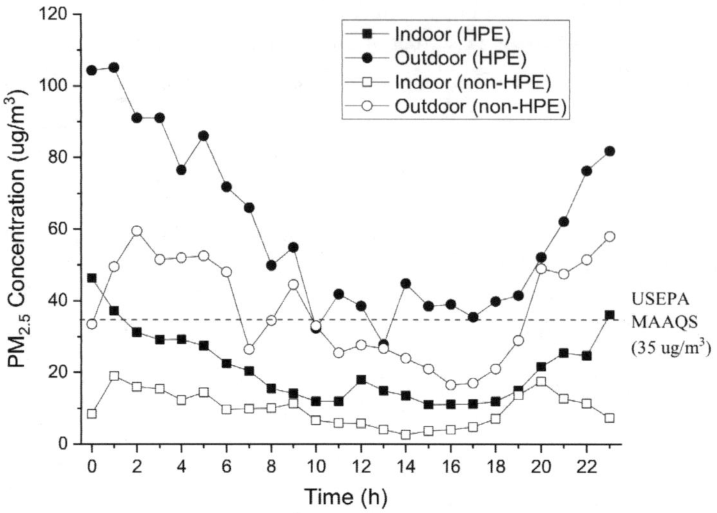

Figure 2. Diurnal variations of $PM_{2.5}$ concentrations during non-HPE and HPE

3.3. Ratio of indoor to outdoor $PM_{2.5}$ concentrations

Table 3 displays the mean I/O ratio of $PM_{2.5}$ concentrations during non-HPE and HPE. The mean I/O ratio for $PM_{2.5}$ during HPE (0.35) was higher compared to the non-HPE (0.26), where both ratios were less than 1.00. This indicates that the outdoor concentrations were very high due to the HPE in addition to the existing biomass burning activities near the school area. [31] found that I/O ratio increases with increasing quantity of people indoor, intensity of activities, and outdoor $PM_{2.5}$ concentration levels. During the non-HPE, the open systems such as windows and doors were opened for ventilation purpose, but this condition did not contribute in the dilution of indoor $PM_{2.5}$ concentration.

Table 3. Mean indoor to outdoor $PM_{2.5}$ concentrations (μgm^{-3}) (I/O) ratio

Measurement Duration	Indoor (mean ± s.d.)	Outdoor (mean ± s.d.)	I/O ratio
Non-HPE (24 hr)	9.70	37.50	0.26
HPE (24 hr)	21.10	60.40	0.35

*Note: HPE: high particulate event; s.d.: standard deviation; I/O: indoor/outdoor ratio

3.4. Pearson correlation of indoor and outdoor $PM_{2.5}$ concentrations and meteorological parameters

Table 4 exhibits the relationships among the indoor and outdoor $PM_{2.5}$ concentrations and other meteorological parameters, where all correlation coefficients (r) were significant ($p < 0.01$). Strong correlations were observed between the $PM_{2.5}$with indoor variables (RH and wind speed (WS)) and outdoor variables ($PM_{2.5}$, RH, Temp., and WS), with r values in the range of 0.70 - 0.79. Positive correlations were recorded between the indoor $PM_{2.5}$ and indoor RH, and also between indoor $PM_{2.5}$ with outdoor $PM_{2.5}$ and RH. Meanwhile, negative correlations were found between $PM_{2.5}$, Temp., and RH, for both indoor and outdoor environments. [40] in their study revealed the rationale behind the negative correlation is due to the mixing height factor, where the dispersion rate of particles towards the upper part of the atmosphere during high temperatures reduces the amount of PM particles.

Table 4. Correlations among indoor and outdoor $PM_{2.5}$ and meteorological parameters

Location	Parameter	Indoor				Outdoor			
		$PM_{2.5}$	RH	Temp.	WS	$PM_{2.5}$	RH	Temp	WS
Indoor	$PM_{2.5}$	1							
	RH	0.70	1						
	Temp.	-0.45	-0.87	1					
	WS	-0.74	-0.93	0.72	1				
Outdoor	$PM_{2.5}$	0.75	0.74	-0.47	-0.83	1			
	RH	0.79	0.94	-0.73	-0.93	0.77	1		
	Temp.	-0.77	-0.93	0.77	0.95	-0.74	-0.99	1	
	WS	-0.76	-0.81	0.58	0.89	-0.69	-0.93	0.93	1

*$PM_{2.5}$: Fine particles; RH: relative humidity; Temp.: temperature; WS: wind speed. Correlation is significant at 0.01 confident level (2-tailed).

4. Conclusion

As a conclusion, the average percentages of indoor and outdoor $PM_{2.5}$ concentration were higher during HPE compared to non-HPE by 54% and 37.9% respectively. In addition, the average outdoor $PM_{2.5}$ concentration after 24 h in HPE exceeded the permissible limit as recommended by the United State of Environmental Protection Agency (USEPA) (35 μgm^{-3}) and Malaysian Ambient Air Quality Standard (MAAQS) (50 μgm^{-3}). The trend for both non-HPE and HPE was similar but different in magnitude. The $PM_{2.5}$ concentration during HPE was higher compared to the non-HPE for both indoor and outdoor environments. Beside the evident HPE effect, an expected increase in biomass burning activities from the existing agricultural area in the vicinity of the selected study area was also considered. The mean I/O ratio for $PM_{2.5}$ concentration during HPE (0.35) was higher compared to the non-HPE (0.26), demonstrating the dispersion of $PM_{2.5}$ from outdoor sources towards the indoor environment. This condition was supported by the strong correlation between indoor and outdoor $PM_{2.5}$ concentrations thus indicating a significant influence from outdoor sources due to HPE and also the existing biomass burning activities near the school area. The findings from this study revealed the infiltration of outdoor $PM_{2.5}$ into indoor environment specifically in naturally ventilated classroom located even at a high floor level. The exceeding concentrations of $PM_{2.5}$ especially in outdoor environment is significantly alarming and immediate enforcement of open burning activities prohibition should be taken by the authorized bodies. Furthermore, the redesigning of classrooms especially the open system (windows and doors) should be considered by architects and engineers in order to provide a conducive study environment particularly for school children.

Acknowledgements

The authors would like to acknowledge the Ministry of Higher Education Malaysia and Universiti Sains Malaysia (USM) for the research funding (203/PAWAM/6071360 and 1001/PAWAM/8014106) provided throughout the completion and publication of this research.

References

[1] Tangang, F., Latif, M.T., Juneng, L. (2010). The roles of climate variability and climate change on smoke haze occurrences in Southeast Asia Region. Climate change: Is Southeast Asia up to the challenge, 36-49.

[2] Radojevic, M. and Hassan, H., (1999). Air quality in Brunei Darussalam during the 1998 haze episode. *Atmospheric Environment*, **33**(22), 3651-3658.

[3] Yusof, N. F. F. M., Ramli, N. A., Yahaya, A. S., Sansuddin, N., Ghazali, N. A., & al Madhoun, W. (2010). Monsoonal differences and probability distribution of PM_{10} concentration. *Environmental monitoring and assessment*, **163**(1-4), 655-667.
[4] Wooster, M.J., Perry, G.L.W. and Zoumas, A. (2012). Fire, drought and El Niño relationships on Borneo (Southeast Asia) in the pre-MODIS era (1980–2000). *Biogeosciences,* **9**(1), 317-340
[5] Awang, M. (2000).Do we have enough clean air to breathe: Syarahan Inaugural. Universiti Putra Malaysia
[6] Quah, E., Johnson, D. (2001). Forest fire and environmental haze in Southeast Asia: using the stakeholders approach to assign costs and responsibilities. *Environmental Management* **63**, 181 – 191
[7] Amil, N., Latif, M.T., Khan, M.F., Mohammad, M. (2016). Seasonal variability of $PM_{2.5}$ composition and sources in Klang Valley urban-industrial environment. *Atmospheric Chemistry and Physics* **16**, 5357 – 5381
[8] Latif, M.T., Othman, M., Idris, N., Liew, J., Abdullah, A.M., Hamzah, W.P., Khan, M.F., Nik Sulaiman, N.M., Jewaratnam, J., Aghomohammadi, N., Sahani, M., Chung, J.X., Ahamad, F., Amil, M., Darus, M., Varkkey, H., Tangang, F., Jaafar, A.B. (2018). Impact of regional haze towards air quality in Malaysia: A review. *Atmospheric Environment* **177**, 28 – 44
[9] Kunii, O., Kanagawa, S., Yajima, I., Hisamatsu, Y., Yamamura, S., Amagai, T., Ismail, I.T.S., (2002). The 1997 haze disaster in Indonesia: its air quality and health effects. *Archive of Environmental Health* **57**, 16 – 22
[10] Liu, Q., Ma, T., Olson, M.R., Liu, Y., Zhang, T., Wu, Y., Schauer, J.J. (2016). Temporal variations of black carbon during haze and non- haze days in Beijing. *Scientific Report* **6**, 1 – 10
[11] DOE (2015). Department of environment, Malaysia. New Malaysian Ambient Air Quality Standard. Department of Environment, Technology and the Environment, Malaysia. Available online at http://www.doe.gov.my/portalv1/wp-content/uploads/2013/01/Air-Quality-Standard-BI.pdf. Accessed on 16th September 2019.
[12] Afroz, R., Hassan, M.N. and Ibrahim, N.A., (2003). Review of air pollution and health impacts in Malaysia. *Environmental research*, **92**(2), 71-77.
[13] Awang, N.R., Ramli, N.A., Shith, S., Zainordin, N.S. and Manogaran, H., 2018. Transformational characteristics of ground-level ozone during high particulate events in urban area of Malaysia. Air Quality, *Atmosphere & Health*, **11**(6), 715-727
[14] Emmanuel, S.C. (2000). Impact to lung health of haze from forest fires: the Singapore experience. *Respirology* **5**, 175 – 182
[15] Mohammadyan,, M., Alizadeh-Larimi, A., Etemadinejad, S., Latif, M.T., Heibati, B., Yetilmezsoy, K., Abdul-Wahab, S., Dadvand, P. (2017). Particulate air pollution at schools: Indoor – outdoor relationship and determinants of indoor concentrations. *Aerosol and Air Quality Research* **17**, 857 – 864
[16] Wichmann, J., Lind, T., Nilsson, M.M., Haines, A. (2010). $PM_{2.5}$, soot and NO2 indoor – outdoor relationships at homes, pre –schools and schools in Stockholm, Sweden. *Atmospheric Environment* **44**, 4536 – 4544
[17] MOE (2019). Ministry of Educational, Malaysia. KM_Laporan _Penutupan_Sekolah_Akibat_Jerebu_di_Selangor_2_ptg_22_Sep_2019.pdf 22nd September 2019 Accessed on 10th October 2019 https://www.moe.gov.my/images/KPM/UKK/2019/09_Sep/KM_ _Laporan_Penutupan_Sekolah_Akibat_Jerebu_di_Selangor_2_ptg_22_Sep_2019.pdf
[18] hmetro (2019). "IPU lebih 200, tutup sekolah" 15th September 2019. Accessed on 10th October 2019 https://www.hmetro.com.my/mutakhir/2019/09/496947/ipu-lebih-200-tutup-sekolah

[19] Stabile, L., Dell'Isola, M., Russi, A., Massimo, A., Buannano, G. (2017). The effect of natural ventilation strategy on indoor quality in schools. *Science of Total Environment* **592**, 894 – 902

[20] Ficco, G., Dell'Isola, M., Vigo, P., Celenza, L. (2015). Uncertainty analysis of energy measurements in natural gas transmission networks. *Flow Measurements Instruments* **42**, 58 – 68

[21] Massimo, A., Del'Isola, M., Frattilo, A., Ficco, G. (2014). Development of a geographical information system (GIS) for the integration of solar energy in the energy planning of a wide area. *Sustainability* **6**, 5730

[22] Selangor Kini (2019). "38 Pembakaran terbuka di Johan Setia sumbang IPU tidak sihat". 9thSeptember 2019 Accessed on 8th October 2019 https://selangorkini.my/2019/09/38-pembakaran-terbuka-di-johan-setia-sumbang-ipu-tidak-sihat/

[23] Li, C.S., Lin, C.H. (2003). Carbon profile of residential indoor PM_1 and $PM_{2.5}$ in the subtropical region. *Atmospheric Environment* **37**, 881 – 888

[24] Huang, H., Lee, S.C., Cao, J.J., Zou, C.W., Chen, X.G., Fan, S.J. (2007). Characteristics of indoor/outdoor $PM_{2.5}$ and elemental components in generic urban, roadside and industrial plant area of Guangzhou City, China. *Journal of Environmental Science* **19**, 35 – 43

[25] Othman, M., Latif, M.T., Mohamed, A.F. (2016). The PM_{10} compositions, sources, and health risks assessment in mechanically ventilated office buildings in in an urban environment. *Air Quality Atmospheric Health* **9**, 597 – 612

[26] Othman, M., Latif, M.T., Matsumi, Y. (2019). The exposure of children to $PM_{2.5}$ and dust in indoor and outdoor school classrooms in Kuala Lumpur City Centre. *Ecotoxicology and Environmental Safety* **170**, 739 – 749

[27] Chen, C. and Zhao, B. (2011). Review of relationship between indoor and outdoor particles: I/O ratio, infiltration factor and penetration factor. *Atmospheric Environment* **45**, 275 – 288

[28] Dominick, D., Latif, M.T., Juahir, H., Aris, A.Z., Zain, S.M. (2012). An assessment of influence of meteorological factors on PM_{10} and NO_2 at selected stations in Malaysia. *Sustainable Environmental Research* **22** (5): 305 - 312

[29] Guildford, J.P. Fundamental Statistics in Psychology and Education (5th ed.). (1973) New York: McGraw-Hill.

[30] USEPA (2015) U.S. Environmental Protection Agency National Ambient Air Quality Standards (NAAQS). Available online at https://www.epa.gov/criteria-air-pollutants/naaqs-table. Accessed on 1st October 2019.

[31] Diapouli, E., Chaloulakou, A. and Spyrellis, N., (2007). Indoor and outdoor particulate matter concentrations at schools in the Athens area. *Indoor and Built Environment*, **16**(1), 55-61

[32] Khan, M.F., Sulong, N.A., Latif, M.T., Nadzir, M.S.M., Amil, N., Hussain, D.F.M., Lee, V., Hosaini, P.N., Shaharom, S., Yusoff, N.A.Y.M. and Hoque, H.M.S., (2016). Comprehensive assessment of $PM_{2.5}$ physicochemical properties during the Southeast Asia dry season (southwest monsoon). *Journal of Geophysical Research: Atmospheres*, **121**(24), 14-589

[33] Khan, M.F., Latif, M.T., Saw, W.H., Amil, N., Nadzir, M.S.M., Tahir, N.M., Chung, J.X. (2016a). Fine particulate matter in the tropical environment: monsoonal effects, source apportionment, and health risk assessment. *Atmospheric Environment* **45**, 597 – 617

[34] Tahir, N. M., Koh, M., Suratman, S. (2013). $PM_{2.5}$ and associated ionic species in a sub-urban coastal area of Kuala Terengganu, Southern South China Sea (Malaysia). *Sains Malaysiana,* **42**(8), 1065-1072.

[35] Rahman, S.A., Hamzah, M.S., Wood, A.K., Elias, M.S., Salim, N.A.A. and Sanuri, E., (2011). Sources apportionment of fine and coarse aerosol in Klang Valley, Kuala Lumpur using positive matrix factorization. *Atmospheric Pollution Research*, **2**(2), 197-206.

[36] Juneng, L., Latif, M.T. and Tangang, F., (2011). Factors influencing the variations of PM_{10} aerosol dust in Klang Valley, Malaysia during the summer. *Atmospheric Environment*, **45**(26), 4370-4378

[37] Dotse, S.Q., Dagar, L., Petra, M.I., De Silva, L.C. (2016). Influence of Southeast Asian haze episode on high PM_{10} concentration across Brunei Darussalam. *Environmental Pollution* **219**, 337 – 352

[38] Khan, M.F., Latif, M.T., Lim, C.H., Amil, N., Jaafar, S.A., Dominick, D., Nadzir, M.S.M., Sahani, M. and Tahir, N.M., (2015). Seasonal effect and source apportionment of polycyclic aromatic hydrocarbons in $PM_{2.5}$. *Atmospheric Environment*, **106**, 178-190

[39] Sahani, M., Zainon, N.A., Mahiyuddin, W.R.W., Latif, M.T., Hod, R., Khan, M.F., Tahir, N.M. and Chan, C.C., (2014). A case-crossover analysis of forest fire haze events and mortality in Malaysia. *Atmospheric Environment*, **96**, 257-265.

[40] Onat, B. and Stakeeva, B. (2013) Personal exposure of commuters in public transport to $PM_{2.5}$ and fine particle counts. *Atmospheric Pollution Research*, **4**; 329-335.

IOP Conf. Series: Materials Science and Engineering **920** (2020) 012008 doi:10.1088/1757-899X/920/1/012008

Physical modelling investigation of a 3-stage landslide prevention system: case study

M A Md Said*, M H Ramli, M Azmi and A S Mohd Azam

[1]School of Civil Engineering, Universiti Sains Malaysia, Engineering Campus, 14300 Penang, Malaysia.

*Corresponding author: ceazlin@usm.my

Abstract: A study was conducted to propose and assess the suitability of a structural design for landslide prevention system consisting of flexible barriers, concrete baffles and rigid barrier. A 1:20 scale model of the landslide system was conducted to observe the impact of a flexible barrier, the concrete baffle configurations and the appropriate rigid barrier height. Soils samples from site were also analysed to identify and characterize the soil properties. The experiments were conducted by varying the soil moisture content of the soils for the flexible barrier and concrete baffles. Results obtained showed that the flexible barrier is permanently deformed when the soil moisture is 26%. The configuration of several types of concrete baffles were studied where for unobstructed flow, the use of two rows of baffles had increased the degree of transverse blockage from 23% to 35% and provided up to 75% run-out distance reduction. A cantilever retaining wall is designed as a rigid barrier to ensure adequate retention and to prevent overflowing if the flexible barrier and concrete baffles failed. The combination of three different barriers was assessed and the result showed that the run-out distance was significantly decreased by 47 percent compared to unobstructed flow.

1. Introduction

Penang Island is a small island situated in northern part of Malaysia, approximately 300 km^2 comprised predominantly of steep topography of more than 50 percent above 60 meter. With a largely urban population, Penang is one of Malaysia's many rapidly industrialized states. In recent decades, efforts have been intensified to industrialize and develop other economic sectors, resulting in greater urbanisation and pressures on flat land [1]. Most of the lowlands have been developed in Penang Island, hills are therefore abundantly available on the island as the low land resources are poor [2]. Construction activities rapidly increase each day in Penang Island, particularly for residential and commercial purposes. Such rapid development jeopardizes the environment through natural disasters such as climatic changes, flood and landslide, has become threats to peace of life and property of locals.

Landslide events are common in Penang Island, particularly during wet seasons between April to May and October to November and in some cases due to slope failures or design failures. The district of Balik Pulau, Jelutong, Air Itam, Tanjung Bungah and Paya Terubong are among the most prominent landslide areas in Penang Island. Whenever heavy rain and flash floods occur, landslides on natural or man - made slopes would almost likely to happen because high amounts of precipitation are the major cause for landslide [3].

Recently, several landslides occurred in Penang which had no casualties such as at Taman Lau Geok Swee in Paya Terubong on 14th July 2017 [4], and at Sungai Kelian in Tanjung Bungah Bungah Hill on 28th May 2018 [5]. Several landslides with casualties had happened at Jalan Bukit Kukus, Paya Terubong, on October 2018 [6,7], and the Granito condominium site in Tanjung Bungah on 22 October 2017 [8].

2. Study Area

The study area shown in Figure 1, is located at a slope owned by Kek Lok Si Temple, Ayer Itam, Pulau Pinang. The Kek Lok Si Temple also generally known as the Temple of Supreme Bliss, is the largest Buddhist temple in Malaysia and the main attraction for Buddhists for other countries in Southeast Asia. The surrounding area are extensively developed, with the huge iconic Pagoda of Ten Thousand Buddha, main access road to Ayer Itam Dam, cable car facilities to carry pilgrims and visitors uphill and road and other facilities at the bottom of the hill.

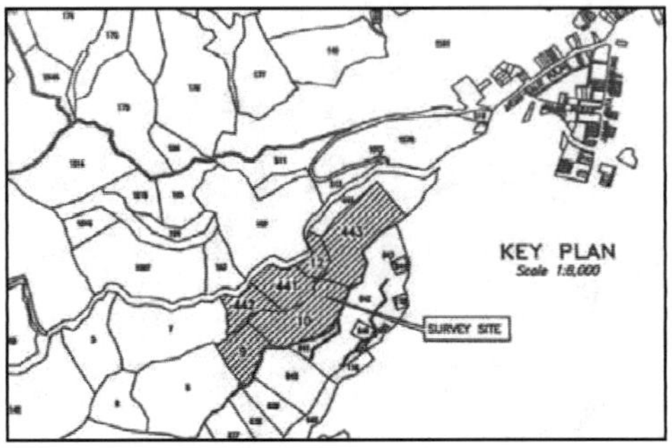

Figure 1. Location of the Study Area

A landslide occurred at Air Itam Dam Road, Penang on 29th October 2016, after heavy rain for more than three hours which had loosened large boulders, trees, and vegetation on a 30-metre section of the hillside [9]. A 350 metre section of the road was closed right after the landslide occurred due to heavy rain causing difficulties for residents, vegetable farmers and tourists.

To prevent further landslides, a counterfort retaining wall was constructed at the location. Although the prevention measure has been taken, there is a still possibility of landslide as the slope above the retaining wall has not been improved or reinforced. Therefore, the retaining walls can be significantly affected if slope failure occurs again [10].

3. Site Survey and Laboratory Tests

A site survey was conducted to determine the elevation, slope angles and information required for the study. The contour of study area is shown in Figure 2 where the elevation varies from 23.16m (maximum) to 8.18m (lowest).

Soil samples were obtained at different locations at the slope and stored in plastic bags for further analysis. Several soil tests were conducted, i.e. the Atterberg limit test, the particle size distribution test, the permeability of the coarse-grained test and the specific gravity test. All the tests followed different standards as shown in Table 1. Standard used for physical properties testing.

Table 1. Standard used for physical properties testing

Physical properties	Testing method and abbreviated reference
Particle size distribution	BS 1377: Part 2: 1990: 9.5 BS 1377: Part 2: 1990: 9.3
Specific gravity, G_s	BS 1377: Part 2: 1990: 8.3
Liquid limit, LL	BS 1377: Part 2: 1990: 4.3
Plastic limit, PL	BS 1337: Part 2: 1990: 5.3
Permeability	BS 1377: Part 5: 1990:5 ASTM D 2434

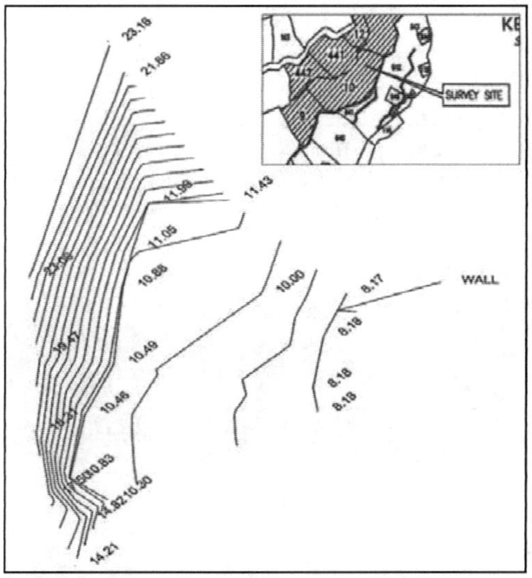

Figure 2. The Elevation of the Study Area

3.1 Physical Model

The physical model is used to understand the behaviour of the landslide prevention system characteristics shown in Figure 3. The landslide prevention conceptual system can be visualized by the physical model. A 1:20 physical landslide modelling was performed using the most critical cross section obtained from the site survey in order to investigate the effect of changes in debris properties on the barrier's structural response subject to debris impact. A 3.5m long with a width of 2m and 1m in height model is used to investigate the influence of baffle arrangement and the suitable height of rigid barrier. The slope inclination is according to site measurements, i.e. 70°. The topsoil is released at the top part of the model to imitate the landslide at the study area.

Figure 3. Physical Model of Landslide Prevention System

3.2 Baffle configurations

A series of 6 different baffles configurations were investigated to measure the run out distances. Two control tests without baffles were conducted as control tests for comparison. Baffles array are separated into 3 degree of transverse blockages and the number of rows is varied from 1 to 2 staggered rows. The degree of transverse blockages is varied as 23%, 26% and 35%. The spacing between each row is varied as 50mm, 75mm and 90mm which show real spacing at site 1m, 1.5m and 1.8m respectively.

Table 2 shows the control test while
Table 3 and ***Table 4*** show the baffle test configurations.

Table 2. Control Test

Test ID	Soil Weight (kg)
CT-1	13
CT-2	20

Table 3. Single Row with constant baffle height of 75mm

Test ID	Soil Weight (kg)	Transverse blockage (%)
R1-13-23	13	23
R1-20-23	20	
R1-13-26	13	26
R1-20-26	20	
R1-13-35	13	35
R1-20-35	20	
R2-20-35	20	

Table 4. Double Rows with constant baffle height of 75mm

Test ID	Soil Weight (kg)	Transverse blockage (%)
R2-13-23	13	23
R2-20-23	20	
R2-13-26	13	26
R2-20-26	20	
R2-13-35	13	35
R2-20-35	20	
R2-13-23	20	

4. Results and Discussion

4.1 Baffle Configuration Results

Three transverse blockages are examined, namely 23%, 36% and 35%. Furthermore, the number of rows investigated ranges from 1 to 2 with spacing between successive rows of 50 mm and constant baffle height of 75mm. Run-out distance results of the testing programme consisting of 14 tests are given in the Table 5 and Table 6 including control test.

Table 5. Control Test Results

Test ID	Soil Weight (kg)	Run Out Distance (cm)
CT-1	13	87
CT-2	20	107

Table 6. Testing Program Results

	Soil Weight (kg)	Transverse blockage (%)	Run out Distance (cm)
R1-13-23	13	23	92
R1-20-23	20		84
R1-13-26	13	26	46
R1-20-26	20		55
R1-13-35	13	35	39
R1-20-35	20		48
R2-13-23	13	23	73
R2-20-23	20		65
R2-13-26	13	26	37
R2-20-26	20		45
R2-13-35	13	35	27
R2-20-35	20		29

Baffles are arranged in a staggered configuration as previous study [11] which disrupts streamlines between successive rows with optimal energy dissipation. Transverse blockage percentages are adopted for ease of applicability to prototype conditions.

4.2 Flexible debris resisting barrier test program

A flexible net with a height of 17.5cm after scaled down was used in physical modelling to investigate its purpose, by minimizing the run-out distances. From the results obtained in Table 7, the flexible barrier undergoes permanent deformation when the moisture content is 26 %. This shows that flexible barriers must be provided with robust foundations and anchors to withstand the impact, as well as the effect of debris weight on the slope [12]. The structure of the flexible barrier may fail due to the high water content in the soil caused by continuous rainfall.

Table 7. Flexible Barrier Test Results Program

Soil Wight (kg)	Water Content (%)	Moisture Content (%)	Run Out Distance (cm)
30	0	0	47
30	20	13%	58
30	40	26%	Failed

Figure *4. Flexible Net Testing Results*
Figure 4 shows flexible net testing results. At some point during heavy rainfall, the weight and forces exceeded the design load of the flexible barrier. The mass of soil with 26% moisture content has

damaged the flexible net as presented in (c) before enabling the energy absorption system to fully operate.

4.3 Combination of Three Phases Results

For ease of interpretation and calibration of theoretical models, dry top soil is used in the test. A 45kg source weight, 35% transverse blockage, two rows of baffles, 17.5cm flexible net were adopted to investigate the effectiveness of the 3-phase combination. Comparing with unobstructed flow, the use of three types of resisting barriers can reduce up to 47% of the debris run out distance from 110cm to 58cm. There is an obvious influence of baffles geometry and flexible net on flow interaction which reduce run-out distance as shown in Table 8.

Table 8. Combination of Three Phases Results

Net Height (cm)	Soil Weight (kg)	Distance Between Baffles (mm)	Transverse Blockage	Run Out Distance (cm)
-	45	-	-	110
17.5	45	50	35%	58

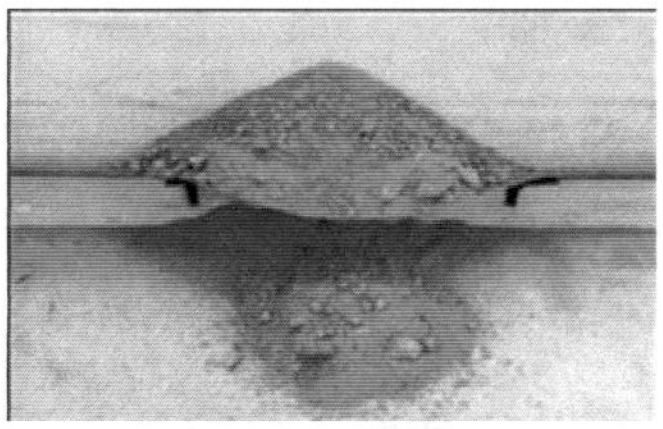

(a) 30kg of soil with 0% moisture content

(b) 30kg of soil with 13% moisture content

(c) 30kg of soil with 26% of moisture content

Figure 4. Flexible Net Testing Results

4.4 Retaining Wall

Rigid barrier for this study are of reinforced concrete construction and is practically the same as the one proposed in Hong Kong. Rigid barrier are designed to withstand the impact force of debris and occasional boulders [12] A cantilever retaining is designed based on BS8100: 1997 where all the geotechnical properties given in Table 9 were obtained from the SI/GI Contractor.

Table 9. Geotechnical Properties

Unit Weight of Soil, ρ	19.64 kN/m^3
Internal Angle of Friction, φ	34°
Bearing Pressure	100 kN/m^2
Coefficient of Friction	0.4
Unit Weight of Concrete, ρ	24 kN/m^3

5. Conclusion

In conclusion, the study was satisfied as the geotechnical and geological parameters for landslide prevention system were identified, three different passive earth retaining system were proposed, the influence of each different type of barriers on flow interaction were assessed and rigid barrier is designed and evaluated using common geotechnical and reinforcement concrete design code.

Three small scale debris resisting barriers including a flexible net, set of baffles and rigid barrier are used to demonstrate and confirm the practicability of the proposed method. The result showed that the usage of a single flexible barrier had failed due to high water content in the soil as it could not withstand the impact and weight of the soil. The flexible barrier undergoes permanent deformation when the moisture content reached 26% only by adopting the flexible barrier. From the findings, an increasing number of rows is the most effective way in reducing the kinetic energy of debris flow compared with unobstructed flow. The use of double rows of baffles can provide more than 50% run-out distance reduction. Furthermore, the arrangement of baffles in staggered position promotes optimal energy dissipation and induced the soil pattern fell more evenly in placed and less dispersed. Moreover, a 45kg source weight, 35% transverse blockage, 2 rows of baffles, 17.5cm flexible net were adopted to investigate the effectiveness of the 3-phase combination.

A 5 meter cantilever retaining is designed based on BS8100: 1997. The height of rigid barrier is designed to ensure adequate retention and avoid overflowing debris. The stability of the whole structure under the service loads including overturning, sliding and bearing failure modes was assessed. The rigid barrier is designed against lateral sliding and overturning and both Factor of Safety (FOS) values exceed 1.5 which achieved the stability.

The combination of three different barriers experiments showed that the run-out distance has reduced up to 47% compared to unobstructed flow.

Acknowledgement

The authors wish to express sincere appreciation for the support and encouragement provided by the Engineering Department, MBPP and School of Civil Engineering, Universiti Sains Malaysia in making this project a success.

References

[1] See Sew G and Chin Y C 2003 The Engineering Aspects of Hill-Site Development Hillside Development – Issues and Challenges (Kuala Lumpur)

[2] Chan N W 1998 Environmental hazards associated with hill land development in Penang Island, Malaysia: some recommendations on effective management *Disaster Prev. Manag.* **7** 305–18

[3] Pradhan B and Lee S 2010 Landslide susceptibility assessment and factor effect analysis: backpropagation artificial neural networks and their comparison with frequency ratio and bivariate logistic regression modelling *Environ. Model. Softw.* **25** 747–59

[4] The Star Malaysia Past lanslide cases reported in Penang Accessed 15/6/2020

[5] The Star Malaysia Huge landslide in Tg Bungah hill | The Star Accessed 15/6/2020

[6] Anon Two confirmed dead, one rescued and 10 missing in Bukit Kukus landslide Accessed 15/6/2020

[7] The Sun Daily Chronology of landslides in Penang since 2017 Accessed 15/6/2020

[8] The Straits Times Three dead, 11 missing in Penang landslide, SE Asia News & Top Stories - The Straits Times Accessed 15/6/2020

[9] Malay Mail Penang water authority closes Air Itam Dam road for safety | Malay Mail Accessed 15/6/2020

[10] Mandali A K 2011 Reliability Analysis of Counterfort Retaining Walls *Electron. J. Struct. Eng.* **11** 15

[11] Ng C W W, Choi C E, Kwan J S H, Shiu H Y K, Ho K K S and Koo R C H 2012 Landslide_Mobility_Analysis_for_Design_of_Multiple.pdf *Proceedings of the One Day Seminar On Natural Terrain Hazard Mitigation Measures* One Day Seminar On Natural Terrain Hazard Mitigation Measures (Kowloon, Hong Kong: HKSAR) pp 16–21

[12] Kwan J S H, Chan S L, Cheuk J C Y and Koo R C H 2014 A case study on an open hillside landslide impacting on a flexible rockfall barrier at Jordan Valley, Hong Kong *Landslides* **11** 1037–50

IOP Conf. Series: Materials Science and Engineering **920** (2020) 012009 doi:10.1088/1757-899X/920/1/012009

On the role of substrate temperature into bonding mechanism of cold sprayed Titanium Dioxide

O Noor irinah[1,2], M Yamada[1], T Yasui[1] and M Fukumoto[1]
[1]Department of Mechanical Engineering, Toyohashi University of Technology, 1-1, Tempaku-Cho, Toyohashi, Aichi,441-8580, Japan
[2]Faculty of Engineering Technology Mechanical and Manufacturing, Universiti Teknikal Malaysia Melaka, 76100, Durian Tunggal, Melaka, Malaysia

Email: irinah_n@isf.me.tut.ac.jp, nooririnah@utem.edu.my

Abstract. Metal oxide, Titanium dioxide, TiO_2 draw a huge attention due to its wide range of applications, chemical stability and low cost. TiO_2 coatings have potential applications in biomedical implants or as photo-catalytic functional systems. Cold spraying is a process where small particles ranging from 1 to 50μm are stimulated by a stream of supersonic gas at a temperature below material melting point and leading to the coating formed in solid state particles. The aim of the present study was to investigate the influenced of once heated substrate properties on bonding mechanism of cold sprayed Titanium dioxide, TiO_2. In this study, tensile adhesion strength, substrate hardness, oxide thickness, oxide composition and interfacial microstructure were evaluated for the coatings. The results of this study indicated that coating adhesion strength reveal a decreased trend from 1.35MPa to 0.49MPa when once heated substrate temperature is increased. Influenced of oxide thickness also will be discussed here as factor that contributed to bonding mechanism between TiO_2 and copper substrate.

1.Introduction

Cold spray that is also named as cold gas dynamic spray has drawn growing attention because of the specific 'cold' feature. In this process, powder particles are accelerated by a supersonic gas stream at a temperature that is lower than the melting point of the material, resulting in a coating formation from particles in a solid state. As a consequence, the cold sprayed coating can effectively eliminate the influence of thermal stress, severe oxidation, thermal phase transformation and other common problems associated with traditional thermal spray processes and distortion [1]. Therefore, cold spray has been broadly implemented for the fabrication of coatings for different applications, as well as a technology of additive manufacturing [2]. Presently, the most applicable theory of bonding mechanism in cold spray of metal is considered as the mechanical interlocking and metallurgical bonding. As a non-chemical reaction, the mechanical interlocking is formed due to intense plastic deformation of material under high-speed collision process [3,4]. The metallurgical bonding is considered a consequence from chemical reaction at the oxide-free interface between particles or particle/substrate [5-7]. Both of them are assumed to make a plastic deformation for both particle and substrate. Based on the bonding mechanisms, it is difficult to fabricate ceramic coating through cold spray process due to lack of its ductility. However, our group succeeded to fabricate Titanium dioxide (TiO_2) thick coating through cold spray process and its deposition mechanism were discussed [8-12].

IOP Conf. Series: Materials Science and Engineering **920** (2020) 012009 doi:10.1088/1757-899X/920/1/012009

Nanostructured TiO_2-anatase coatings draw a huge attention due to have an extensive active surface and chemical stability, and are at a comparatively reasonable cost, they are utilized as functional material. The crystalline framework of TiO_2 has a major impact upon its photocatalytic performance where TiO_2 in anatase stage supplies greater photocatalytic activity than in its rutile stage. At temperatures exceeding 900˚C, which is above the melting point of TiO_2 (1908˚C), the anatase stage irreversibly changes into the rutile stage. At temperatures above the melting point of TiO_2, the deposition of molten or semi-molten droplets form thermal impossible to avert the phase transformation of TiO_2 in thermal spray procedures [13] Otherwise, it is not necessary to melt the material in order to apply cold spraying. Consequently, such technology can avert the unwanted phase transformation [14]. It was enabled to fabricate TiO_2 ceramic coating by cold spray, the details of bonding mechanism have not been clarified yet. Our previous study described that the possibility of chemical bonding between the fractured agglomerated particles and the substrate [15]. K.Schmidt et al claim the plastic deformation of the substrate lead to a large continuous contact zone between TiO_2 particle and substrate and induced to a durable bonding[16]. M.Gardon et al reveal that titanium sub-oxide coating provided a necessary surface geometry and composition that favored the deposition of anatase particles, bonded by chemical interactions with the substrate [13]. Nevertheless, the role of substrate properties toward the bonding mechanism has not been discussed yet. The substrate properties such as hardness, roughness and oxide formation can be controlled by heat treatment of the substrate. The importance of substrate temperature on the cold spraying of metallic coating has already been reported, but the details of the mechanism even in the metal materials has not been clarified [17]. In this study, the effect of once heated pure copper substrate properties from room temperature to 400ºC were carefully investigated toward the bonding mechanism of cold sprayed TiO_2.

2. Methodology

2.1 Cold spray process

We applied a commercially available cold-spray apparatus (CGT KINETIKS 4000; Cold Gas Technology, Ampfing, Germany) to all of the coating experiments. We utilized high-pressure nitrogen gas for the powder carrier gases as well as for the procedure itself. A De-Laval 24TC nozzle was used to accelerate the process gas and to inject the feedstock powders. In order to investigate the influence of the substrate conditions, we used same spray conditions for all experiments. Table 1 shows the cold spray conditions and figure 1 shows a schematic diagram of cold spray process.

Table 1. Cold Spray Conditions

Working gas	Nitrogen
Gas pressure	3 MPa
Gas temperature	500˚C
Spray distance, mm	20
Traverse speed, mm/s	10

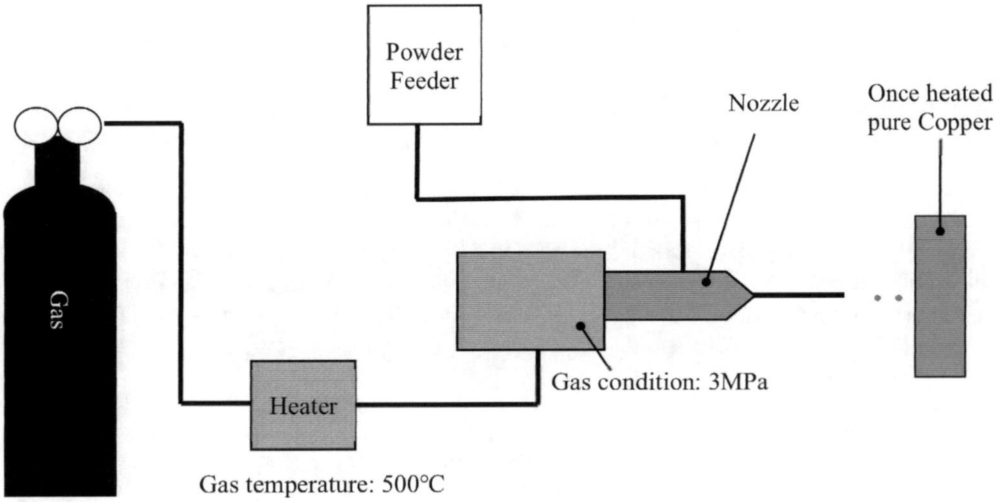

Figure 1. Schematic diagram of Cold spray process

2.2 Materials

As a feedstock, we applied TiO_2 powder (TAYCA, Okayama, Japan) which has been succeeded to fabricate thick coatings by cold spray in our previous studies [8-12]. Figure 2 shows the morphology of the TiO_2 powder particles. The average particle size (D_{50}) is 7μm measured by a particle size analyser (SALD). For a substrate, we utilized pure copper, C1020 having a dimension of Ø25mm x 10mm. We used an electric furnace in air environment to preheat the grit-blasted substrate to four different temperatures: 100°C, 200°C, 300°C and 400°C respectively, the increment being 15°C/5 minutes, and after a five-minute soaking it was cooled in the furnace to room temperature. The substrates were rinsed with acetone before they were sprayed.

Figure 2. Morphology of the TiO_2 feedstock powder

2.3 Coating evaluations

A scanning electron microscope (SEM: JSM-6390, JEOL, Tokyo, Japan) was used to observe the coating's cross-sectional microstructures.

2.4 Adhesion Strength

In accordance with JIS H 8402, specimens of dimensions of Ø25mm x 10mm assessed the coating's adhesion strength which was given as the fracture load's value as measured by a universal testing machine (Auto graph AGS-J 10kN Shimadzu) and subsequently divided by the cylindrical coating area. The epoxy-based strong adhesive bonded the pin and coating (Huntsman Advanced Materials Company: Aralkyl Standard). We conducted the strength test subsequent to curing the glue for a period of 24h. We acquired the adhesion strength with an average of five specimens for each of the spray conditions.

2.5 Substrate evaluations

An X-ray Photoelectron Spectroscopy (XPS: Quantera SXM-CI, ULVAC-Phi,Inc.) was used to measure compositional and chemical state analysis for 5 different substrates. The substrate conditions are room temperatures and once heated temperatures of 100°C, 200°C, 300°C and 400°C. We used an HMV-G micro-Vickers hardness tester (Shimadzu) to measure the substrates' micro-hardness. We acquired the micro-hardness value by an average of five points per substrate conditions.

3. Results

3.1 Interfacial coating microstructure

Figure 3 depicts the TiO_2 coating cross-section area on pure copper, C1020 for various substrate temperatures. Almost every figure depicted a dense coating with a thickness in the range of 200μm to 300μm, showing that a critical velocity has been reached. Figure 2(a) to 2(e) show some degree of porosity was observed in pure copper substrate as well as crack between the coating at substrate especially on once heated substrate at 300°C and 400°C, shown by figure 2(d) and 2(e).We can categorize the cold-spray procedure into two stages. The adhesion or interface formation between substrate and particle is the first stage. The once-heated substrates can apparently implement this stage, which forms the first coating layer. The cohesion between the particles and coating growth upon the first layer is the second stage. The once-heated temperature substrate in this study was below the process gas temperature. It appears that the substrate conditions like temperature had minimal impact on the cohesion procedure.

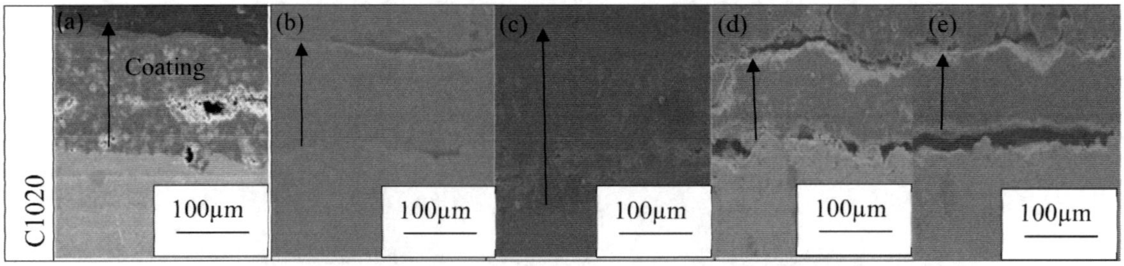

Figure 3. Cross section microstructure of TiO_2 coatings on different once heated substrate temperature : (a) RT, (b) 100°C, (c) 200°C (d) 300°C (e) 400°C

IOP Conf. Series: Materials Science and Engineering **920** (2020) 012009 doi:10.1088/1757-899X/920/1/012009

3.2 Adhesion strength

Figure 4 shows the adhesion strength of the coatings deposited onto different substrate conditions. The coatings which are sprayed upon the substrate at a temperature of 400°C in the course of specimen preparation peel off automatically; consequently, at this temperature the evaluation was not succeeded. The result indicates that the adhesion strength showed a reduced trend from 1.35MPa to 0.49MPa. Therefore, increasing the once heated temperature reduce the adhesion strength of cold-sprayed TiO_2 coatings on the copper substrate. Figure 5 shows fracture picture of cold sprayed TiO_2 coating onto once heated C1020 substrates after adhesion strength testing. It is confirmed the interface fracture occur between coating and substrate for pure copper substrates in all conditions as reveal by figure 5 (a) to 5(d).

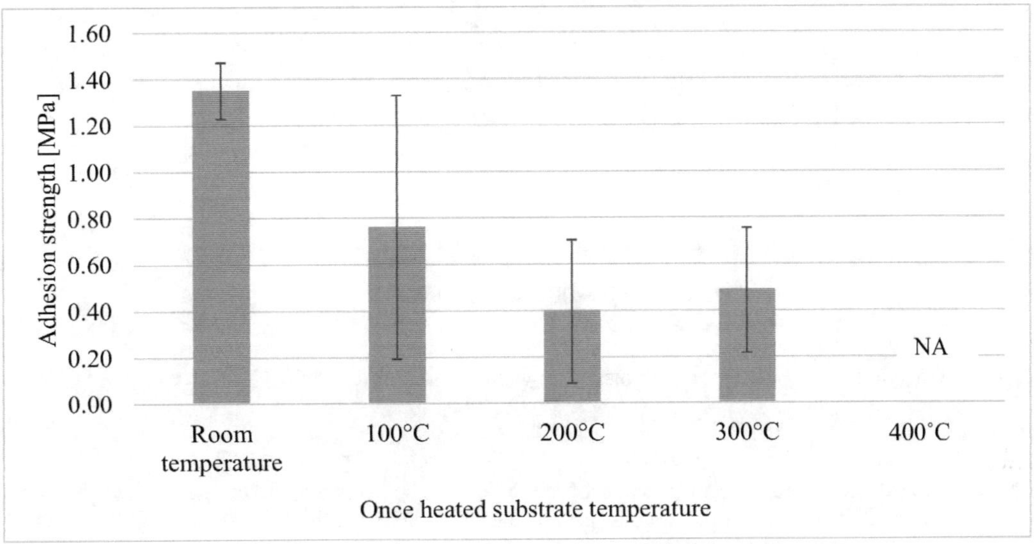

Figure 4. Adhesion strength of cold sprayed TiO_2 coating onto once heated C1020 substrates

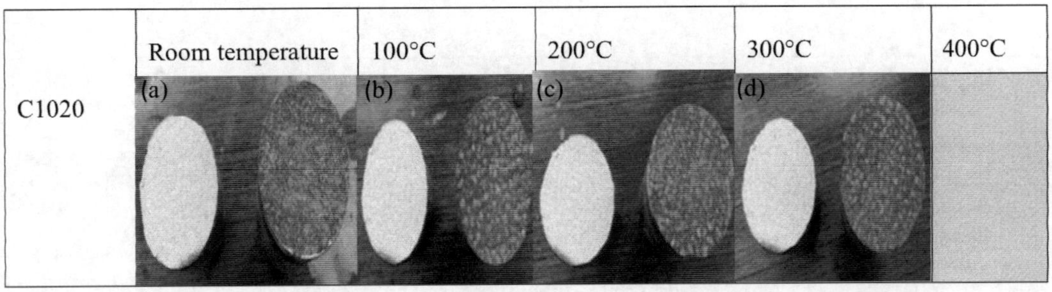

Figure 5. Fracture picture of cold sprayed TiO_2 coating onto once heated C1020 substrates after adhesion strength testing of (a) room temperature (b) 100°C (c) 200°C and (d) 300°C

3.3 Substrate hardness

Substrate hardness for room and once heated conditions are depicted in Figure 6. The hardness from room temperature to 300 ˚C was almost similar. On the other hand, the substrate once heated with 400˚C exhibited lower hardness compared with other substrates. Thermal re-crystallization could cause the apparent reduction of the hardness. It is known that re-crystallization temperature about 0.4 times the melting point (copper is 1080 ˚C). Grains become coarser when they are heated to this temperature, and slow cooling in the furnace enables the alloy to develop larger grains, thereby supporting the reduction of hardness.

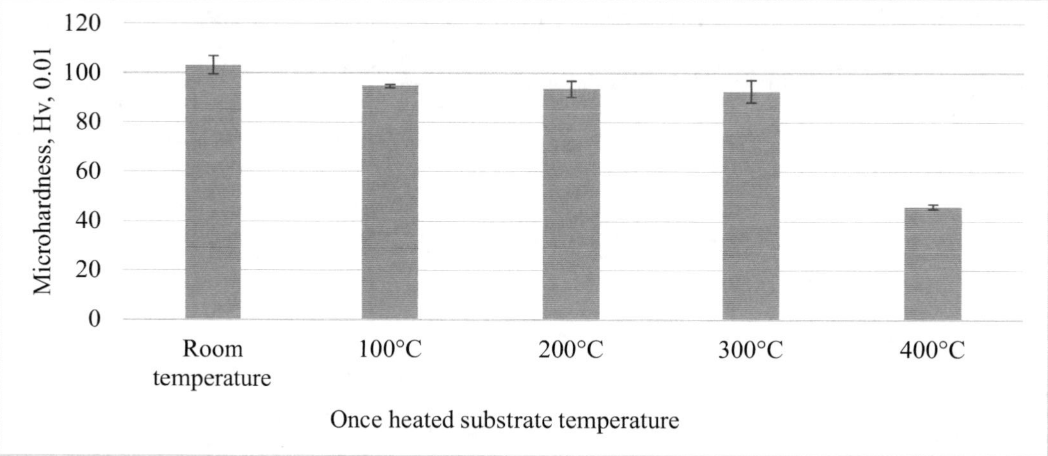

Figure 6. Once heated pure Copper hardness from room temperature to 400˚C

3.4 Oxide thickness

Figure 7(a) to (e) shows the concentration as a function of depth of oxygen and copper in the substrates. Oxygen atomic concentration in the deepest part of the oxide layer, in the substrates, C1020, increases significantly as the annealing temperature increases from room temperature to 400°C . This indicate that the oxide layer of pure copper grows thicker as the annealing temperature of substrate increases where oxide thickness for pure copper at room temperature is approximately 20nm and increased to more than 90nm of annealed temperature 400°C.

C1020	Room temperature	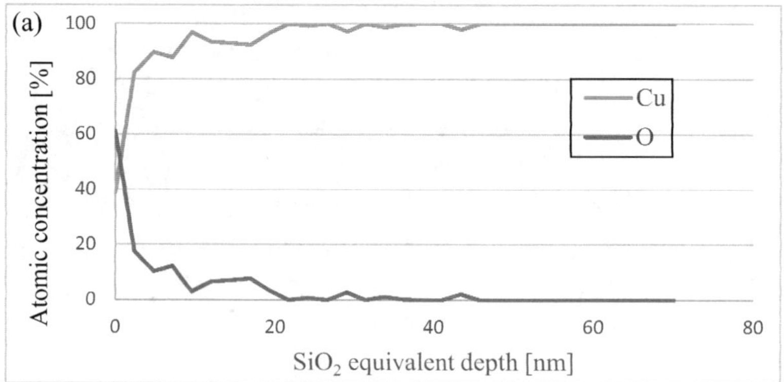

<table>
<tr><td></td><td>100°C</td><td>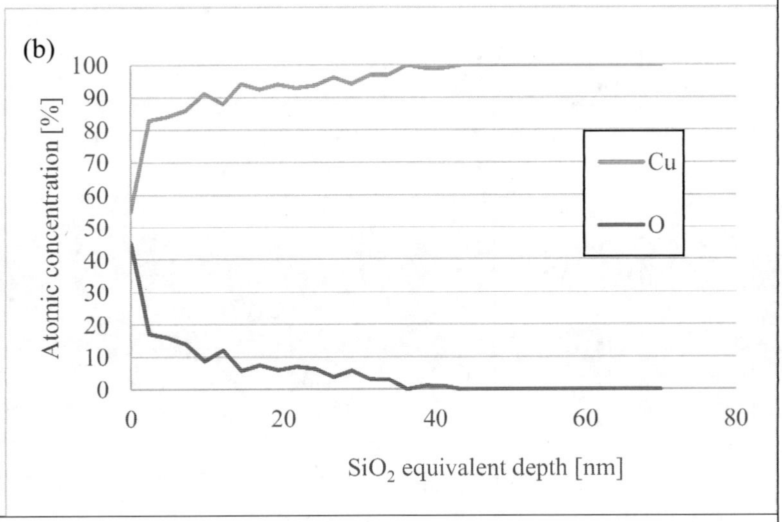
</td></tr>
<tr><td></td><td>200°C</td><td>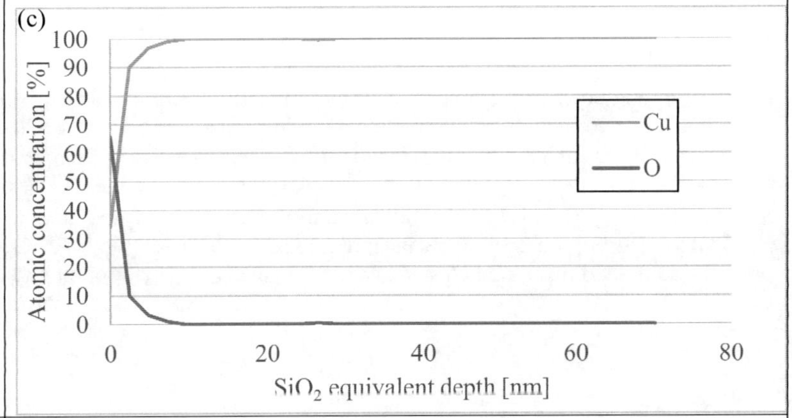
</td></tr>
<tr><td></td><td>300°C</td><td>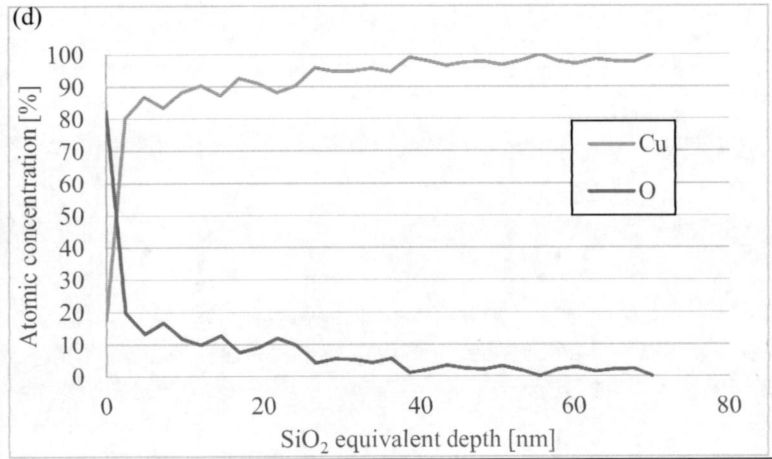
</td></tr>
</table>

	400°C	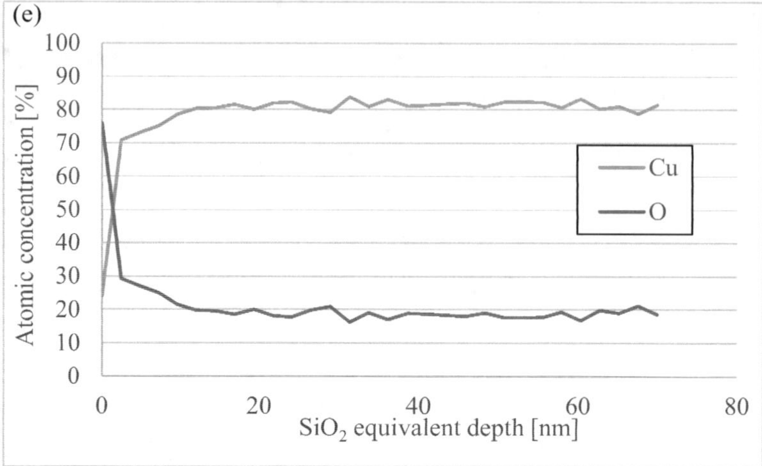

Figure 7. Depth profile analysis of pure copper, C1020 result from X-Ray photoelectron spectroscopy

3.6 Oxide composition

Result of through depth direction analysis by utilizing XPS for substrates is indicated in figure 8(a) and (b). Oxide layer formed on pure copper shown in figure 8(a) and (b) is Cu_20 for once heated substrate temperature 100°C and 200°C, while Cu0 for once heated substrate temperature 300°C and 400°C. The peak for Cu 2p3/2 in this study is 932.5eV, indicated the presence of Cu_20. The O1s peaks shown in figure 8 (b) are 530.3,531.3 eV indicated the presence of Cu_2O and CuO and hydroxide species physically adsorbed on the Cu surface. The peak at 529.3eV was the CuO sub-peak, when the Cu substrate contained two oxidized states, each Cu peak and corresponding O peak were detected [18].

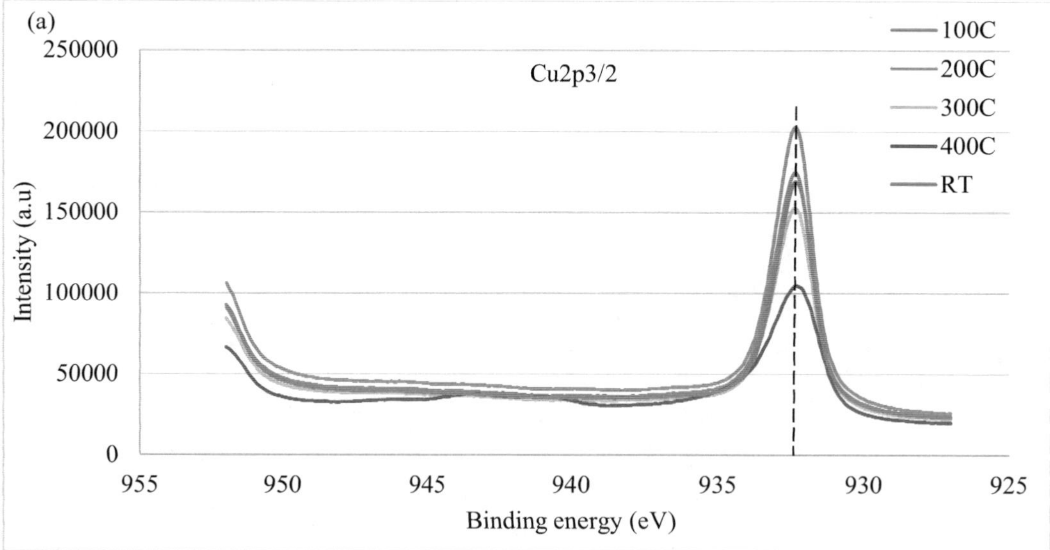

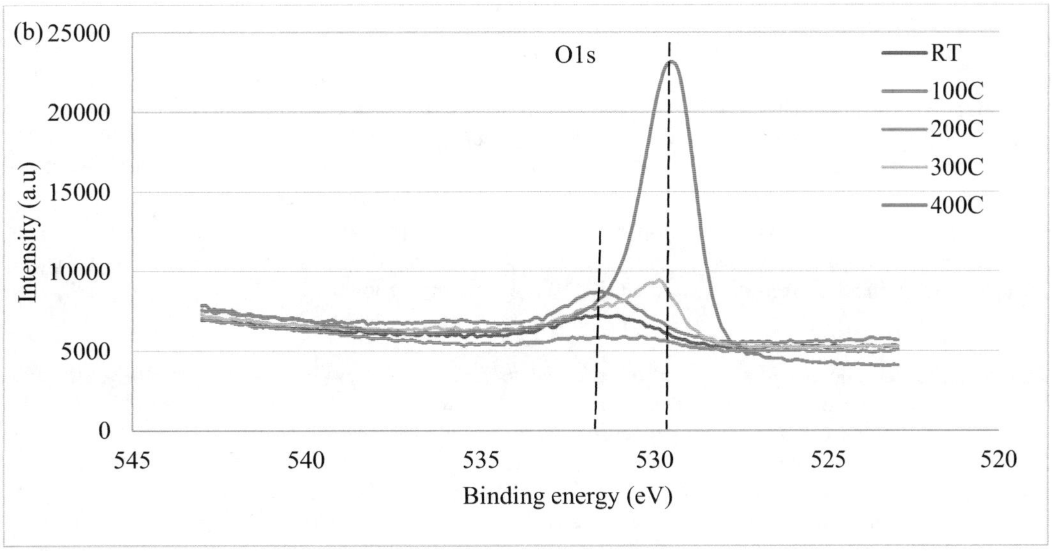

Figure 8. XPS spectra of (a) Cu 2p3/2 and (b) O1s for C1020 from room temperature to 400°C

4. Discussion

Cold-sprayed TiO_2 coatings on the pure copper, (C1020) substrates revealed a decreased trend in adhesion strength from 1.35MPa to 0.49MPa for room temperature substrate to once heated substrate of 400°C.The adhesion strength value represent the adhesion bonding occurring between the first layer of TiO_2 coating and the surface of copper substrate. Referring to factors that we studied; substrate hardness, oxide thickness and oxide composition, some of these factors may pronounce to bonding mechanism of cold-sprayed TiO_2 onto pure copper substrate. Once heated substrate of pure copper from room temperature to 400°C revealed a decreasing trend of hardness due to re-crystallization effect by the heat treatment. In this situation, substrate become softer and easier to deform due to kinetic impact energy of the sprayed particles. If the main deposition factor is the mechanical anchoring, softer substrate should have an advantage for the adhesion strength. Nevertheless, the adhesion strength was decreased at higher heated temperature. It means that the mechanical anchoring is not the main bonding mechanism between the cold-sprayed TiO_2 coating and the copper substrate.

According to K.H.Ko et al the increasing of quantity of high oxidation state in copper feedstock can produce large amounts of rebounding particles which do not accumulate but tamping of the surface of the coating[18]. This condition maybe one of the explanation due to thicker oxide thickness at once heated substrate of 400°C and SEM images of picture 2(e) reveal a crack between particle/substrate. Thick oxide formation on the copper substrate prevented the bonding between the particle and the substrate. Figure 7(a) to (e) shown depth profile analysis of pure copper, C1020 result from X-Ray photoelectron spectroscopy. Oxide thickness is increased from 20nm on surface of room temperature C1020 to more than 90nm on surface of once heated 400°C . According to W.Ya Li et.al stated that the increasing of oxide film thickness, it will need more kinetic energy to break up and extrude the oxide film, thus a higher particle velocity is needed for bonding. In other words, the effective bonding area is decreased under the same particle impact conditions [19]. It could explain the decreased trend of adhesion strength of once heated substrate of pure copper from room temperature to 400°C, because the particle velocity was constant in all condition. Therefore, a chemical bonding between newly- formed pure copper and cold-sprayed TiO_2 particle can be the main factor on this process.

There is another possibility for the chemical bonding, which is cold-sprayed TiO_2 and surface oxide of copper substrate. The composition of copper oxide films that coexist two phases, cuprous oxide, Cu_2O and cupric oxide, CuO. According to G.Papadimitropoulos et al claim that at temperature up to 225°C, Cu and Cu_2O is formed while above this temperature only CuO forms[20]. Both Cu_2O and CuO phases existed on our substrate and it has been supported by O1s peak shown in figure 8(b). F-H.Lu stated for solid state reaction between CuO-TiO_2, phase diagram of the CuO-TiO_2 system, the eutectic temperature is 900°C in air, therefore the sample will react each other due to eutectic reaction [21]. Referring to our spray conditions that involved gas temperature of 500°C and pressure at 3MPa, this condition is not reached up to eutectic temperature. Therefore, the oxide composition may not be the main factor that influences the adhesion strength in cold sprayed TiO_2 onto pure copper.

The metallurgical bonding is considered a consequence from chemical reaction at the oxide-free interface between particles or particle/substrate [5-7]. According to the results we obtained in this study, chemical bonding between cold-sprayed TiO_2 particle and newly formed copper substrate is considered be the main factor. Once heated temperature of 400 °C decreased the hardness due to the recrystallization, however, thick oxide layer which prevent the bonding was formed and reduced the adhesion strength. Hence the bonding mechanism between cold-sprayed TiO_2 onto pure copper substrate can be explain with the chemical reaction which is similar with the cold spraying of metallic materials. Figure 9 shown a schematic diagram of factor that influenced toward adhesion strength of TiO_2 coating on substrate.

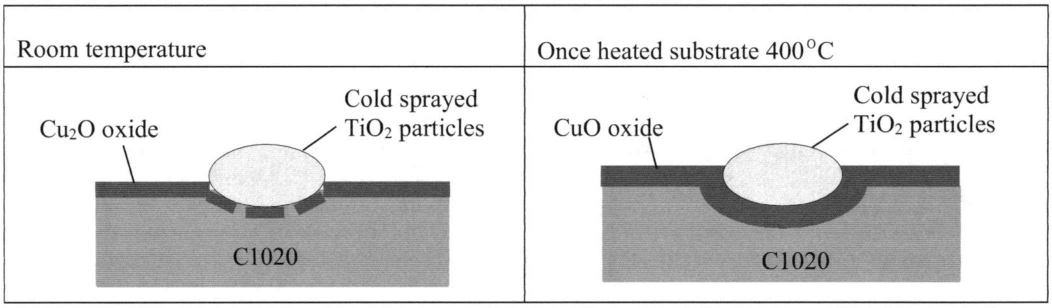

Figure 9. Factor that influenced toward adhesion strength of TiO_2 coating on pure copper

5. Conclusions

The present work demonstrates factors that influenced toward bonding mechanism of cold sprayed titanium dioxide, TiO_2 onto soft material, pure copper, C1020. We summarize the outcome attained in this study as follows:

1. Thick TiO_2 coatings could be depositied onto all substrate conditions, though some cracks were observed in the substrates once heated at 300°C and 400°C.

2. Increasing the heating temperature decreased the coating adhesion strength and substrate hardness. On the other hand, oxide thickness on the substrate was increased with the heating temperature.

3. CuO and Cu_2O were formed on the substrate by the heat treatment. These oxydes did not contribute the chemical reaction with the cold-sprayed TiO_2 particles.

4. The major bonding mechanism is considered to be the chemical bonding between the cold-sprayed TiO_2 particle with newly-formed copper substrate.

Acknowledgments
This work was supported by JSPS KAKENHI Grant Number JP17K06857 and this work was performed, in part, at the Cooperative Research Facility Center at Toyohashi University of Technology. We also acknowledge the Interface & Surface fabrication lab, ISF, Majlis Amanah Rakyat, MARA and Universiti Teknikal Malaysia Melaka, UTeM for Noor Irinah's PhD scholarship.

References
[1] H.Assadi, H.Kreye, F.Gartner and T.Klassen 2016 *Acta Mater* **116** 382
[2] J.Pattison,S.Celotto,R.Morgan,M.Bray and W.O`Neil 2007 *Int.J.Mach Tool Manu* **47** 627
[3] Shuo Yin, Xiaofang Wang, Xinkun Suo, Hanlin Liao, Zhiwei Guo, Wenya Li and Christian Coddet 2013 *Acta Mater.* **61** 5105
[4] M.Grujicic, J.R.Saylor, D.E.Beasley,W.S.DeRosset and D.Helfritch 2003 *Appl.Surf.Sci* **219** 211
[5] G.Bae, J-I Jang and C.Lee 2012 *Acta Mater* **60** 3524
[6] Y. Ichikawa, K. Ogawa 2015 J. Therm. Spray Technol. **24** 1269
[7] K. Ito, Y. Ichikawa, K. Ogawa 2016 Mater. Trans. **57** 525
[8] M. Yamada, Y. Kandori, K. Sato and M. Fukumoto 2009 J. Solid Mche. Mater. Eng. **3** 210
[9] M. Yamada, H. Isago, H. Nakano and M. Fukumoto 2010 J. Therm. Spray Technol. **19** 1218
[10] N. Tjitra Salim, M. Yamada, H. Nakano, K. Shima, H. Isago and M. Fukumoto 2011 Surf. Coat. Technol. **206** 366
[11] A.R. Toibah, M. Sato, M. Yamada and M. Fukumoto 2016 Mater. Manuf. Process. **31** 1527
[12] A.R. Toibah, K. Takahashi, M. Yamada and M. Fukumoto 2016 Mater. Trans. **57** 1345
[13] M. Gardon, C. Ferna ´ ndez-Rodrı ´ guez, D. Garzo ´ n Sousa, J.M. Don˜a-Rodrıguez, S. Dosta, I.G. Cano, and J.M. Guilemany 2014 *J. Therm. Spray Technol.* **23** 1135
[14] V. K. Champagne 2007 *The cold spray materials deposition process* (Cambridge: Woodhead Publishing Limited)
[15] M. Yamada, H. Isago, K. Shima, H. Nakano, M. Fukumoto, T. J, and G.M. B.R. Marple, A. Agarwal, M.M Hyland, Y.-C. Lau, C.-J. Li, R.S. Lima (Ed.) 2010 *Proc.Int.Conf.Therm. Spray Glob. Solut. Futur. Appl.,Eds.* (*Singapore*) p 172
[16] K.Schmidt ,S. Buhl, N.Davoudi , C.Godard ,RolfMerz , I.Raid ,E.Kerscher, M.Kopnarski, C. Müller-Renno, S.Ripperger and J.Seewig et.al 2017 *Surf.Coat technol.***309** 749
[17] M. Fukumoto, H. Wada, K. Tanabe, M. Yamada, E. Yamaguchi, A. Niwa, M. Sugimoto and M. Izawa 2007 J. Therm. Spray Technol. **16** 643
[18] K.H. Ko, J.O. Choi, H. Lee and B.J. Lee 2012 *Powder Technol* **218** 119
[19] W-Ya Li,Chang-Jiu Li and Hanlin Liao 2010 *Appl.Surf.Sci.* **256** 4953
[20] G Papadimitropoulos, N Vourdas, V Em Vamvakas and D Davazoglou 2005 *J. Phys.: Conf. Ser.***10** 182
[21] F-Hsing Lu, F-X Fang and Y-S Chen 2001 *J.Eur.Ceram.Society* **21** 1

IOP Conf. Series: Materials Science and Engineering **920** (2020) 012010 doi:10.1088/1757-899X/920/1/012010

State-of-the-art for Magnetorheological Elastomers Fabrication and Hydro-tribological Characterisation

R MD Nasir[1,*], C F Ru[1], R K Shuib[2]

[1]School of Mechanical Engineering, Engineering Campus, Universiti Sains Malaysia, 14300 Nibong Tebal, Pulau Pinang, Malaysia
[2]School of Material and Mineral Resources Engineering, Engineering Campus, Universiti Sains Malaysia, 14300 Nibong Tebal, Pulau Pinang, Malaysia

*Corresponding author: ramdziah@usm.my

Abstract. Magnetorheological effect on mechanical properties of its specific applications has been studied. Mechanical properties such as hardness, tribology characteristic and salt water and water absorption with hygrothermal effect will be further studied on wear performance of MRE on marine equipment applications. Isotropic, anisotropic and anisotropic with added N330 carbon black filler of Ni Zn Ferrite Magnetorheological Elastomers (MREs) were prepared with Ni Zn Ferrite filler, $Ni_{1-x}Zn_xFe_2O_4$ particles weight fraction of 70%. Images of Scanning Microscopy Electron (SEM) and Energy Dispersive X-Ray Analysis (EDX) are identified with the presence of the structure of Ni Zn Ferrite in the natural rubber matrix. Anisotropic Ni Zn Ferrite MREs with added carbon black filler given highest hardness value of 90.78; least effected on salt water and water absorption and hygrothermal effect; better wear performance with a low average specific coefficient of friction(COF), specific wear rate and least loss of wear volume. Therefore, it is most suitable use for future application in a wet environment for a long period.

Introduction

Magnetorheological elastomers (MREs) are classified as smart materials that have the capability to alter their mechanical, electrical, magnetic properties with external factors applied such as temperature, magnetic or electrical field, stress and moisture. Combination of rubber matrix (non-magnetic) with ferrite particles (magnetically permeable) in the scale of a micron or nano size with external magnetic field applied on the magnetic mold during curing process provide elastomers to enhance its rheological properties. The reasons of chosen ferrite particles as the main component of MREs as they have high permeability and low hysteresis loss compared to other magnetical particles [1] .

There are two types of solid-state Magnetorheological elastomers (MREs) which are isotropic MREs and anisotropic MREs. With the discovery of phenomena structure of Isotropic MREs and anisotropic MREs by Khimi et al. shown in Fig. 1[2].Structure of isotropic MREs shows a noticeable structure of a uniform distribution of ferrite particles inside the rubber matrix. Anisotropic MREs give special chain-like structures of ferrite particles linked arrangement in rubber matrix from the process of curing matrix when applied the magnetic field.

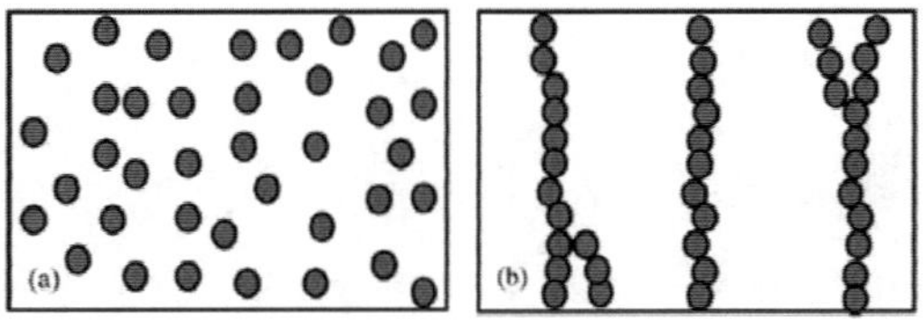

Fig. 1 Structure of MREs (a) Isotropic (b) Anisotropic[2]

The chosen of ferrite particles are Ni Zn Ferrite (NZF) $Ni_{1-x}Zn_xFe_2O_4$, where x>1 with particle weight fraction is 70% inside a base natural rubber matrix and applied magnetic field with 230 mT. Particles size of NZF in between 10 to 30 μm with irregular shape form which added to rubber matrix during compounding stage. Advantages of input NZF particles used in rubber matrix as formed of MREs provide high saturation magnetization, high electrical resistivity, enhance mechanical hardness, tensile strength and cost-effectiveness [3, 4]. Magnetorheological elastomers (MREs) consists of liquid-state and solid-state material[5-6]. The advantages of providing MR fluid (MRF) are it can change its rheological properties frequently and provides better mechanical, electrical, damping properties compare to the conventional rubber products application. However, for liquid-state MREs has occurred a problem of particle sediment which can be solved by solid-state MREs. Solid-state MREs are formed with the rubber-matrix such as natural rubber as the major component to provide excellent mechanical performances without disturbing the performance of MRF effect [7].Natural Rubber used was under the Standard Malaysia Rubber Grade L (SMRL). Natural rubber also is known as the ecological friendly natural product that provides less pollution and waste to the environment [8-15].

The aim of this research is to study the relationship between water and salt water absorption properties along with hygrothermal effecting tribological performance on isotropic, anisotropic and anisotropic with added Carbon black(CB) filler NZF MREs prepared. Additional of CB as a secondary fillers could increase viscosity, the rate of vulcanisation, hardness, tan δ, and the tensile strength of the finished MRE's compound[16]. In this research, solid-state MRE is chosen with added NZF filler as the experimental specimen. Physical properties of Ni Zn Ferrite Magnetorheological Elastomers (NZF MREs) such as mechanical hardness and tribology studies on wear and friction characteristic will be studied with specially applied of the external factor such as water and salt water absorption at room temperature and a high temperature about 60°C.

Experiment

Three types of magnetorheological elastomer (MRE) compounds which natural rubber filled with NZF particles sized between 10 to 30 μm and comprising with particles weight fraction 70%. Compound 1 is an isotropic compound without applied magnetic field. Compound 2 and 3 are an anisotropic compounds with the applied magnetic field 230mT whereby Compound 2 without added filler of N330 carbon black and Compound 3 with an added filler of N330 carbon black with particles sized around 300μm. These three compounds are formulated using a conventional two roll mill machine (model XK 160). Cure times were determined using rheometer MDR 2000. The three compounded MREs are cured using hot press under temperature of 150 °C. N330 Carbon Black filler is chosen as another anisotropic compound of MRE with particle weight fraction of 50 wt.% for better wear performance. Basic formulation is shown in Table 1.

Table 1 Formulation of Ni Zn Ferrite MREs

Materials	Isotropic MRE (phr)	Anisotropic MRE (phr)	Anisotropic MRE CB (phr)
SMRL grade NR	100	100	100
ZnO	5	5	5
Stearic acid	1	1	1

IOP Conf. Series: Materials Science and Engineering **920** (2020) 012010 doi:10.1088/1757-899X/920/1/012010

Paraffin oil	3	3	3
NiZn Ferrite	70	70	70
CBS	2	2	2
TMTD	1	1	1
Sulphur	1.5	1.5	1.5
IPPD	1.5	1.5	1.5
Carbon Black N330	-	-	50
Total Materials (phr)	**185**	**185**	**235**

The fabrication of Isotropic NZF MREs consists of two major steps which are mixing and curing without the applied magnetic field. While fabrication of anisotropic NZF MREs consists of three major steps which are mixing, pre-alignment and curing. The mixing process for both isotropic and anisotropic are using a conventional laboratory two roll mill with model XK 160 according to ASTM D3184-80. The total cycle time of mixing is tabulated in Table 2 [16].

Table 2 Total cycle time of mixing process

Sequence of Materials compounding	Cycle Time (in minutes)		
	Isotropic MRE	Anisotropic MRE	Anisotropic MRE CB
SMRL grade Natural Rubber	3	3	3
Zinc Oxide	2	2	2
Stearic Acid	2	2	2
Ni Zn Ferrite with paraffin oil	5	5	5
Carbon Black with paraffin oil	-	-	7
CBS	2	2	2
TMTD	2	2	2
IPPD	2.30	2.30	2.30
Sulpur	2	2	2
Total Cycle Time	**20.3**	**20.3**	**27.3**

The cure time of three MREs compounds were determined by using MDR Rheometer with model 2000 according to ASTM D 2084. The optimum cure time (t_{90}) is the time required for the torque to reach 90 % of the maximum achievable torque. The curing characteristic was determined at 150°C and a vulcanization curve was plotted from the graph of torque versus time to obtain t_{90} [16]. The value of t_{90} in units of m.m represented as required minutes of cure per millimeter of wall thickness. Therefore, the appropriate cure time of MRE compounds should multiply with the thickness of the MREs fabricated using equation (1).

$$Appropriate\ Cure\ Time = t90(m.m)x\ thickess\ of\ the\ frame \quad (1)$$

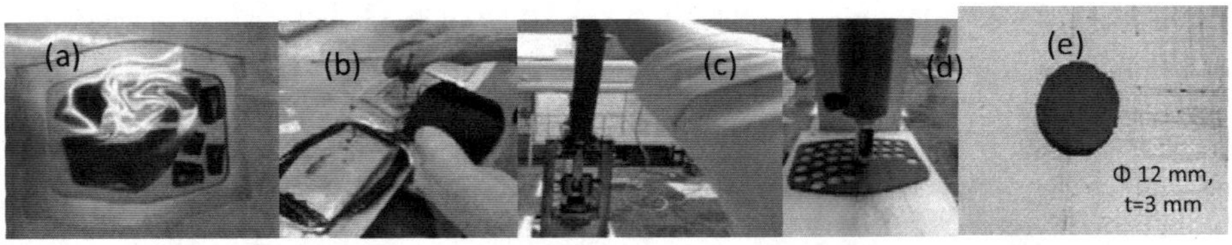

Fig. 2 (a) Compounded rubber sample weighted in frame mould (b) Fabricated Anisotropic NZF (c) and (d) manual punching with 12 mm puncher (e) punched sample 12 mm diameter with 3 mm thickness

Compounded rubber samples were weighted 35 g to 40 g and then placed in a frame mould with dimension 288 x 240 mm and covered by as shown in Fig.2 (a).For compounded rubber to fabricate as isotropic NZF MRE were cured in compression frame mould at 150 °C under pressure 10 MPa by using hot press machine with model 120T. For anisotropic NZF MREs were fabricated with external magnetic mould 230 mT. The external magnetic mould needs to preheat at 80 °C before compounded rubber sample was put between the external magnetic mould to undergo pre-alignment of NZF particles using manual hot press machine at 80 °C for 30 minutes under pressure 12 MPa. Laser thermometer was used to check the temperature frequently. After the pre-alignment process, the compounded rubber samples undergoes post cure at 150 °C under pressure 10 MPa using hot press machine with model 120T. The fabricated NZF MRE sample was taken out and removed the excessive rubber compound as shown in Fig.2(b). Treatment of cooling fabricated NZF at room temperature to re-orientate the magnetic dipoles after post cure [5]. After 24 hours of cooling treatment, NZF MRE can be punched out using manual puncher with the 12mm diameter puncher to a sample with dimension 12 mm diameter x 3 mm thickness as shown in Fig.2(c)-(e).

The degree of hardness of isotropic, anisotropic and anisotropic with added of carbon black NZF MREs samples were measured using TECLOCK GS-706G Type A Durometer according to ASTM D2240. A base of durometer was in contact with NZF MREs samples by applied pressure force manually. These tests were repeated on three times for each compound.

The density of isotropic, anisotropic and anisotropic with added of carbon black MREs sample were determined by using indirectly method from Archimedes' Principle. Initial mass of dry and clean volumetric cylinder was measured using a digital analytic balance with precision 0.0001 g and recorded as M_a. The NZF MRE sample in air was weighed as M_b. Distilled water was filled in the volumetric cylinder and a sinker was put into water and read from the scale level from volumetric cylinder and recorded as V_a. The NZF MRE sample with a sinker was submerged in the volumetric cylinder with containing distilled water. The volume of water displaced from submerged NZF MRE sample was read on the scale of volumetric cylinder and recorded as V_b. The density of NZF MRE sample was calculated using equation (2).

$$Density, g/cm^3 = \frac{M_b}{(V_b - V_a)} \quad (2)$$

Morphology of microstructures of isotropic, anisotropic and anisotropic with added of carbon black NZF MREs were observed using Hitachi S-3400 Scanning Electron Microscope (SEM). The samples were coated with a thin layer of Au using Quorum SC7620 Sputter Coater with applied pressure 2 MPa and time 90 seconds. Elemental composition of NZF MREs samples were analyzed using Energy Dispersive X-Ray Analysis (EDX). Source of X-Ray has interacted with points of samples surface that selected and each element has a unique atomic structure with the presence of peak on its electromagnetic emission spectrum.

Wear and friction characteristic of isotropic, anisotropic and anisotropic with added of carbon black NZF MREs were studied by DUCOM Pin On Disc tester with model TR-20 according to ASTM G99. For the preparation of the testing sample, a MRE sample with diameter 12 mm and thickness 3 mm was attached to the bottom end of a metal pin by using super glue and dried using air dryer around 10 minutes. The MRE sample pin was weighted using a digital analytical balance with precision 0.0001 g and recorded as initial mass before wear test, M_1. Sandpaper with grade 1000 was stick on top the steel rotating disc and wear track diameter was adjusted to 50 mm and screwed tightly. The MRE sample pin was held against the counterface of the rotating steel disc and ensure the surface of the MRE sample pin has an effective contact of the flat surface with the sandpaper on rotating disc. The MRE sample pin was loaded 5 N against the disc through a dead weight loading system. Tribology testing for all MRE samples were conducted under sliding speed of 30 rpm, 50 rpm and 70 rpm and normal loads of 5 N and 10 N. By setting up of WINDUCOM 2010 software in desktop, filename, sample id and remarks were entered and rotary mode was selected before beginning the test. Sliding speed (RPM), load applied (in kg), wear track diameter (mm) and test time for 30 minutes on each sample testing were acquired. The reference point needs to be set zeroing before the test started.

IOP Conf. Series: Materials Science and Engineering **920** (2020) 012010 doi:10.1088/1757-899X/920/1/012010

After reaches elapsed time, MRE sample pin was ejected and measured mass after wear test, M_2. Wear volume loss was calculated by using equation (3) and specific wear rate, k was calculated by using equations (4) and (5) [17].

$$Volume\ loss, mm^3 = \frac{Mass\ loss\ (M_1 - M_2)}{Density} x1000 \quad (3)$$

$$Specific\ Wear\ rate, k\ (mm^3/Nm) = \frac{Volu \quad loss}{Load\ x\ Sliding\ distance} \quad (4)$$

$$Sliding\ distance, m = \pi\ x\ D(m)\ x\ Sliding\ Speed(\frac{m}{s})\ x\ Time(s) \quad (5)$$

Fig. 3 MRE samples immersed in water at room temperature

The behavior of isotropic, anisotropic and anisotropic with added carbon black filler NZF MREs were studied by immersing the samples with the dimension of 12 mm diameter and thickness 3 mm in salt and distilled water separately according to ASTM D570. At first, the samples were dried in an oven at 70 °C about 16 hours and weighed of the conditioned samples were taken before immersion in salt water and water. The water absorption experiments were carried out for 14 days at room temperature as shown in Fig.3. During the period of 14 days immersion, the weight of wet samples were taken daily to calculate the percentage increase in weight during immersion by using equation (6). The samples need to be wiped with absorbent paper before weighing using a digital analytical balance with precision 0.0001 g. After weighing process, the samples need to immerse back into the water immediately. At the end of 14 days of water absorption test, tribology characteristic testing was being studied on the NZF MREs samples to investigate the effect of water absorption on these three compounds.

$$Increase\ of\ weight, M = \frac{M_w - M_c}{M_c} x100 \quad (6)$$

where M_w the weight of is wet sample after immersion to a certain period and M_c is the weight of conditioned dried sample before immersion in water and in salt water.

RESULTS AND DISCUSSION

In Fig. 4 (a), the graph of torque against time, a vulcanization curve was shown to determine T_{90} value. The curves consist of two parts which are maximum torque and minimum torque. Maximum torque represented as the shear modulus of a fully vulcanized compound at the 150 °C while minimum torque is used to measure the viscosity of NZF MRE compound as shown in Fig. 4 [17-19].

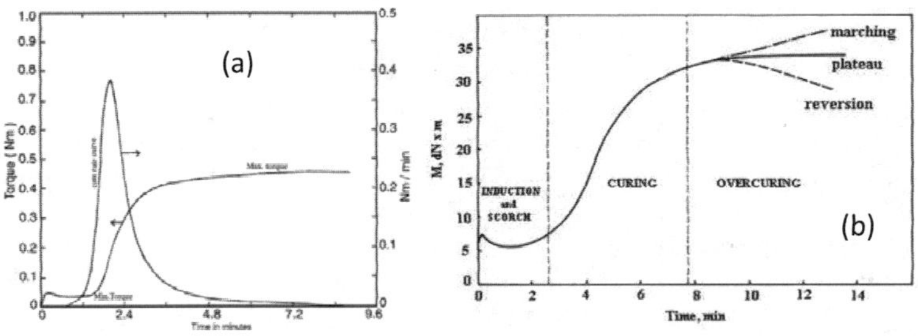

Fig. 4 (a) Vulcanization curve [20] (b) stages of vulcanization curve[21]

There are 3 stages of curing conditions as represented by a vulcanization curve shown in Fig. 4(b). As the first stage, a slow chemical reaction between rubber and the additives when compound was heated under pressure. The second stage is when curing of rubber compound started which lead to torque increases with time. Scorch time (t_{s2}) was determined as induction time where the time crosslinking starts. The slope will continue to increase until it reaches the steady peak value of torque and plateaus out as the third stage (overcuring) [19-20]. T_{90} value is determined when the torque reaches 90% from the maximum torque. This is also represented as the time required for optimum vulcanization of NZF MRE. Comparative table of determine cure time and scorch time for three compounds was tabulated as shown in Table 3 which can be referred to Figs. 5(a) –(b) and (c).

Table 3 Cure time and scorch time determined for three compounds

Compounds	Isotropic	Anisotropic	Anisotropic with added carbon black
T_{90} (m.m)	3.12	3.58	2.80
t_{s2} (m.m)	1.30	1.64	1.12

Each compound of NZF MREs has its own unique cure time and scorch time due to different mixing time when under fabrication of rubber compounding. Lower cure time and scorch time for anisotropic with added carbon black are because of the cross-linking happened between natural rubber and others additives with carbon black [21].

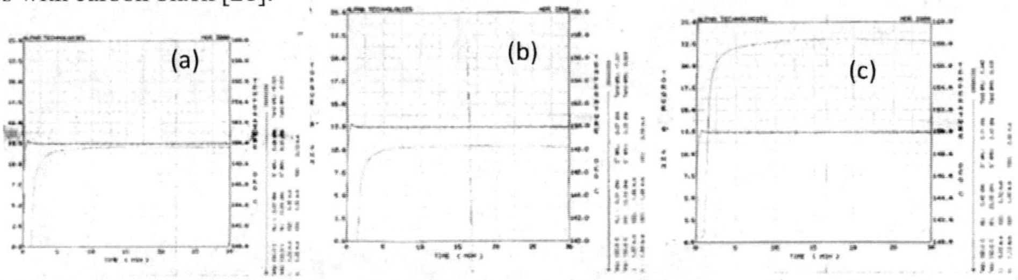

Fig.5 (a) Isotropic compound cure time determined (b) Anisotropic compound cure time determined (c) Anisotropic with added carbon black compound cure time determined

Average hardness value against all type of MRE and density was plotted as shown in Fig. 6(a) and (b) respectively.

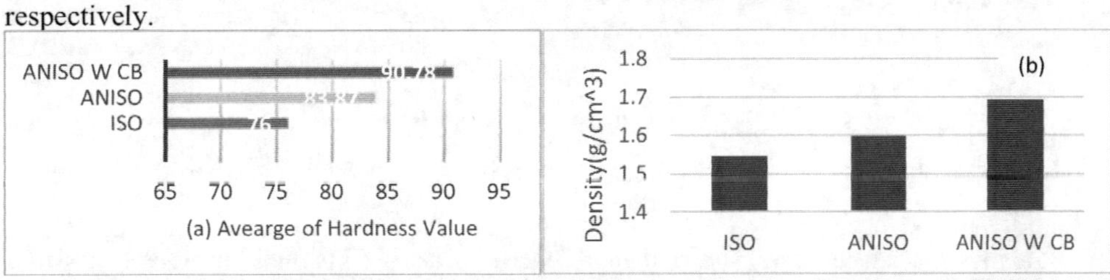

Fig. 6 (a) Average Hardness Value (b) Density specification for all types of NZF MRE

Isotropic NZF MRE shows the lowest hardness compare to anisotropic and anisotropic with added carbon black. Isotropic NZF MRE does not undergo special treatment on applying the magnetic field during curing. For anisotropic and anisotropic with added carbon black with applied magnetic field have higher hardness value which are 83.87 and 90.78 respectively than the isotropic non-applied magnetic field which is 76. The applied magnetic field during curing of NZF will change the stiffness and strength of NZF MRE due to the formation of the chain-like structure of magnetic particle within the rubber matrix. Moreover, additional N330 carbon black filler reinforced NZF MRE by enhancing the mechanical properties of NZF MRE and modify the stiffness, hardness, thermal stability and shrinkage of anisotropic NZF MRE [22].

The result of density specification of NZF MRE was plotted as shown in Fig.6(b). Anisotropic with carbon black NZF MRE shows the highest density (1.6927 g/cm^3). Additional of material 50 phr N330 carbon black in rubber matrix causes more material particles concentrated to the NZF MRE volume given. Next, anisotropic NZF MRE has advantages of alignment of ferrite particles with the applied magnetic field during curing. Therefore, the NZF particles will packed together tightly that lead to increase the density of the material (1.5991 g/cm^3). For isotropic NZF, NZF particles are not packed together and distributed uniformly through the volume space of NZF which causes it has lesser dense (1.5474 g/cm^3) compare to the others. SEM images of isotropic, anisotropic and anisotropic with added carbon black of NZF MREs are captured and shown in Fig. 7 respectively. It can be clearly seen that isotropic NZF MRE had uniform NZF particles distribution in rubber matrix without any particles packed together in Fig.7(a). Both anisotropic NZF MREs in Fig.7(b) and 7(c) shows that the NZF particles are aligned and formed chain-like columnar structure due to the cured under applied magnetic field during fabrication. In Fig.7(c), N330 carbon black particles will be distributed evenly in the rubber matrix. The carbon black particles have restricted the movement of NZF particles during curing under the applied magnetic field. Therefore, alignment of NZF particles are shorter in anisotropic NZF with added carbon black as agreed by [23].

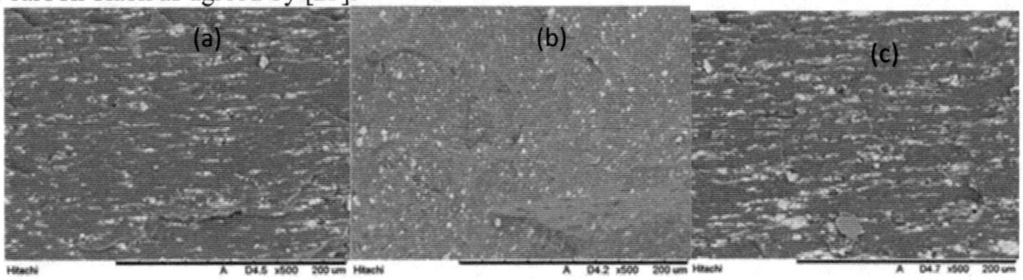

Fig. 7 SEM images of (a) isotropic NZF MRE (b) anisotropic NZF MRE (c) anisotropic NZF MRE with added carbon black at 500 X magnification

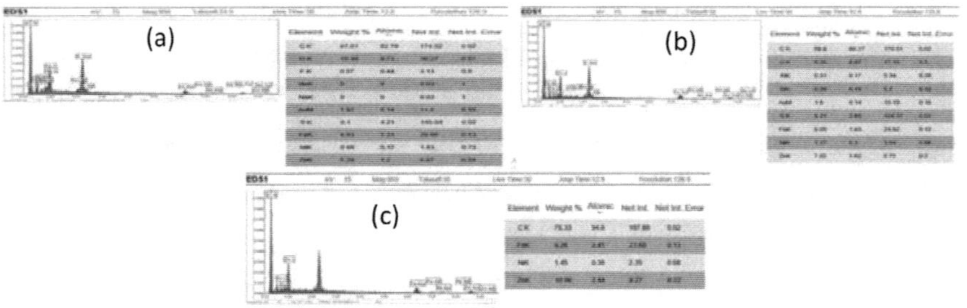

Fig.8 (a)EDX analysis for isotropic (b) anisotropic NZF MRE (c) anisotropic NZF MRE with added carbon black

From the EDX analysis, it can be clearly shown that presence of peaks NZF particles in all type of MREs as shown in Figs. 8(a)-(c) accordingly. The percentage weight of Nickel (Ni) shows between 0.68 % - 1.45 %. The percentage weight of Zinc (Zn) element shows between 5.09 % - 10.96 %. The percentage weight of ferrite (Fe) element shows between 4.93 % - 9.26 % while the percentage weight of carbon content for isotropic and anisotropic NZF MRE shows almost the same which are 67.01 % and 68.8 %. Yet, for anisotropic NZF MRE with added carbon black has the highest percentage weight amount of carbon content which is 78.33 % (Fig. 8c). These have confirmed the presence of added carbon black filler has contributed to the percentage weight of carbon content in the rubber matrix. After immersion of salt water and water in 14 days (336 hours) for all type of NZF MREs at room temperature, a graph of percentage of moisture absorption against immersion for 14 days was plotted as shown in Fig.9(a) and (b).

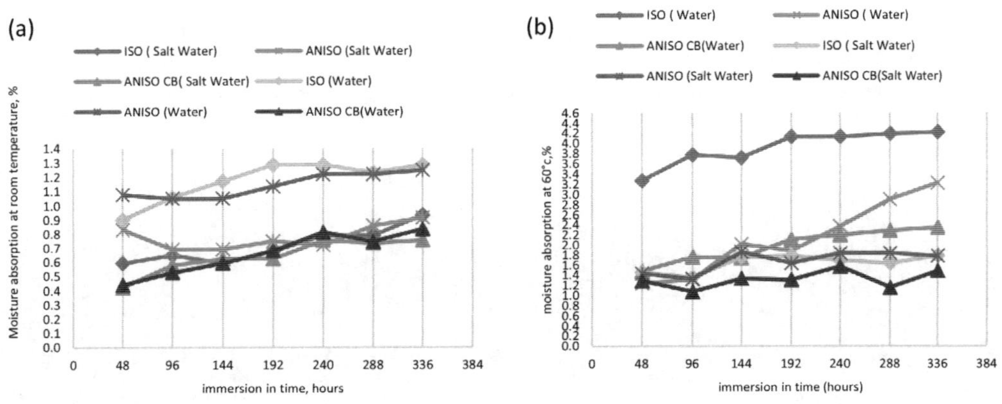

Fig.9 (a)Salt water and water absorption at room temperature (b) Hygrothermal effect at 60 °C upon immersion for 14 days

From the result, it can be clearly shown that low moisture absorbed at weight fraction in salt water and water for all type of NZF MREs at room temperature. For isotropic NZF MRE, moisture absorbed at weight fraction up to 0.9301 % (salt water) and 1.2850 % (water); for anisotropic NZF MRE only 0.9124 % (salt water) and 1.2465 % (water); anisotropic with added carbon black NZF MRE only 0.7545 % (salt water) and 0.8333 % (water). This can be explained by free volume approach which defined as the available volume of the material being unoccupied by the molecules [24]. With the proper curing, uniform dispersion of NZF nanoparticles filler and N330 carbon black filler can reduce the segmental movement of the polymer chain and improve cross-linking in a rubber compound that highly reduces free volume inside a material.

Carbon black filler able to control a large extent of water soluble residues and known as low absorption group [22]. Presence of sodium chloride (NaCl) in salt water solution has prevented the growth of microcavities therefore the percentage of moisture absorption for all NZF MREs are lesser than water absorption [10]. 35% of high concentration salt particles NaCl inside the NZF MREs will create an osmotic pressure which prevents the elastomer to absorb further water ingression [24]. For hygrothermal effect at temperature 60 °C, amount of salt water and water absorption of all NZF MREs had been increased as shown in Fig.9(b). For isotropic NZF MRE, moisture absorbed at weight fraction up to 1.7771 % (salt water) and 4.2162 % (water); for anisotropic NZF MRE only 1.7697 % (salt water) and 3.2172 % (water); anisotropic with added carbon black NZF MRE only 1.4769 % (salt water) and 2.3432 % (water). This has proven that NZF MREs have a weak resistance to temperature effect that soften the elastomer surface leads to higher moisture absorbed inside material when temperature increases [11].

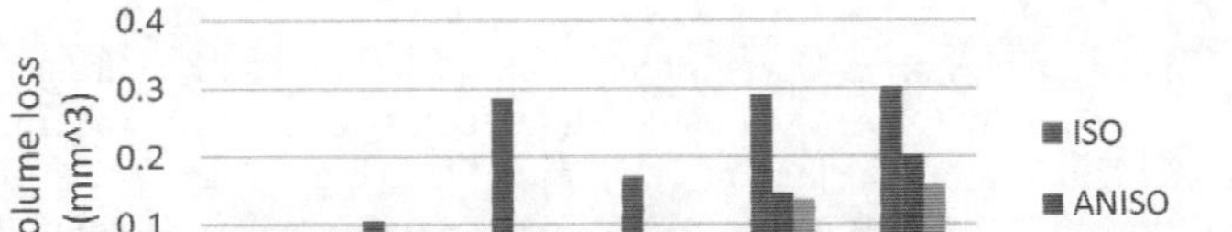

Fig.10 Volume loss for all types of NZF MREs

A combination of sliding speed and load applied increases during wear test, wear loss volume will increases as shown in Fig.10. More wear particles are dispersed out from the elastomer surface causes high loss of mass during high sliding speed and high load applied. Isotropic NZF MREs has the highest wear loss volume in all states of sliding speeds and loads applied followed by anisotropic and anisotropic with added carbon black. This shows that contribution of carbon black filler and applied the magnetic field during curing have significantly changed hardness, stiffness and strength of the elastomer cause anisotropic NZF MREs shows lower wear loss volume. The high hardness of anisotropic with added carbon black shows the lowest wear loss volume and fewer wear particles generated[18].

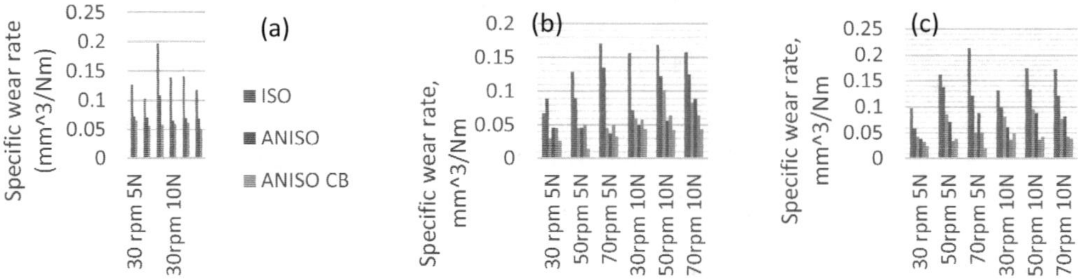

Fig.11 Specific wear Rate for (a) dried NZF MREs (b) Moisture absorption at room temperature (c) Hygrothermal effect upon moisture absorption at 60 °C

Anisotropic with added carbon black NZF MRE shows the lowest specific wear rate compared to the other two compounds at the initial dry condition as in Fig.11a. The reinforced carbon black filler contributed to the hardness of anisotropic NZF MRE which led to better mechanical properties and less deformation of elastomer composites [17]. In Figs. 11b-c, isotropic NZF MREs always shows the highest specific wear rate in all states of environmental conditions applied and different sliding speeds and loads applied on wear tests. This is because the low hardness of isotropic NZF MREs compound has contributed to the low shear strength and high wear loss volume without treatment of alignment of NZF particles through the applied magnetic field. For anisotropic and anisotropic with added carbon black, with the treatment of water and salt water absorption has enhanced wear performance on these two compounds which shows the reduction of specific wear rate compared to the initial dry condition shown in Fig. 12 except on the high sliding speed 70 rpm. Less NZF particles have been washed away by water and salt water absorption because of the NZF particles has been properly aligned by applied magnetic field[14].

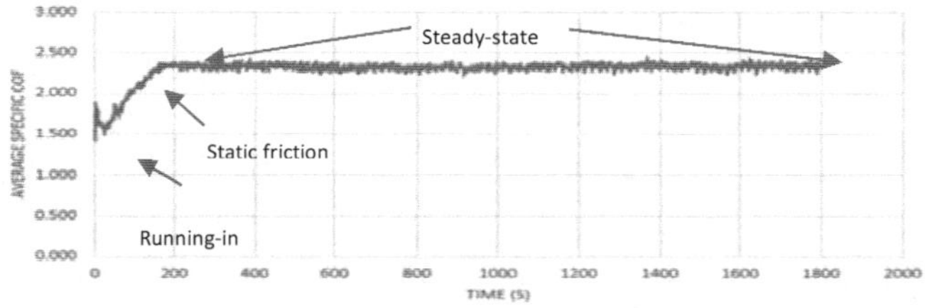

Fig.12 Isotropic compound at sliding speed 50rpm and load applied 10N at initial dry condition

Friction caused by shearing elastomer due to adhesion and defined as resistance to sliding. Friction caused by adhesion when actual contact area, the molecular interaction will form adhesion bonds between the elastomer and smooth surface of sandpaper [13]. Fig. 13(a) shows the result of average specific friction coefficient for all types of NZF MREs with applied of different sliding speed and load applied during wear and friction test. The average specific COF was calculated from the graph of friction coefficient range generated on pin-on-disc machine. For all states of sliding speeds with 10 load applied have shown the reduction of the COF. This result opposed the law of friction where load increases the average specific COF will also increase. This is because the wear debris on the surface causing surface roughness of NZF MRE formed a thin layer of lubricant effect which lead to the decrease of average specific COF when increase of load applied [13]. With the moisture absorption of water and salt water at room temperature, the results was shown in Figs. 19-20.

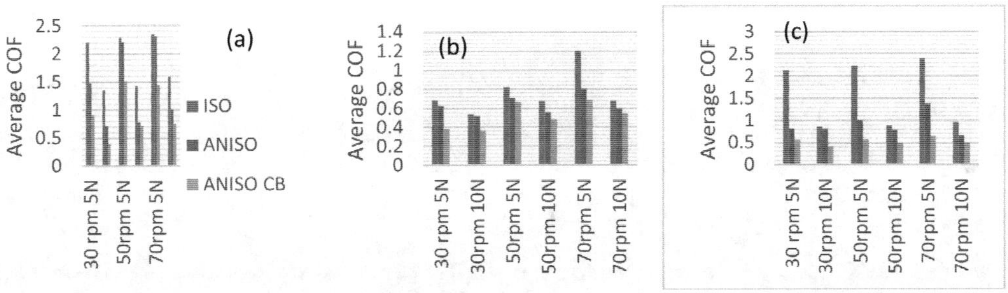

Fig.13 (a) Average COF for all types of NZF MREs (b) with water absorption at room temperature (c) with salt water absorption at room temperature

With the absorption of water and salt water can cause a decrease in the average specific coefficient for all types of NZF MREs compared to the initial dry condition as referred to Figs. 14b-c. The decrease of coefficient depends on the thickness of water film, nature of elastomer formed and smoothness of surface material and presence of foreign particles such as NaCl in salt solution. Increase water and salt water temperature at 60°C in water bath are shown in Fig. 14(a). In the case when there is absent of stick-slip phenomena, the average specific COF with the effect of water absorption at 60 °C for isotropic and anisotropic compounds are slightly increased compared to with room temperature as referred to Fig.14(b). This is because of increase in temperature leading to reduce the hardness of MREs due to the surface soften [12]. At salt water absorption at 60 °C, overall average specific COF for all types of NZF MREs compound are slightly reduced compared to the salt water at room temperature as in Fig. 15(b). The soften surface has been enhanced by penetrated salt particles on the surface leads to reduction of average specific COF.

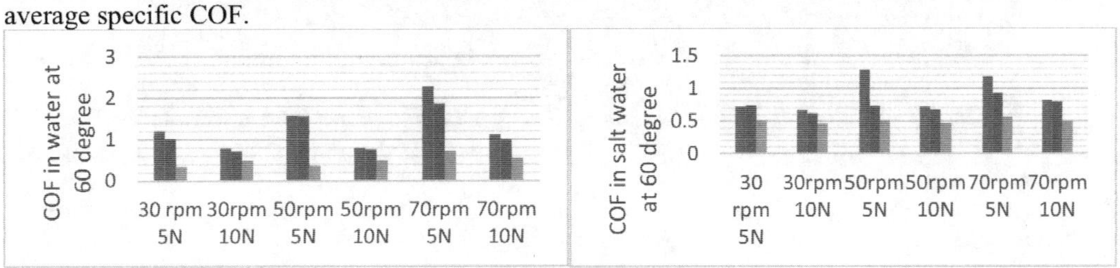

Fig.14 Average COF for all types of NZF MREs (a)with water absorption at 60°C (b)with salt water absorption at 60°C

Surface morphology analysis for all types of NZF MREs were captured using SEM and width of abrasive wear debris formed were measured after wear and friction test with sliding speed 70 rpm and 10 N load applied as shown in Fig. 15. At the high sliding speed and load applied, the abrasive wear debris formed obviously. Anisotropic with added carbon black NZF MRE have less wear debris formed and width of wear debris around 0.082 mm to 0.172 mm. This is can be proven with its good mechanical properties such as high hardness, low wear rate and small and stable friction coefficient. Therefore, anisotropic with added carbon black compound shows relatively slight wear on the surface and highest wear resistance compared to the two compounds. For isotropic compound, long and large size of wear debris formed and width of wear debris around 0.362 mm to 0.558 mm. For anisotropic compound, wear debris formed shorter and smaller size than isotropic compound. Width of wear particles is between 0.181 mm to 0.204 mm. From the SEM observation on surface morphology on water and salt water absorption of NZF MREs, there is no any worn holes or obviously differences on wear surfaces which shows that moisture absorption treatment have reduced wear rate and better wear performance as shown in Figs. 16. With the water and salt water absorption at 60 °C for all types of NZF MREs still showing the similar pattern of wear debris formed as water absorption at room temperature as shown in Fig. 16. However, for high temperature effect at 60 °C on water absorption, isotropic compound started produces worn holes as shown in Fig. 17. This is because small size of sample isotopic NZF MRE started being wear off at the end of the sample with high sliding speed and high load applied. The isotropic NZF MRE has weak resistance to temperature effect, high specific wear rate, low wear performance and low hardness.

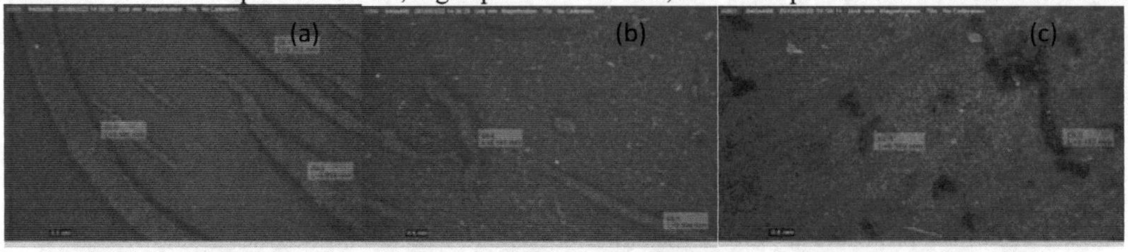

Fig. 15 SEM of worn surface for (a) Isotropic (b) Anisotropic and (c) Anisotropic with C-black

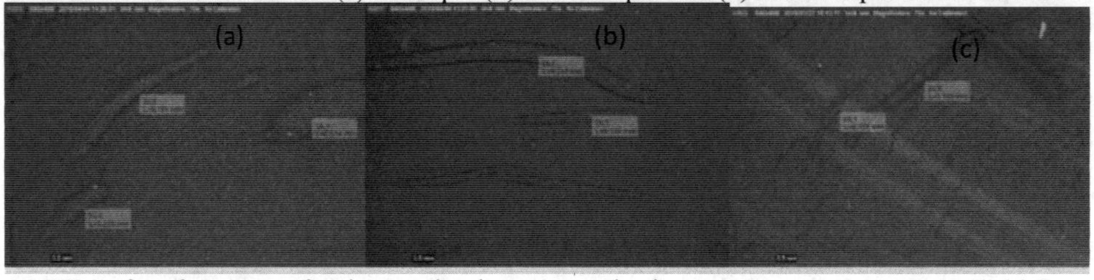

Fig. 16 SEM of surface state after immersion in water and salt water at room temperature) (a) Isotropic (b) Anisotropic (c) Anisotropic with C-black

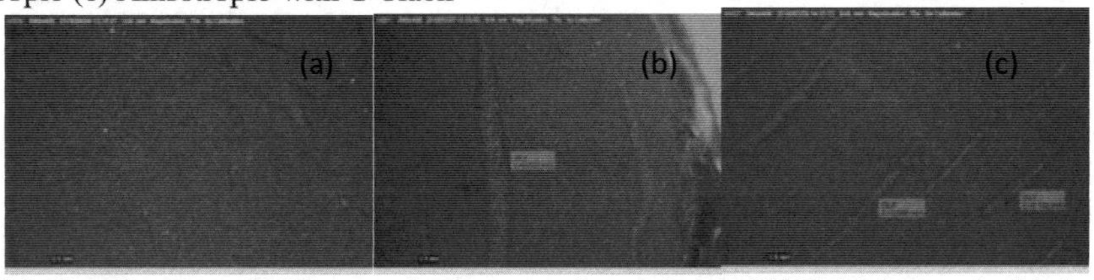

Fig. 17 SEM of surface state in Water and salt water absorption at 60°C) (a) Isotropic (b) Anisotropic (c) Anisotropic with C-black

Conclusion

Wear and friction properties inversely proportional to the hardness of NZF MREs. The high hardness of anisotropic with added carbon black leads to low wear loss volume, low specific wear rate, low average specific COF. Low hardness of isotropic NZF MREs has shown an increase of wear rate, wear loss volume and COF.With the effect of different environmental conditions such as water and salt water absorption at room temperature and at 60 °C, wear and friction characteristic have changed for all types of NZF MREs. Isotropic compound shows slight increase on specific wear rate and average specific COF because of the low hardness nature of isotropic NZF MREs on weak resistance to temperature effect and corrosion of NZF particles in both solutions due to improper alignment of particles during curing without applied magnetic field. It also shows the highest intake of moisture absorption compared to the other compounds. For anisotropic compounds, specific wear rate and average specific COF have slightly decreased when applied to the lubricating effect on water and salt water absorption and does not have significantly increase when effect of high temperature. With the advantages of added carbon black in anisotropic compound has the lowest specific wear rate, average specific COF and least moisture absorption when immersed in solutions. It shows the high resistance to any external factors on applied different environmental conditions. In a nutshell, based on all states of testing conditions such as low or high sliding speed and load applied; temperature effect at room temperature and at 60 °C; different testing solution such as water and salt water, anisotropic with added carbon black compound is the best durability, better wear resistance and always shows steady and stable COF without effect of tested environment conditions.

References

1. Ruddy, C., Ahearne, E., Byrne, G. A Review Of Magnetorheological Elastomers: Properties and Applications. Advanced Manufacturing Science (AMS) Research. 2012
2. Khimi, S.R., K.L. Pickering, and B.R. Mace, Dynamic properties of magnetorheological elastomers based on iron sand and natural rubber. Journal of Applied Polymer Science, 2015. 132(8).
3. Flaifel, M.H., et al., NiZn ferrite filled thermoplastic natural rubber nanocomposites: Effect of low temperature on their magnetic behaviour. Cryogenics, 2012. 52(10): p. 523-529.
4. Sarangi, P.P., et al., Synthesis and characterization of pure single phase Ni–Zn ferrite nanopowders by oxalate based precursor method. Powder Technology, 2010. 203(2): p. 348-353.
5. Khimi, S.R. and K.L. Pickering, Comparison of dynamic properties of magnetorheological elastomers with existing antivibration rubbers. Composites Part B: Engineering, 2015. 83(Supplement C): p. 175-183.
6. Lian, C., K.-H. Lee, and C.-H. Lee, Friction and Wear Characteristics of Magneto-Rheological Elastomers Based on Silicone/Polyurethane Hybrid. Journal of Tribology, 2015. 137(3): p. 031607-031607-7.
7. Chen, L., et al., Investigation on magnetorheological elastomers based on natural rubber. Journal of Materials Science, 2007. 42(14): p. 5483-5489.
8. P. H. Mott, C.M.R., Aging of Natural Rubber in Air and Seawater. Polymer Physics, 2000. Code 6120, Naval Research Laboratory, Washington DC 20375: p. 1- 10
9. J.R. Vinson, R.B.P., W.J. Walker, D.R. Ulrich (Eds.),, The Effect of Relative Humidity and Elevated Temperature on Composite Structures. Center for Composite Materials, University of Delaware (1976), 1976. (3rd Edition), AFOSR TR-77-0030.
10. I., I.K., P.R. A., and A. Stanley, Hygrothermal aging of rubber modified and mineral filled dicyandiamide cured digylcidyl ether of bisphenol A epoxy resin. I. Diffusion behavior. Journal of Applied Polymer Science, 2001. 82(14): p. 3468-3476.
11. Hiratsuka, K.i. and K. Hosotani, Effects of friction type and humidity on triboelectrification and triboluminescence among eight kinds of polymers. Tribology International, 2012. 55: p. 87-99.
12. Lian, C., K.-H. Lee, and C.-H. Lee, Effect of Temperature and Relative Humidity on Friction and Wear Properties of Silicone-Based Magnetorheological Elastomer. Tribology Transactions, 2018. 61(2): p. 238-246.
13. Lee, D.-W., et al., A Study on the Tribological Characteristics of a Magneto-Rheological Elastomer. Journal of Tribology, 2012. 135(1): p. 014501-014501-5.
14. Shuib, R.K. and K.L. Pickering, A new method to predict optimum cure time of rubber compound using dynamic mechanical analysis. JOURNAL OF APPLIED POLYMER SCIENCE, 2013. 131(6): p. n/a-n/a.
15. Hani Aziz Ameen, K.S.H., Ethar Mohamed Mhdi Mubarak, Effect of loads, sliding speeds and times on the wear rate for different materials. American Journal Of Scientific And Industrial Research, 2011: p. 99-106.
16. Malini, K.A., P. Kurian, and M.R. Anantharaman, Loading dependence similarities on the cure time and mechanical properties of rubber ferrite composites containing nickel zinc ferrite. Materials Letters, 2003. 57(22): p. 3381-3386.
17. Raa, K.S. and P.K. L., A new method to predict optimum cure time of rubber compound using dynamic mechanical analysis. Journal of Applied Polymer Science, 2014. 131(6).
18. SALLEH, M.W.B.M., The Effect of Carbon Black Loadings On The Tensile Strength of Rubber Vulcanizates in Faculty of Chemical and Natural Resources Engineering. APRIL 2010, Universiti Malaysia Pahang. p. 52.
19. Bal, S. and S. Saha, Effect of sea and distilled water conditioning on the overall mechanical properties of carbon nanotube/epoxy composites. International Journal of Damage Mechanics, 2015. 26(5): p. 758-770.
20. G.J.Briggs, D.C.E., E.B Storey, Water Absorption of Elastomers Polymer Corporation Limited Research and Development Division Sarnia, Ontario 1967.

21. Song Jun Ping , M.L.X., Contribution of Carbon Black to Thermal Conductivity of Natural Rubber Key Engineering Materials, 2013. 561(© 2013 Trans Tech Publications, Switzerland): p. 158-163.
22. H.W.Kummer, W.E.M., Rubber and Tire Friction. Engineering Research Bullentin B-80, 1967(Automotive Safety Project, Department of Mechanical Engineering, Pennsylvania State University): p. 108.
23. Chowdhury, M.A., et al., The Effect of Sliding Speed and Normal Load on Friction and Wear Property of Aluminium. Int J Mech & Mechatron Eng, 2011. 11: p. 45-49.
24. Frank L.Roth, R.L.D., William L.Holt, Frictional Properties of Rubber. Journal of Research of the National Bureau of Standards, April 1942. 28: p. 440-461.

Acknowledgement:
Special thanks to the Bridging Grant 304.PMEKANIK.6316322 USM

IOP Conf. Series: Materials Science and Engineering **920** (2020) 012011 doi:10.1088/1757-899X/920/1/012011

Experimental investigation on the pull through failure of roof cladding to purlin connection considering overhang roof

T A Majid*, S S Zaini, M K A Muhammad and N A S Nik Ahmad
School of Civil Engineering, Universiti Sains Malaysia, Engineering Campus, 14300 Nibong Tebal, Penang, MALAYSIA

*Corresponding author: taksiah@usm.my

Abstract. The study on the effects wind loading to the roof cladding and purlin connection has raised a great interest among researchers. In some cases, the failure of roof cladding initiated at the overhang roof was due to the development of uplift force created by the strong wind. However, these studies are limited to the non-engineered building without considering overhang roof. This paper studies the effect of roof cladding types, overhang length and cladding thickness in term of pull through capacity of the overhang roof using simple pull through tests. The tests were conducted on two types of claddings namely, corrugated and trapezoidal roof cladding. In the case of corrugated cladding with 0.28 mm thickness, the specimen with 200 mm overhang length shows the highest pull through capacity with 1.534 kN followed by 300 mm overhang length with 1.083 kN. While, for trapezoidal cladding with 0.35 mm thickness, the 300 mm overhang length recorded the highest pull through capacity at 1.163 kN compared to 200 mm overhang length (1.068 kN). It was noted that shorter length of overhang developed stronger roof cladding to purlin connection. The most efficient connection for roof cladding to purlin connection was found to be 0.23 mm thickness of corrugated roof cladding with 200 mm overhang length.

1. Introduction

The wind is caused by different atmospheric pressure where air is moving from higher to lower pressure resulting in winds of various speeds. Malaysia experiences two monsoon seasons namely the Southwest Monsoon from late May to September and Northeast Monsoon from November to March where these seasons nominate wind and rainfall [1]. The structural response due to fluctuating nature of wind loads can cause large scale disaster across community [2]. During high wind events, light-gauge steel roofing system is highly susceptible to premature failures [3]. The roof of a house is usually the most vulnerable part to experience such large wind loading. The interaction between wind flow and building surfaces will give a significant effect on wind loads of a roof of a low rise building as stated by [4].

Uniquely, in the Northern region of Peninsular Malaysia, a study done by Majid *etal.* (2016) found that most of the rural houses are designated by having overhang roof [5] as shown in Figure 1. This part of roof is located at the edge of the roof and protruding outwards where its function is to protect the wall façade during heavy rain. However, due to inadequate support to hold the roofing system from uplift might initiate damage at overhang especially during strong wind events [6]. Inadequate connection strength, insufficient strength of structural member and fluctuation of high wind speed can be the reasons that cause damage to the roof system. Most of the houses in rural areas are non-engineered low rise buildings. This type of house is easily damaged during strong wind events [7]. Hence, this issue should be further investigated focusing on the connection at the roof overhang and purlin connection. This study aims to compare pull through capacity of roof cladding to purlin connections assembled from different cladding thickness, roof cladding type, and overhang length.

Figure 1. Example of overhang steel roof cladding on the rural houses roofing system

Extreme wind events can cause extensive damage to residential buildings due to failure of the roof connection [8]. The worst failure of the roof connection occurred in roofing system might leads to accelerated damage to the whole building [9]. Generally, the roof connection failure can be split into two types namely pull through failure and pull out failure as shown in Figure 2. The pull through failures occurred when the fastener head is pulled through the thin cladding and pull out failure occurred when the fastener pulled out from the purlin as reported by Sivapathasundaram and Mahendran [10]. In addition, the study conducted by Maharaachchi and Mahendran [11] showed that the pull through failure of the connection system was initiated by the presence of large stress concentration due to wind loading around the fastener hole. As a result, the cladding failed locally within the vicinity of the fastener thus forming a sizeable hole that allows the fastener to pull through the cladding [12].To address the drawback, this paper presents an experimental investigation on pull through failure of the steel roof cladding under simulated wind uplift loading considering overhang roof.

(a) (b)

Figure 2. Local failure of the steel roof cladding (a) pull through failure (b) pull out failure

2. Methodology

The methodology in this study consists of two parts namely, rural house survey and experimental work. The rural house survey was carried out to obtain useful information pertaining to the roof dimensions especially on overhang roof length, roof cladding types and purlin spacing. This information was used in shaping the experimental work. The rural house survey covers several districts in Kedah, Penang and northern part of Perak because this region showed to be among the highest recorded house damage due to strong wind event [13].The survey in this study involved 64 number of non-engineered rural houses. The data was collected from 21st January 2019 until 3rd February 2019.

In the case of the experimental work, two types of roof cladding was used namely corrugated and trapezoidal steel roof cladding. Table 1 shows a summary of the overall specimen. Umbrella head nails were used to secure the tested steel roof cladding to the timber purlin. The diameter of umbrella head nail, the shaft and length are 16 mm, 3.90 mm and 50.8 mm length, respectively. The type of timber purlin used in this experimental work is Dark Red Meranti from Strength Group 5 based on MS544 Part 2.This type of wood is common for the construction of roof truss in the rural area of Malaysia due to the relatively low cost compared to metal purlin.

Table 1. Summary of specimen used for experimental work.

Material	Type	Dimension
Roof cladding	Corrugated	Thickness: 0.18 mm, 0.23 mm
	Trapezoidal	Thickness: 0.23 mm, 0.35 mm
Timber purlin	Dark Red Meranti (SG5)	50.8 mm × 50.8 mm × 914 mm
Fastener	Umbrella head nail	Head diameter: 16 mm Shaft diameter: 3.90 mm

The steel roof cladding was fastened to the two timber purlins (50.8 mm × 50.8 mm) and adjusted to achieve a 750 mm span length. The roof cladding was secured using umbrella head nails at the crest and inversely assembled to the steel frame. This type of arrangement is adopted in order to simulate the uplift wind loading acting on roof cladding as shown in Figure 3(a) and the schematic diagram of the experiment is shown in Figure 3(b).The reaction force of the central nail fastener was measured using 5kN load cell attached to the central umbrella head nail at the center timber purlin. Moreover, the deflection of the roof cladding overhang was measured using three Lateral Vertical Displacement Transducer (LVDT). In this case, two were located at unscrewed crest and one was located at screwed crest. The loading pad was designed such that it is able to distribute uniform load from the hydraulic testing machine to the steel roof cladding. The test was proceeding until pull through failure occurred where the roof cladding pullover from the fastener head.

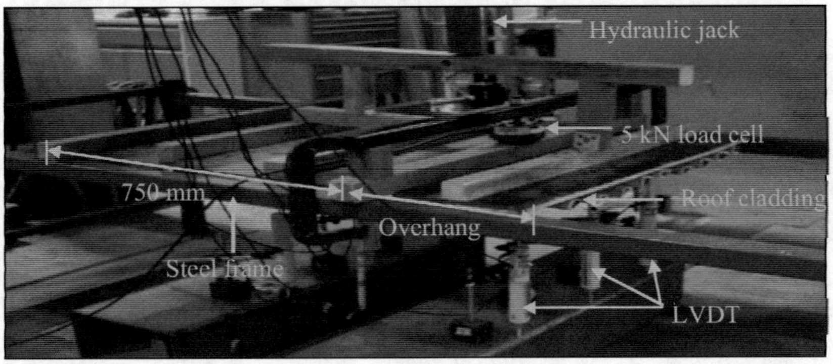

(a)

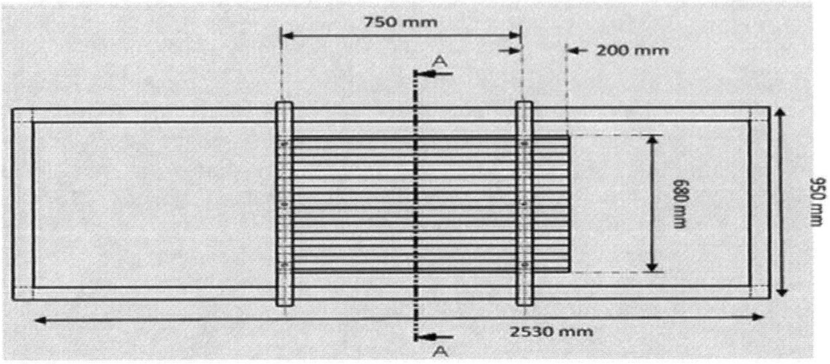

(b)

Figure 3. Pull through test experimental set up showing (a) test set up and (b) schematic diagram

3. Results and discussions

3.1. The rural house survey result

The rural house survey activity involving a total of 64 houses measured the important parameters such as types of claddings, purlin spacing and steel roof cladding overhang length. Figure 4.1 shows the percentage distribution for two types of steel roof cladding found for the damaged rural houses. A number of 36 houses (56%) used corrugated and 28 houses (44%) used trapezoidal as the roof cladding. The result indicates all surveyed houses used steel roof cladding compared to tiles or other types of roofing materials. This finding is particularly true because this material is cheap and requires less number of purlins for the roof assembly. However, the lightness of thin steel cladding compared to tiles or asbestos makes this type of roof material prone to be uplifted during strong wind.

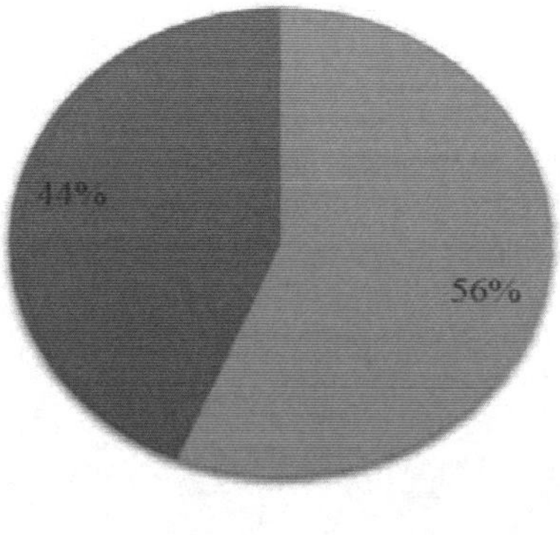

Figure 4. Types of steel roof cladding for rural houses

The variation in terms of the purlin spacing of rural houses is shown in Figure 5. The result showed that 34 houses (53%) were built with purlin spacing between 600 mm to 900, 19 houses (30%) with purlin spacing more than 900 mm and 11 houses (17%) used purlin spacing of less than 600 mm. It is thought that the variation of purlin spacing observed in this survey is merely due to the general practice of the traditional tradesman and has no technical relationship. However, it is well understood that higher purlin to purlin span can result in larger tributary area to the cladding connection hence imposing more wind load to the connections. As a result, the risk of connection failure is increased. As for the experimental set up, a value of 750 mm was selected (average length of 600 mm and 900 mm) to be the span length of the roof cladding for the pull through test. The main function of purlins is to carry roof cladding.

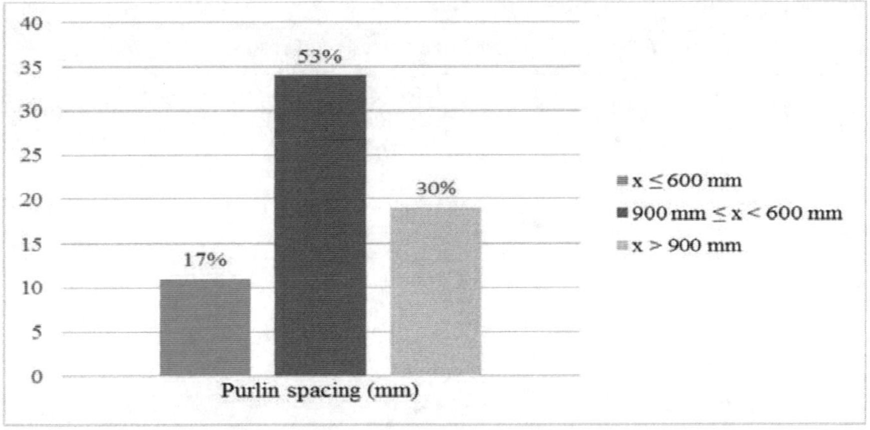

Figure 5. Number of rural houses respected to the purlin spacing

Figure 6 shows the overhang length of the steel roof cladding for the surveyed rural houses. It is apparent from the bar chart that 53 houses (83%) adopted the length of overhang roof between 200 mm to 300 mm. In addition, 6 houses (9%) used the length of less than 200 mm and 9 houses (8%) constructed the overhang roof having more than 300 mm length. The selection of the overhang cladding length in the experimental work done by choosing the dominant length based on-site survey data.

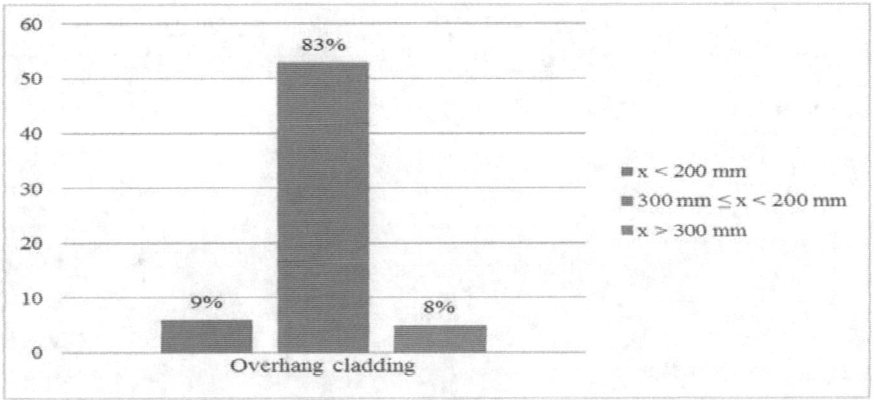

Figure 6. Number of rural houses according to the length of overhang cladding

3.2. Pull through test results

The result for the pull through test was determined based on the average value of the maximum pull through capacity and maximum applied load. The pull through capacity was recorded at the central fastener using 5 kN load cell and the maximum applied load was obtained from the hydraulic jack. The summary of the experimental results is shown in Table 2. It can be seen that, with the same cladding overhang length, the highest thickness of the steel roof cladding shows the increased of pull through capacity and maximum applied load. The results are particularly true for both types of steel roof cladding.

Table 2. Summary of experimental results.

Type of cladding	Thickness (mm)	Cladding overhang length (mm)	Pull through capacity (kN)	Max. applied load (kN)
Corrugated	0.18	200	0.764	2.270
		300	0.548	2.350
	0.23	200	1.534	3.391
		300	1.083	2.830
Trapezoidal	0.23	200	0.944	3.578
		300	0.951	3.444
	0.35	200	1.068	3.738
		300	1.163	3.578

3.3. Effect of type of roof cladding

Figure 7 shows the relationship between the pull through capacity and the displacement of corrugated and trapezoidal roof cladding for 0.23 mm thickness and 200 mm roof cladding overhang length. It can be seen that the pull through capacity of corrugated profile is higher compared to trapezoidal profile. The maximum value of pull through capacity for the corrugated and trapezoidal roof cladding was recorded at 1.534 kN and 0.944 kN, respectively. The difference in pull through capacity is calculated to be approximately 62.5 %.

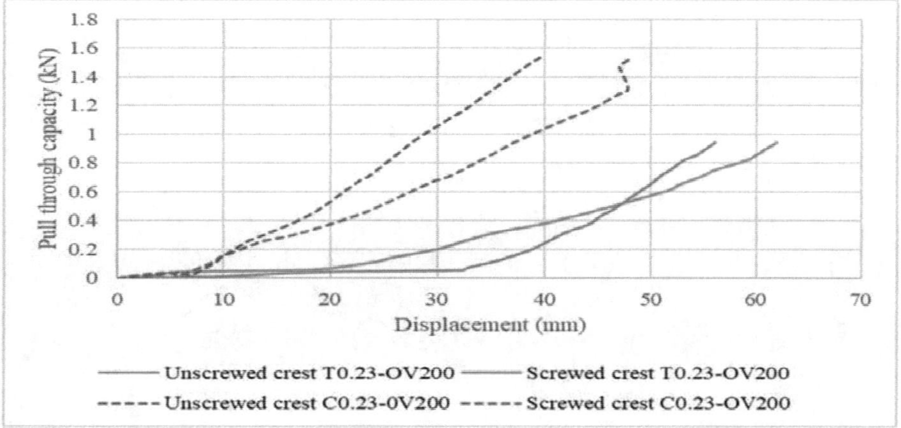

Figure 7. Relationship between pull through capacity and the overhang displacement for the corrugated and trapezoidal profile

3.4. *Effect of length of overhang cladding*

This section discusses the effect of roof cladding overhang length based on pull through capacity and maximum applied load. Figure 8 shows the relationship of pull through capacity with respect to the recorded deformation for 0.23 mm thickness corrugated roof cladding with different overhang cladding length. In this case, the maximum pull through capacity of the specimen with 200 mm and 300 mm overhang length was recorded to be 1.534 kN and 1.083 kN, respectively. This shows that corrugated roof cladding with 200 mm overhang cladding length has better connection strength compared to 300 mm overhang cladding length. In this study, it can be seen that, by reducing 33% of the overhang length, the pull through capacity can be increased up to 41.6%.

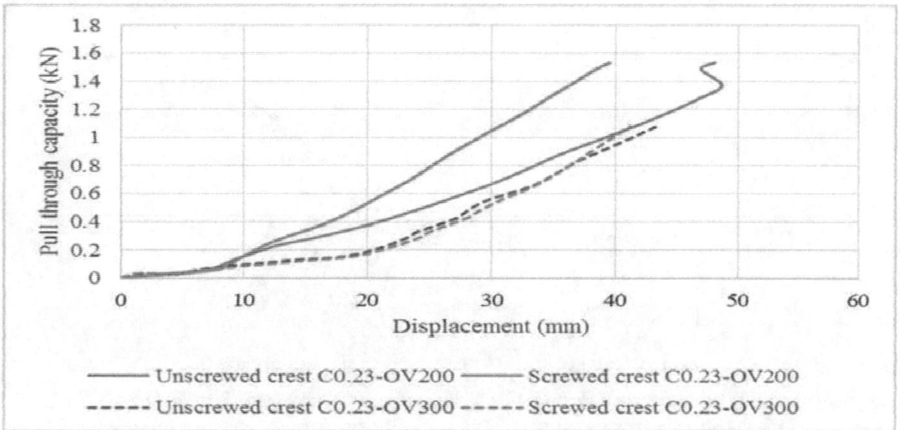

Figure 8. Effect of roof cladding overhang length for corrugated steel roof cladding

Figure 9 shows the results of applied load against displacement with different overhang length for the corrugated roof cladding with 0.23 mm thickness. Similar to the above finding, the magnitude of the maximum applied load for corrugated roof cladding with 200 mm overhang cladding length was found to be higher than corrugated roof cladding with 300 mm overhang cladding length. The maximum applied load for 200 mm overhang length was recorded to be 3.391 kN compared to 2.83 kN for the 300 mm overhang length. It can be seen that by reducing the overhang length 100 mm, the resistant is able to withstand an additional of 19.8% prior to failure.

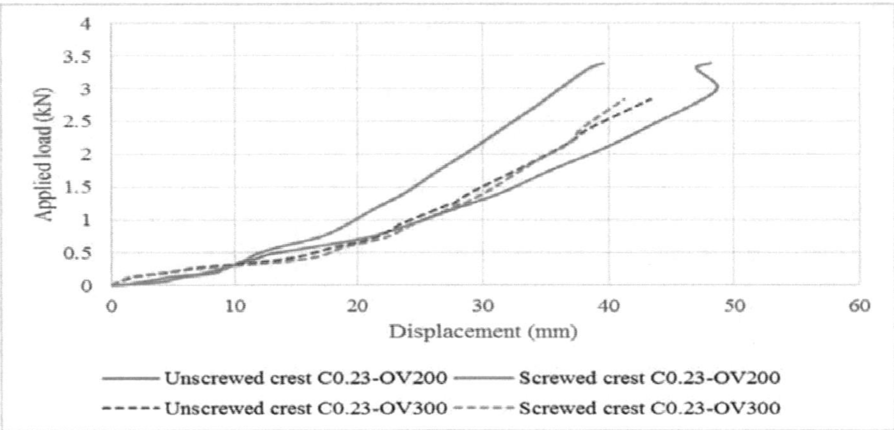

Figure 9. Load-displacement curve for 0.23 mm corrugated steel roof cladding with different roof overhang length

Figure 10 shows the effect of roof cladding overhang length on the pull through capacity for trapezoidal roof cladding with 0.23 mm thickness. The results show that the pull through capacity for the 300 mm overhang length is 0.951 kN and 0.944 kN for the specimen with 200 mm overhang length. Unlike the trapezoidal cladding, the magnitude of the pull through capacity for 200 mm overhang length was found to lower than the 300 mm overhang length. However, the difference was calculated to be 0.74% and can be classified as insignificant.

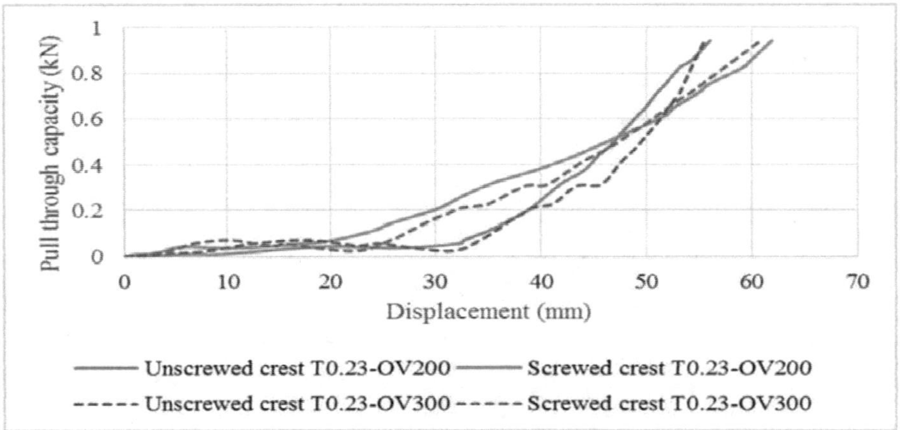

Figure 10. Pull through capacity-deformation curve for 0.23 mm trapezoidal steel roof cladding with different roof cladding overhang length

Figure 11 shows the effect of roof overhang length to the applied load for trapezoidal roof cladding with 0.23 mm thickness. The graph shows that the maximum applied load for the specimen with 200 mm overhang cladding length is marginally greater than 300 mm overhang cladding length. The maximum applied load for the 200 mm and 300 mm overhang length is recorded to be 3.578 kN and 3.445 kN, respectively. The difference in the maximum applied load is calculated to be approximately 3.86%. Theoretically, the overhang cladding length contributes to the maximum applied load where the shorter the overhang length, the higher the maximum applied load will be.

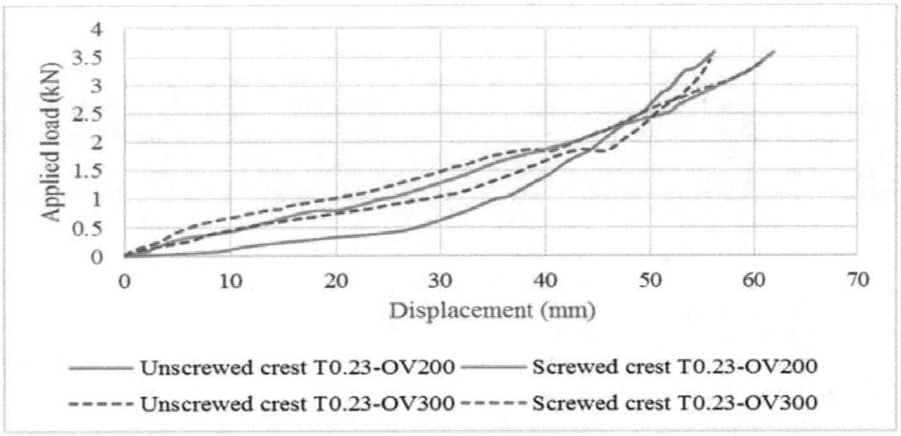

Figure 11. Applied load-deformation curve for 0.23 mm trapezoidal steel roof cladding with different roof cladding overhang length

3.5 Effect on the thickness of the roof

Figure 12 shows the pull through capacity and displacement of corrugated roof cladding with 0.18 mm and 0.23 mm thickness, both having 300 mm roof overhang length. It can be seen that the corrugated roof cladding with 0.23 mm thickness is significantly higher than 0.18 mm thickness corrugated roof cladding. The value of pull through capacity for 0.23 mm thickness corrugated roof cladding is recorded at 1.083 kN while for 0.18 mm thickness is 0.548 kN. The difference between the two values is calculated approximately 97.7%.

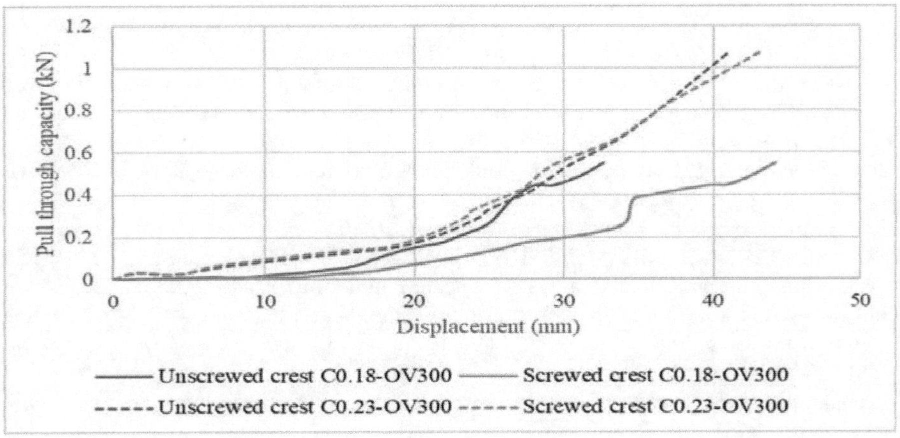

Figure 12. Comparison on roof cladding thickness for pull through capacity-displacement relationship

4. Conclusion

Rural house survey was carried out to determine the length of the overhang roof typically used in northern region of Peninsular Malaysia and the experimental work was carried out to determine the maximum pull through capacity for two overhang length (200 mm and 300 mm) of roof cladding to purlin connection. The tests were carried out using in-house modified frame and hydraulic jack machine. Based on the experimental results, the following conclusion can be drawn: -

- The length of the roof overhang produced significant effect on the roof cladding to purlin connection. Shorter overhang length is able to resist higher wind load and manifested via the increase in the pull through capacity and applied load compared to the longer overhang length.
- The highest pull through capacity and applied load was shown by the 0.35 mm thickness trapezoidal roof cladding with 200 mm overhang length. However, in the case of roof cladding thickness and overhang length, the 0.23 mm thickness corrugated roof cladding with 200 mm overhang length produced the highest value. The pull through resistant can be significantly contributed by the cladding thickness.
- Shorter overhang improved the pull through capacity by increasing the stiffness and reducing the tributary area that accumulate the wind pressure.

5. Acknowledgments

The research team would like to acknowledge Universiti Sains Malaysia for the financial support received from the RUI Grant (1001/PAWAM/8014023).

6. References

[1] Satari S, Zubairi Y, Hussin A and Hassan S 2015 Some statistical characteristic of Malaysian wind direction recorded at maximum wind speed: 1999–2008 *SainsMalaysiana*, **44**, 1521-1530.

[2] Jayasinghe N C, Ginger J D, Henderson D Jand Walker, G. R. 2018 Distribution of wind loads in metal-clad roofing structures *Journal of Structural Engineering* **144**.

[3] Sivapathasundaram M and Mahendran M 2018 Development of suitable strengthening methods for thin steel roof battens subject to pull-through failures *Journal of Architectural Engineering*, **24**, 04018004.

[4] Mahaarachchi D 2003 *Behaviour and design of profiled steel cladding systems subject to pull-through failures* Queensland University of Technology Australia.

[5] Majid T A, Zaini S S, Ismail M A, Deraman S N C and Poon J L 2016 Numerical investigation on the effect of overhang roof around rural house *Advances in Civil, Environmental and Materials Research* (ACEM16).

[6] Zaini S S, Majid, T A, Deraman S N C, Chik F A Wand Muhammad M K A 2017 Post windstorm evaluation of critical aspects causing damage to rural houses in the northern region of peninsular Malaysia. *Matec Web of Conferences*, 2017.

[7] Sabah, S.H.A., Ramli, N.I., Majid, T.A., Zaini, S.S. 2020 Development of site-specific wind hazard map for peninsular Malaysia via spatial modelling, *Energies* **13**(3),729

[8] Chowdhury A G, Canino I, Mirmiran A, Suksawang N and Baheru T 2012 Wind-loading effects on roof-to-wall connections of timber residential buildings *Journal of Engineering Mechanics* **139** 386-395.

[9] Mahendran M. 1995 Wind-resistant low-rise buildings in the tropics *Journal of Performance of Constructed Facilities* **9** 330-346.

[10] Sivapathasundaram M and Mahendran M 2016 Development of suitable test methods for the screw connections in cold-formed steel roof battens *Journal of Structural Engineering* **142**.

IOP Conf. Series: Materials Science and Engineering **920** (2020) 012011 doi:10.1088/1757-899X/920/1/012011

[11] Mahaarachchi D and Mahendran M 2009 A strain criterion for pull-through failures in crest-fixed steel claddings. *Engineering Structures* **31** 498-506.
[12] Yang Nand Bai F. 2017 Damage analysis and evaluation of light steel structures exposed to wind hazards *Applied Sciences* **7** 239.
[13] Wan Chik F A, Che Deraman S N, Ramli N I, Muhammad M K A, Majid T A and Zulkarnain, N Z 2014 Development of windstorm database system for wind damages in Malaysia *Journal of Civil Engineering Research* **4** 214–217.

Assessment of Seismic Scenario- Structure Based Limit State Criteria for a Reinforced Concrete High-Rise Building

Puteri Nihal Che Kamaludin[1], Moustafa Moufid Kassem[1], Ehsan Noroozinejad Farsangi[2], Chee Ghuan Tan[3], and Fadzli Mohamed Nazri[1*],

[1]School of Civil Engineering, Universiti Sains Malaysia, Engineering Campus, 14300 Nibong Tebal, Penang, Malaysia.

[2]Faculty of Civil and Surveying Engineering, Graduate University of Advanced Technology, Kerman, Iran.

[3]Department of Civil Engineering, Faculty of Engineering, University of Malaya, 50603 Kuala Lumpur, Malaysia

*Email: cefmn7@gmail.com/cefmn@usm.my

Abstract. Issues regarding reinforced concrete (RC) high-rise buildings involve different seismic scenarios, such as near- and far-field earthquakes, which can result in different levels and complex seismic scenario structures related to global damage. This study aims to evaluate the seismic scenario structures of a 23-storey RC high-rise building based on different damage measures and to develop a fragility curve with different limit state criteria. Six ground motions are selected to represent two seismic scenarios. Seismic scenario-based global response of the building at increasing earthquake intensity measures is adopted using incremental dynamic analysis (IDA). IDA curves and interstorey drift are used in the parametric study. The structural performance and damage measure of RC high-rise buildings are evaluated with performance-based seismic design limit state. The four performance levels proposed by ATC-43 are operational performance, immediate occupancy, life safety and collapse prevention. The IDA curves showed that near-field effect has a high frequency that gives impact at early intensity measures on building collapse compared with far-field effect. Meanwhile, interstorey drift result indicates that the near-field effect has a larger effect on building damages compared with the far-field effect. Based on the fragility curves, near-field earthquakes have a larger effect towards structural damages than far-field earthquakes.

1. Introduction

At present, the number of high-rise buildings has expanded rapidly because of the changing socioeconomic conditions, rapid population growth and urbanisation. The expansion of high-rise buildings has spread to cities that are subjected to multiple-scenario earthquake-prone regions. The development of commercial high-strength concrete and new advances in construction technologies has influenced the developments of RC high-rise buildings. This development increases exposure to seismic risks that need to be accurately quantified, and thus, the hazard and vulnerability of RC high-rise buildings should be appropriately mitigated [1-4]. Issues regarding RC high-rise buildings involve different seismic scenarios, such as near- and far-field earthquakes, and can result in different levels and complex seismic scenario structures related to global damage. Engineering demand parameters (EDPs) for buildings include interstorey drift ratios and inelastic component structure deformations, such as plastic hinge rotations and floor accelerations, which can be defined from inelastic time history. The structural element performance results from analyses have been used to link EDP to damage measure (DM). DMs describe the physical condition of a structure, including effective descriptions of damage to characterise the performance limit state criteria as reported [5-8]. Noting that, the observation is given to the maximum interstorey drift ratio as a structural damage measure, to evaluate the seismic performance of buildings. Instead of directly considering the distributions of maximum interstorey drift ratio, fragility curves are developed for three different limit states, namely, immediate occupancy (IO), life safety (LS) and collapse prevention (CP). These states will provide a framework to examine the sensitivity of seismic structural responses to ground motion duration on a wide variety of earthquake intensity levels.

Therefore, the behaviour of the structure of RC high-rise buildings must be studied where suitable DMs need to be considered. Two of the major objectives of this research are to evaluate the seismic scenario structure for a 23-storey RC high-rise building based on different damage measures and to develop a fragility curve for the building with different limit state criteria.

2. Methodology

Figure 1 shows the flowchart of the research methodology.

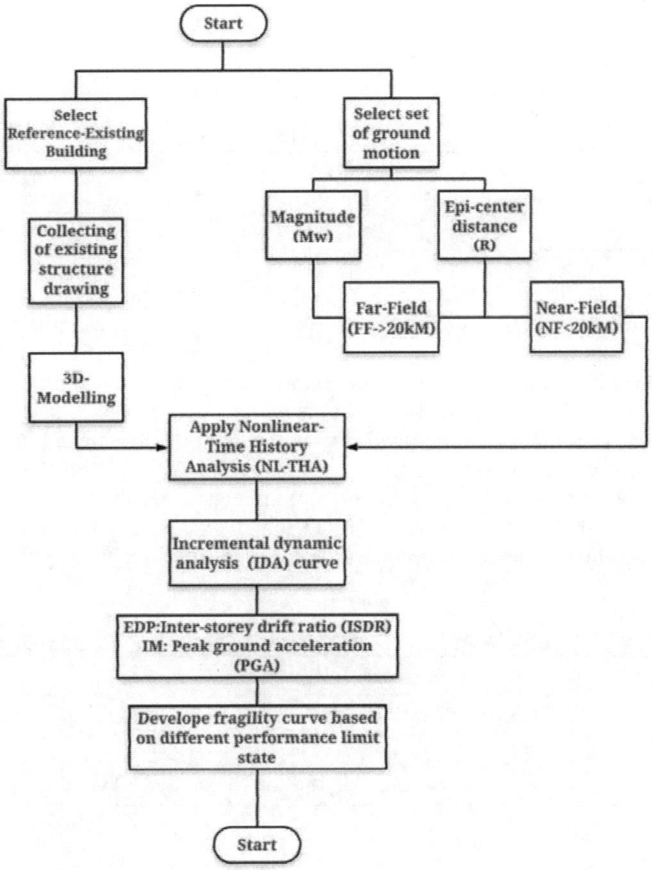

Figure 1. Flowchart of the research methodology

This study focuses on the reinforced concrete of an RC high-rise building. A 23-storey, 83.95 m in height, RC structure is selected as a representative sample building to evaluate seismic scenario structures. The building consists of a basement, 3 storeys for car parking, 16 typical storeys for apartments, a storey for roof floor and 2 storeys for motorised lift. The site class of the building is assumed as site class A (hard rock). Table 1 shows the material properties used to model the building. Table 2 shows the size of slabs, shear walls, columns and beams used to model the building. Table 3 shows the design loads used to define the loads in SAP2000 software. A brief summary of the structure details are presented as follows:

Table 1. Material properties

Material Name	Strength (MPa)
Concrete Compressive Strength, f'c	40
Yielding Tensile Steel Strength fy	420

Table 2. Section properties

No.	Section	Size (mm)
1	Slab	250 mm
2	Shear Wall	200 mm
3	Column	400 x 400 mm
4	Beam	400 x 200 mm

Table 3. Design loads

No.	Design Load	kN/m²
1	Dead Load	2
2	Live Load	4

Earthquake scenarios are a major source of determining seismic demand. According to Oyguc et al. [9], the distances from the considered stations to the epicentre, soil properties and peak ground acceleration (PGA) values are the key parameters for selecting the records. In addition, liquefaction effects are neglected. In this study, seismic scenarios, which are near- and far-field ground motions, are recorded. Three sets of near- and far-field records with magnitudes ranging from 5 to 7 are selected for conducting IDA. The records are downloaded from the Pacific Earthquake Engineering Research Centre website. Selection of the ground motion records is conducted based on the method of Vahdani et al. [10]. Table 4 shows the list of far-field ground motion records selected for this study. Table 5 shows the list of near-field ground motion records selected for this study. Each of the ground motion records is applied to the building by using SAP2000 under time history analysis until the building collapses or reaches the drift limits of 2.5%.

Table 4. Far-field ground motion records (FF record set)

No	Earthquake	Magnitude (M)
1	Imperial Valley-01, 6th of June 1938	5.00
2	Northridge-01, 1st of July 1997	6.69
3	Northern Calif-06, 18th of December 1967	5.20

Table 5. Near-field ground motion records (NF record set)

No	Earthquake	Magnitude (M)
1	Northwest China-03, 11th of April 1997	6.10
2	Parkfield-01 CA, 28th of September 2004	6.00
3	Taiwan SMART1 (40), 20th of May 1986	6.32

A 3D analytical tool is utilised to create a nonlinear model of the sample building as shown in Figure 2. IDAs are performed using six records of earthquake ground motion records. To construct IDA curves that can represent the variable state of the response of the structure related to different earthquake records with increasing intensities, a suitable ground motion intensity measure needs to be defined. Peak ground acceleration (PGA) have been used to be the most accepted intensity measure[11, 12]. A nonlinear time history analysis (NL-THA) is adopted. Ground motion intensity will increase until the building collapses or reaches the drift limit of 2.5%. Based on the analysis, the displacement for each of the storey can be obtained. This finding will help plot the IDA curves, interstorey drifts and fragility curves.

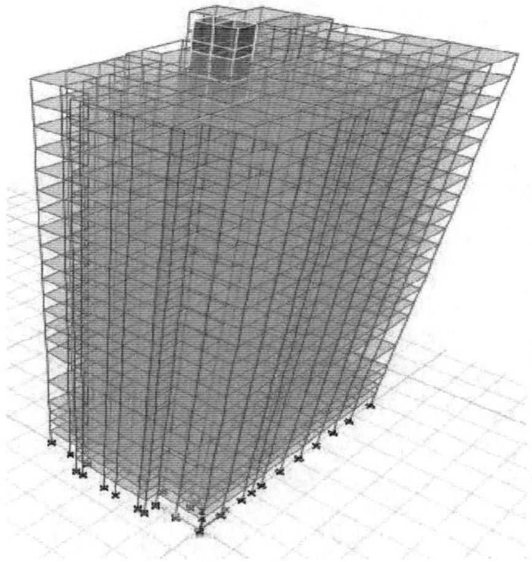

Figure 2. 3D- model of 23-storey RC high-rise building

3. Results and Discussion

The analysis of a 23-storey RC high-rise building is performed based on incremental dynamic analysis (IDA), and the development of the fragility curves for different performance levels OP, IO, LS and CP.

3.1. Incremental dynamic analysis (IDA) curves

As shown in Figure 3, the building experienced operational performance (OP) limit state (drift=0.5%) according to Imperial Valley, Northern California and Northridge ground motions when the PGA reached 0.387g, 0.099g, and 0.055g, respectively. In addition, the building experienced IO limit state (drift=1%) when the PGA reached 0.773g, 0.198g, and 0.111g, respectively. Nevertheless, the building experienced LS limit state (drift=1.5%) when the PGA reached 1.159g, 0.297g, and 0.166g, respectively. Lastly, the building experienced CP limit state (drift=2.5%) when the PGA reached 1.932g, 0.495g, and 0.277g, respectively.

As shown in Figure 4, the building experienced OP limit state (drift=0.5%) according to Northwest, Taiwan and Parkfield seismic records when the PGA reached 0.129g, 0.156g, and 0.066g, respectively. Moreover, the building experienced IO limit state (drift=1%) when the PGA reached 0.257g, 0.310g, and 0.131g, respectively. As well, the building experienced LS limit state (drift=1.5%) when the PGA reached 0.386g, 0.464g, and 0.197g, respectively. Lastly, the building experienced CP limit state (drift=2.5%) when the PGA reached 0.643g, 0.772g, and 0.328g, respectively.

On the basis of the results of the seismic scenarios, by comparing far- and near-field ground motions, the near-field effect experiences of the CP limit states are faster compared with the far-field effect. According to the result, the highest PGAs from the three records of ground motion under far- and near-field cases are 1.932 and 0.772 g, respectively. This finding indicates that the near-field effect affects the ground motion at the site through path attenuation and induced high-frequency filtration. Moreover, the value depends on the characteristics of the site soil strata, where the seismic wave is either amplified or dissipated while moving from the bedrock to the ground level. The modification in the seismic wave impacts the dynamic behaviour of high-rise buildings [2].

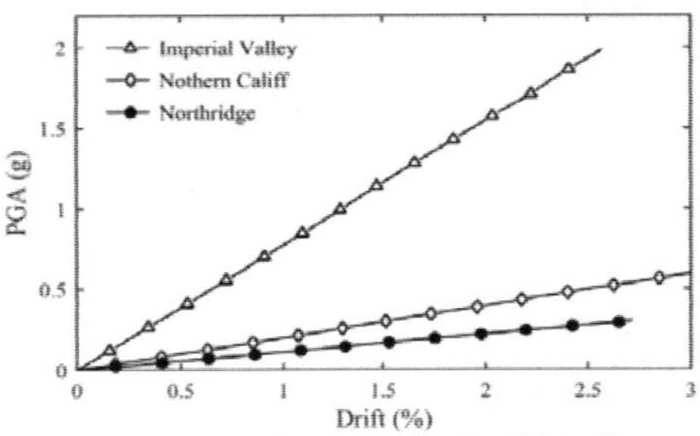

Figure 3. IDA curve for far-field effect

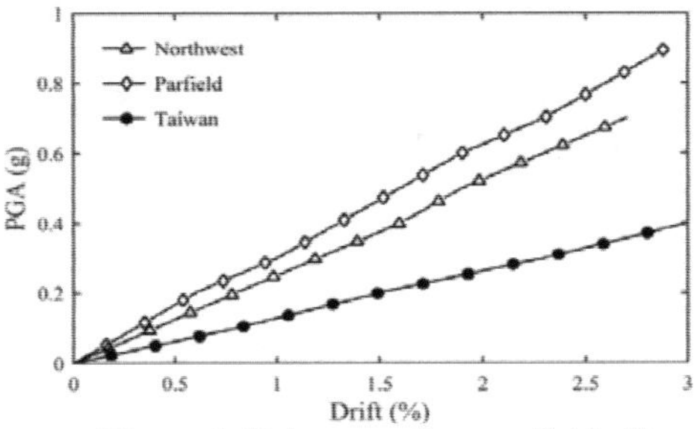

Figure 4. IDA curves for near-field effect

3.2. Fragility Curves

Based on the fragility curve in Figure 5, under far-field effect, the OP limit state achieved 100% of damage when PGA is equal to 0.2g. The IO limit state achieved 100% of damage when PGA is equal to 0.4g. The LS limit state achieved 100% of damage when PGA is equal to 0.55g. Lastly, the CP limit state achieved 100% damage when PGA is equal to 0.95g.

Based on the fragility curve in Figure 6, under near-field effect, the OP limit state achieved 100% of damage when the PGA is equal to 0.25g. The IO limit state achieved 100% of damage when the PGA is equal to 0.5g. The LS limit state achieved 99% of damage when PGA is equal to 1.0 g. Lastly, the CP limit state achieved 81% damage when the PGA is equal to 1.0g.

The far-field effect has a larger effect on the structural damage compared with the near field effect. However, this conclusion is only valid for the present case study. This argument may not be true for all case studies because each has different ground motion that depends on frequency, soil strata and seismic waves.

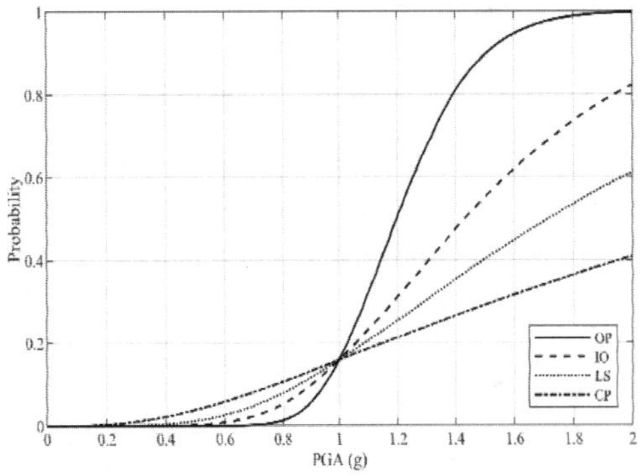

Figure 5. Fragility curve for far-field effect

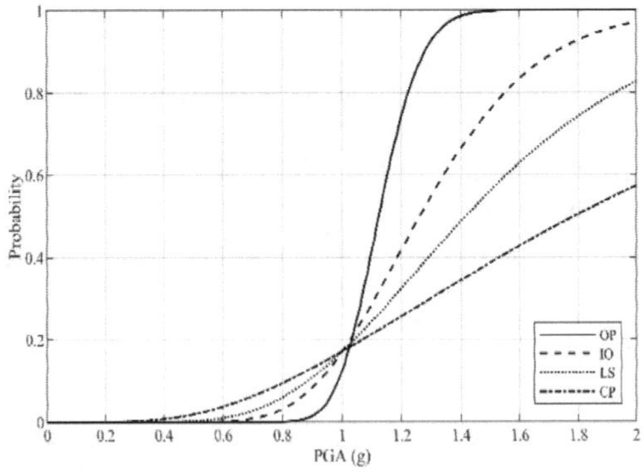

Figure 6. Fragility curve for near-field effect

4. Conclusion

This study evaluates a 23-storey RC high-rise building under different seismic scenarios. Damage measures of the 23-storey RC high-rise building were evaluated with different seismic scenarios, and the fragility curves were developed with different limit criteria. From the numerical comparisons, the following findings can be concluded:

1) In IDA curves, near-field effect produces high frequency that gives impact at early intensity measures on building collapse compared with the far-field effect. Meanwhile, in interstorey drifts, the near-field effect has large effects towards the building damage compared with the far-field effect. This finding indicates that the near-field effect affects the ground motion at the site through path attenuation and induced high-frequency filtration. The effect also depends on the characteristics of the site soil strata. The seismic wave is either amplified or dissipated while moving from the bedrock to the ground level. The modification in the seismic wave impacts the dynamic behaviour of wide-frequency sensitive high-rise buildings.

2) Regarding fragility curves, near-field effect has a larger effect towards the structural damage compared with the far-field effect. However, this conclusion is only valid for the present case study. This argument may not be true for all case studies because each has different ground motion that depends on frequency, soil strata and seismic waves.

Acknowledgement
This research was funded under the Research University Individual (RUI) Grant Scheme (8014080) by the Universiti Sains Malaysia.

References

[1] Alwaeli W, Mwafy A, Pilakoutas K, Guadagnini M 2014. Framework for developing fragility relations of high-rise RC wall buildings based on verified modelling approach. *The Proceedings of the Second European Conference on Earthquake Engineering and Seismology*;.
[2] Alwaeli W, Mwafy A, Pilakoutas K, Guadagnini M 2017. A methodology for defining seismic scenario-structure-based limit state criteria for rc high-rise wall buildings using net drift. *Earthquake Engineering & Structural Dynamics*.;**46**(8):1325-44.
[3] Alwaeli W, Mwafy A, Pilakoutas K, Guadagnini M 2017. Multi-level nonlinear modeling verification scheme of RC high-rise wall buildings. *Bulletin of Earthquake Engineering*;**15**(5):2035-53.
[4] Ji J, Elnashai AS, Kuchma DA 2009. Seismic fragility relationships of reinforced concrete high-rise buildings. *The Structural Design of Tall and Special Buildings*;**18**(3):259-77.
[5] Cornell C 2004. Hazard, ground motions and probabilistic assessment for PBSD. Performance based seismic design concepts and implementation PEER report. **5**:39-52.
[6] Dantanarayan H, MacRae G, Dhakal R, Yeow T, Uma S 2012. Quantifying building engineering demand parameters in seismic events.
[7] Kiani J, Camp C, Pezeshk S 2018. Role of conditioning intensity measure in the influence of ground motion duration on the structural response. *Soil Dynamics and Earthquake Engineering*;104:408-17.
[8] Kassem MM, Nazri FM, Wei LJ, Tan CG, Shahidan S, Zuki SSM 2019. Seismic fragility assessment for moment-resisting concrete frame with setback under repeated earthquakes. *Asian Journal of Civil Engineering*;**20**(3):465-77.
[9] Oyguc R, Toros C, Abdelnaby AE 2018. Seismic behavior of irregular reinforced-concrete structures under multiple earthquake excitations. *Soil Dynamics and Earthquake Engineering* ;**104**:15-32.
[10] Vahdani R, Gerami M, Razi M 2017. Seismic vulnerability assessment of steel moment-resisting frames based on local damage. Journal of Earthquake and Tsunami;11(05):1750016.

[11] Vamvatsikos D, Cornell CA 2002. Incremental dynamic analysis. Earthquake *Engineering & Structural Dynamics*;**31**(3):491-514.
[12] Whittaker A, Deierlein G, Hooper J, Merovich A, Rojahn C, Hamburger RO, et al. *ATC-58 Project Task Report*.

Water quality assessment at selected sites in Perak and Kinta River

S A M Wakif, N Sabtu* and I Abustan

School of Civil Engineering, Universiti Sains Malaysia 14300 Nibong Tebal, Malaysia

*Corresponding author: nsabtu@usm.my

Abstract. This paper describes the water quality study which was carried out at five (5) sampling sites along Sg. Perak and Sg. Kinta respectively. The objective of the study was to examine the chemical, physical and biological characteristics of the river and to determine the level and classification at each site based on the Water Quality Index (WQI) and National Water Quality Standards (NWQS). Water samples from the river were subjected to in-situ and laboratory water quality analyses and focused on six (6) important parameters based on WQI which are Dissolved Oxygen (DO), Biochemical Oxygen Demand (BOD_5), Chemical Oxygen Demand (COD), Total Suspended Solid (TSS), Ammoniacal Nitrogen (NH_3-N), and pH. All testing procedures used in this study adhered to the American Public Health Association (APHA) testing procedures. Based on the results obtained, the quality of water for four (4) sampling stations along Sg. Perak can be concluded as CLASS III with only one station concluded as CLASS IV according to the National Water Quality Standards (NWQS). For Sg. Kinta, four sampling stations were classified as CLASS III whereas only one station falls into CLASS II. It is important to determine the WQI at different areas of along the river in order to predict (possible) points of pollution for monitoring purposes.

1. Introduction

Perak River (Sg. Perak) is the second longest river in Peninsular Malaysia after Sungai Pahang. Sg. Perak is approximately 400 km long with a basin size of 14,900 km^2. The river extends from the mountainous Perak-Kelantan-Thailand border of the Belum Forest Reserve and ends at the Straits of Malacca. As the second longest river in Peninsular Malaysia, Sg. Perak is an important asset to the state of and people of Perak. It serves as the primary source of raw water and contributes to (approximately) 70 percent of raw water in Perak that is channelled to the population. This amounts to about 1300 million litres of water that is distributed to 2.5 million residents in Perak daily [1]. This number is expected to increase in the future as the population of Perak increases. Apart from this, the river also serves as one of the main sources of livelihood for fishermen and is also a tourist attraction by providing tourist activities such as kayaking and canoeing. The river also contributes in many other contexts in terms of benefit, profit and serviceability.

Kinta River (Sg. Kinta) on the other hand is also one of the most significant rivers in Perak as it is one of the main tributaries of Sg. Perak. The length of the river itself extends to approximately 100 km long with a catchment basin of 2500 km^2. Sg. Kinta originates from Gunung Korbu and the downstream end of Sg. Kinta adjoins into Sg. Perak. Sg. Kinta main function is for the provision of water supply, but it is also a critical resource for industrial/production in Ipoh [2,3].

An increase of urbanisation, industrialisation, and agricultural developments along the river basins over the years contributes to the higher levels of river pollution. Household sewage; fertilisers; herbicides

and pesticides; sand mining activities; waste and sewage from heavy industries such as iron and steel industry; food industry; pharmaceuticals and others are released into the rivers through drainage systems thus contaminating and deteriorating the quality of rivers. The release of untreated organic waste into these rivers also reduces the availability of oxygen in the river hence affecting the ecosystem [4].
In order to address the issue of river pollution, a river water quality-monitoring programme was started in 1978 by the Ministry of Natural Resources and Environment in order to establish the baseline as well as to detect the water quality of rivers in Malaysia. This programme has also extended to identifying the sources of pollution and attempts to continuously monitor the changes in river water quality. The programme collects water samples at regular intervals from designated stations for in-situ and laboratory analyses in order to determine its physico-chemical and biological characteristics [5]. Continuous monitoring of river water quality is vital for early detection of pollution influx.
River water quality in Malaysia has always been a cause for concern for various local authorities, government agencies as well as the public at large. This study was therefore performed with the aim of assessing river water quality of two important rivers at different locations/sites where continuous monitoring is not (yet) conducted so that the (possible) sources of pollution may be identified. Ultimately, the outcome of this study would increase awareness on the importance of clean rivers for the ecosystem, to reduce the cost of water treatment for water supply, and the preservation of raw water supply for the future.

2. Materials and Methods

2.1. Study Area

Sg, Perak and Sg. Kinta river were selected in this study as both rivers are important rivers in the state of Perak Darul Ridzuan. Many industrial, domestic, and agricultural activities occur along these rivers. For this study, five (5) sampling stations were selected for each river. The water quality monitoring in Sg. Kinta and Sg. Perak were conducted to obtain qualitative information as well as to examine the characteristics (chemical, physical, and biological) of the river. Figure 1 shows the location of samples collected along Sg. Perak. Exact location and details for the sampling locations for both Sg. Perak and Sg. Kinta are as shown in Table 1(a) and (b) respectively.

Figure 1. Location of sampling sites along Sg. Perak

Table 1(a). Sampling location and coordinates (Perak River).

Station	Site Location	Coordinates	
		Latitude (N)	Longitude (E)
K1	Sg. Kinta	4°05′48.9″	101°00′48.8″
K2	Muara Sg. Kinta	4°05′30.3″	101°00′40.4″
K3	Sg. Kinta – Sg. Bidor (Middle)	4°04′17.9″	101°00′41.8″
K4	Muara Sg. Bidor	4°02′25.6″	101°01′40.1″
K5	Muara Benteng Beras Basah	3°59′26.0″	100°45′36.7″

Table 1(b). Sampling location and coordinates (Kinta River).

Station	Site Location	Coordinates	
		Latitude (N)	Longitude (E)
S1	Kg. Harewood	4°26′34.35″	101°3′0.57″
S2	Taman Senawi	4°19′3.03″	100°43′48.75″
S3	Biovalley	4°18′35.69″	101°4′40.22″
S4	Jong Baruk	4°6′57.17″	101°1′4.75″
S5	Kg. Pelanduk	5°8′24.83″	100°37′28.36″

Water samples were obtained from the river and in-situ test and laboratory experiments were conducted in order to determine the river water quality. In-situ parameters such as pH, Dissolved Oxygen (DO) and temperature were determined using YSI Quatro equipment whereas other parameters such as Biological Oxygen Demand (BOD_5), Chemical Oxygen Demand (COD), Total Suspended Solid (TSS) and Ammoniacal-Nitrogen (NH_3-N) were determined in the laboratory in accordance to the specific American Public Health Association (APHA) standards for each parameter. All laboratory analyses were conducted at the Environment Laboratory of School of Civil Engineering, Universiti Sains Malaysia. Table 2 shows all of the parameters tested and the methods/standards referred to in this study. Only six (6) parameters were measured/determined as all of these parameters are significant in order to determine the Water Quality Index.

Table 2. Water quality parameters and the related methods used.

Parameters	**Method**	**Unit**
Temperature	In-situ	°C
pH	In-situ	-
DO	In-situ	mg/l
BOD	APHA 5210B	mg/l
COD	APHA 5220B	mg/l
TSS	APHA 2540D	mg/l
NH_3-N	APHA 4500	mg/l

2.2 Water Quality Index

Water Quality Index (WQI) is a rating tool that summarizes six important water quality parameters into one concise and objective value representing the state of water quality in a river. The resulting value of WQI determines the status of the river water quality based on National Water Quality Standards Malaysia. The status of the rivers are divided into six (6) classes ranging from Class 1 to Class V. Each class would relate to the appropriate/allowable uses of the river. Each of the parameters may also be

scored individually (based on the sub index) and the corresponding status (clean, slightly polluted, or polluted) may be determined. The advantages of using WQI is that it is an most effective tools to communicate information on the quality of any water body [6] and may be used for comparison purposes (where one river may easily be compared to another). However, there are limitations of using WQI as it only takes into consideration six parameters. In reality, there are various other parameters that may be polluting the river that is not measured or considered.
WQI is calculated using the following equation where SI is a sub-index value for each designated parameter mentioned previously. The WQI score obtained should be in the range of zero (0) to one hundred (100). A score of hundred would effectively indicate that the river is pristine and the lower the score the more polluted the river is.

$$WQI = (0.22 \text{ X SIDO}) + (0.19 \text{ X SIBOD}) + (0.16 \text{ X SICOD}) + (0.15 \text{ X SIAN}) + (0.16 \text{ X SISS}) + (0.12 \text{ X SIpH})$$

where,
SIDO = Sub Index DO (% saturation)
SIBOD = Sub Index BOD_5
SICOD = Sub Index COD
SIAN = Sub Index NH_3-N
SISS = Sub Index SS
SIpH = Sub Index pH

Table 3 shows the sub index (SI) and WQI range and its corresponding classification whereas Table 4 shows the classification of water quality based on the National Water Quality Standard (NWQS, 2018) [7].

Table 3. Sub Index and Water Quality Index range and its corresponding classification

SUB INDEX & WATER QUALITY INDEX	INDEX RANGE		
	Clean	Slightly Polluted	Polluted
Biochemical Oxygen Demand (BOD)	91-100	80-90	0-79
Ammoniacal Nitrogen (NH_3-N)	92-100	71-91	0-70
Suspended Solids (SS)	76-100	70-75	0-69
Water Quality Index (WQI)	81-100	60-80	0-59

Table 4. Classification of Water Quality based on Water Quality Index [7]

PARAMETER	UNIT	KELAS				
		I	II	III	IV	V
Ammoniacal Nitrogen	mg/L	< 0.1	0.1 – 0.3	0.3 – 0.9	0.9 – 2.7	> 2.7
Biochemical Oxygen Demand	mg/L	< 1	1 – 3	3 – 6	6 – 12	> 12
Chemical Oxygen Demand	mg/L	< 10	10 – 25	25 – 50	50 – 100	> 100
Dissolved Oxygen	mg/L	> 7	5 – 7	3 – 5	1 – 3	< 1
pH	-	6.5-8.5	6 –9	5 – 9	5 – 9	-
Total Suspended Solid	mg/L	< 25	25 – 50	50 – 150	150 – 300	> 300
Indeks Kualiti Air (IKA)		> 92.7	76.5 – 92.7	51.9 – 76.5	31.0 – 51.9	< 31.0

3. Results and Discussion

Table 6 show the WQI results obtained for all of the sampling sites for Perak River (Sg. Kinta – Sg. Bidor). Based on these results, the quality of water for sampling stations - Muara Sg. Kinta, Sg. Kinta - Sg. Bidor (M), and Muara Beteng Beras Basah can be concluded as CLASS III whereas Sg. Kinta can be concluded CLASS IV according to the National Water Quality Standards [7]. This classification is based on Classification of Water Quality Based on WQI as shown in Table 4.

However, on the basis of the Water Quality Index itself, samples obtained from station Muara Sg. Kinta, Sg. Kinta and Sg. Bidor (M) and Muara Benteng Beras Basah falls into the slightly polluted category. Samples obtained from station Sg. Kinta and Muara Sg. Bidor falls into the polluted category. This classification is based on Water Quality Index category as shown in Table 3.

Table 5. Water Quality Index (Perak River).

Station	Water Quality Sub Index (SI)						WQI	Class	Classification
	DO	BOD	COD	AN	SS	pH			
K1	7	71	79	0	72	97	51	IV	Polluted
K2	50	90	94	17	78	97	70	III	Slightly Polluted
K3	18	79	88	25	76	100	61	III	Slightly Polluted
K4	18	73	86	0	86	96	57	III	Polluted
K5	63	71	64	0	65	99	60	III	Slightly Polluted
Clean			Slightly Polluted			Polluted			

Table 7 show the WQI results obtained for the sampling sites for Kinta River. Five (5) sampling sites were chosen – which were Kg. Harewood, Taman Senawi, Biovalley, Jong Baruk and Kg. Pelanduk. Based on the results obtained, the quality of water for sampling stations Kg. Harewood, Taman Senawi, Jong Baruk and Kg. Pelanduk is categorized as CLASS III whereas Biovalley falls into CLASS II according to the National Water Quality Standards (NWQS). Figure 2 represents the classification results of water quality in (WQI) for Perak River from upstream to downstream. It can be seen that the water quality from the downstream of Sg. Kinta (K1) is polluted – but as the river merges into Sg. Perak, the pollution is perhaps diluted. The water quality at Muara Sg.Kiinta (K2) is only slightly polluted. As it travels downstream, the river water quality starts to deteriorate as it accumulates more pollutants.

Table 6. Water Quality Index (Kinta River).

Station	Water Quality Sub Index (SI)						WQI	Class	Classification
	DO	BOD	COD	AN	SS	pH			
S1	58	67	44	0	76	99	57	III	Polluted
S2	78	80	87	0	24	96	62	III	Polluted
S3	89	97	92	0	86	98	78	II	Slightly Polluted
S4	53	95	78	0	24	99	58	III	Polluted
S5	49	82	89	0	72	99	64	III	Slightly Polluted
Clean			Slightly Polluted			Polluted			

IOP Conf. Series: Materials Science and Engineering **920** (2020) 012013 doi:10.1088/1757-899X/920/1/012013

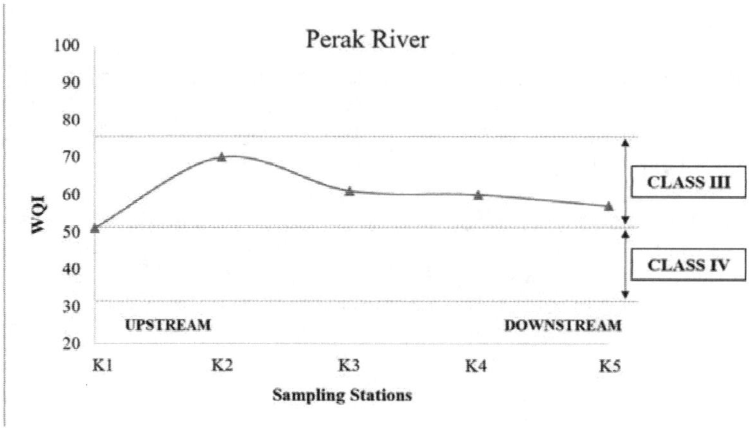

Figure 2. Water Quality Index for Perak River (from upstream to downstream)

3.1. Water Quality Results

3.1.1 ***DO*** *Concentration, mg/L*

Based on the Department of Environment (DOE) limit, most of the parameters sampled were within the DOE limit for Class III river as shown in Table 6. However, based on the Water Quality Index, samples obtained from station Muara Sg. Kinta, Sg. Kinta and Sg. Bidor (M) and Muara Benteng Beras Basah falls into the slightly polluted category. Samples obtained from station Sg. Kinta and Muara Sg. Bidor however falls into the polluted category. Based on Figure 3(a), it can be seen that all of the samples fall within the DOE limit which is 3-5 mg/L for Class III river. The DO concentration at Sg Kinta, Sg Kinta – Sg Bidor (M) and Muara Sg Bidor are 1.35 mg/L, 2.27 mg/L and 2.24 mg/L respectively which is within the limit of Class IV.

Based on Figure 3(b), it can be seen that the results obtained from Taman Senawi and Biovalley falls within the DOE limit which is 5-7 mg/L for a Class II river. DO concentration for Kg Pelanduk, Jong Baruk, and Kg Harewood, all falls within the Class III river category with the lowest DO concentration found at Kg. Pelanduk with 4.1 mg/L of DO.

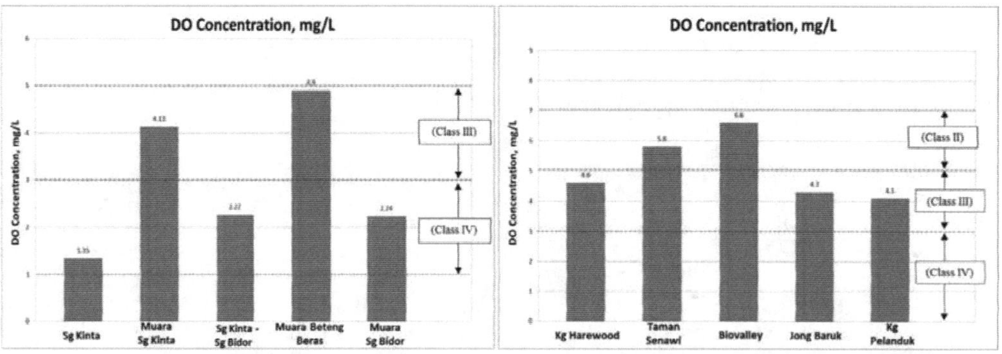

Figure 3(a). Perak River **Figure 3(b).** Kinta River

Figure 3(a),(b) – DO (mg/L) concentration at each sampling location

3.1.2 $\boldsymbol{BOD_5}$ *Concentration, mg/L*

Figure 4(a) shows the results obtained for Biological Oxygen Demand (BOD_5) in mg/L for all of the sites. From this figure, it is observable that the concentration of BOD_5 for Sg. Kinta, Muara Beteng Beras and Muara Sg. Bidor falls within the DOE limit, which is 6-12 mg/L for Class IV. The results for

BOD_5 for Sg. Kinta – Sg. Bidor (Middle) falls within the limit of Class III. The lowest BOD_5 concentration is at Muara Sg Kinta which is 2.5 mg/L and falls within the range of a Class II river. Based on Figure 4(b), it can be seen that the concentration of BOD_5 for Kg. Harewood and Kg. Pelanduk falls into Class IV. The BOD_5 concentration for Taman Senawi falls into Class III whereas the concentration of BOD_5 for Jong Baruk and Biovalley falls within Class II and I respectively. Biovalley shows the lowest concentration of BOD_5 with only 0.78 mg/L.

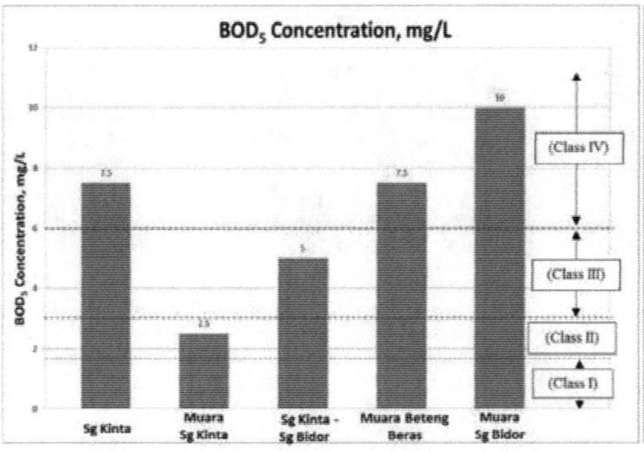

Figure 4(a). Perak River

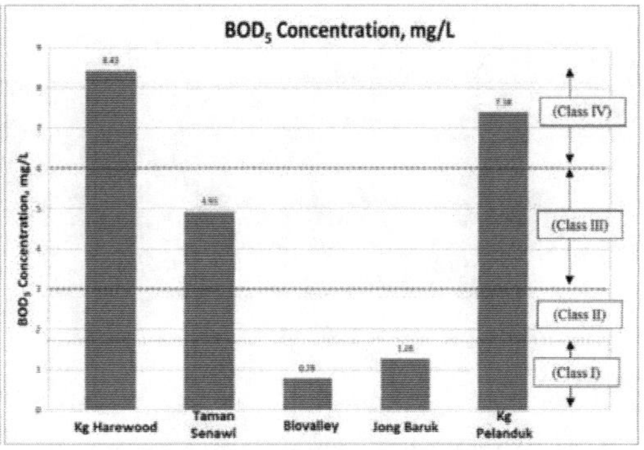

Figure 4(b). Kinta River

Figure 4(a),(b) – BOD_5 (mg/L) concentration at each sampling location

3.1.3 ***COD*** *Concentration, mg/L*

Figure 5(a) shows the analysis for Chemical Oxygen Demand (COD) in mg/L for those sampling stations. Based on this figure, it is perceptible that the concentration of COD for Muara Benteng Beras is 29 mg/L and falls into the Class II river. The concentration of COD at Sg. Kinta falls within Class II whereas all other sampling stations have COD concentrations that fall within Class I river. The lowest COD concentration was found at Muara Sg Kinta which is 4 mg/L.

Based on Figure 5(b), it can be seen that the concentration of COD for samples taken at Kg. Harewood falls in DOE limit of a class IV river. The results obtained for COD at Jong Baruk falls within the DOE limit of a class II river whereas other sites – Taman Senawi, Biovalley and Kg. Pelanduk all has concentrations of COD that falls within Class I. Biovalley and Kg. Pelanduk has the lowest concentrations of COD which is 5 mg/L.

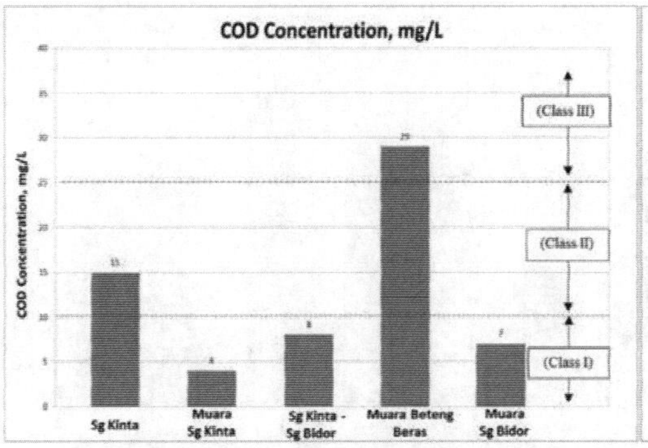

Figure 5(a). Perak River

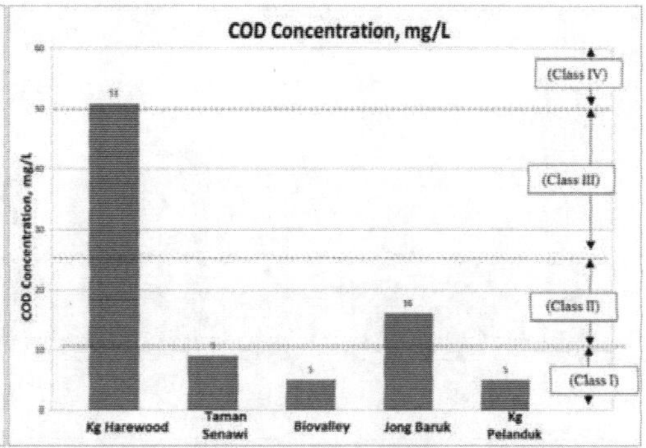

Figure 5(b). Kinta River

Figure 5(a),(b) – BOD_5 (mg/L) concentration at each sampling location

3.1.4 ***TSS*** *Concentration, mg/L*

Figure 6(a) shows the analysis for Total Suspended Solid (TSS) in mg/L for all of the stations sampled. From this figure, it is shown that the Total Suspended Solid (TSS) for Sg. Kinta and Muara Beteng Beras

IOP Conf. Series: Materials Science and Engineering **920** (2020) 012013 doi:10.1088/1757-899X/920/1/012013

falls within the DOE limit, which is 50-150 mg/L for Class III. TSS reading for Muara Sg. Kinta and Sg. Kinta- Sg. Bidor (M) falls within the DOE limit of Class II. The lowest TSS reading is at Muara Sg Bidor which is 21 mg/L and falls within the limit of Class I river.

Based on Figure 6(b), it can be seen that the Total Suspended Solid (TSS) for Taman Senawi and Jong Baruk is 160 mg/L and falls within the DOE limit of Class IV. The TSS concentration for Kg Harewood and Kg Pelanduk falls within the DOE limit of Class II. The lowest TSS concentration was found at Biovalley which is 20 mg/L and is within the limit of a Class I river.

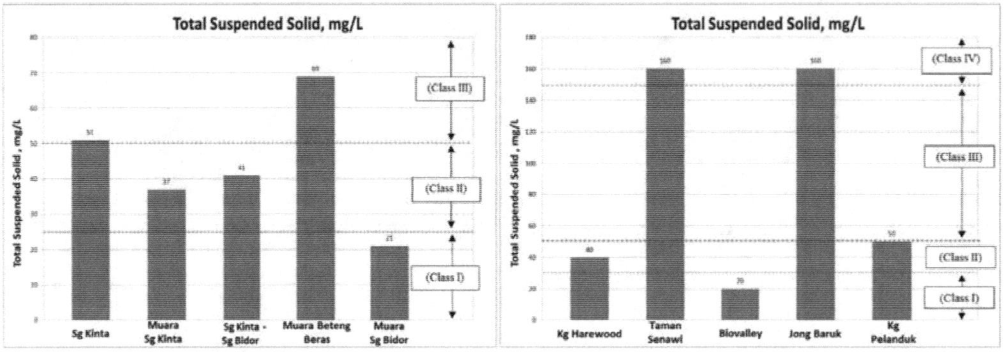

Figure 6(a). Perak River **Figure 6(b).** Kinta River

Figure 6(a),(b) – TSS (mg/L) concentration at each sampling location

3.1.5 NH_3-N Concentration, mg/L

Figure 7(a) shows the analysis for Ammoniacal-nitrogen (NH_3-N), in mg/L for all of the sites. Based on this figure, it can be said that the NH_3-N for Sg. Kinta, Muara Beteng Beras and Muara Sg. Bidor falls into Class V river whereas Muara Sg. Kinta and Sg. Kinta - Sg. Bidor (M) falls into Class IV river. The lowest NH_3-N reading is at Sg. Kinta – Sg. Bidor (Middle) which is 2.24 mg/L. Therefore, it is likely that the high concentration of ammoniacal nitrogen contributed in lowering the total WQI obtained. Lower WQI value indicates higher pollution in the river.

Based on Figure 7(b), it can be seen that the NH_3-N for all of the samples from all of the sites falls within the DOE limit of a Class V river. The lowest concentration NH_3-N was found at Kg. Harewood of 15.68 mg/L. This still far exceeds the maximum allowable limit of NH_3-N for a Class IV river of 2.7 mg/L.

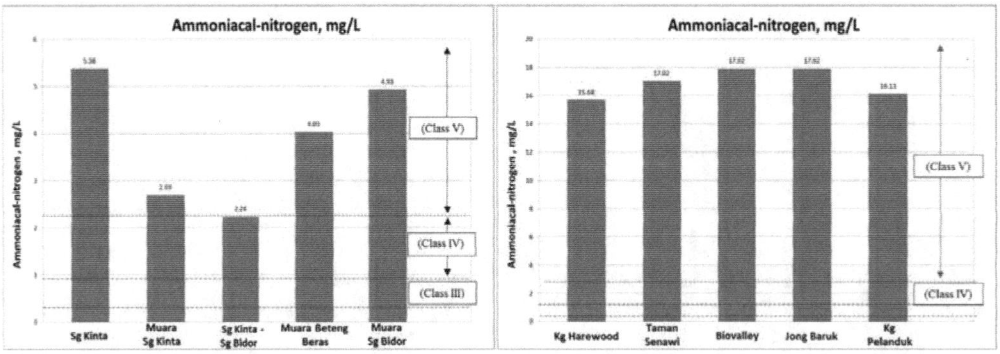

Figure 7(a). Perak River **Figure 7(b).** Kinta River

Figure 7(a),(b) – TSS (mg/L) concentration at each sampling location

Based on the results obtained, it may be concluded that the primary contributing factor to polluting the river is the high content of ammonia from all of the sites sampled. This is evident based on the high concentration of ammoniacal-nitrogen found on all of the sites especially at the downstream of Sungai

Kinta. Sungai Kinta is classified as Class II river [2,8] and Sungai Perak is also classified as Class II river [8].

Ammonia is a colourless gas that is highly soluble in water and is harmful to aquatic environment. The occurrence of ammonia in the rivers are the result of anaerobic decomposition of nitrogen containing compounds or from sewage flow containing ammonia. This may be the result of agricultural activities such as application (fertilisers) which leads to the slow release of inorganic nitrogen into the rivers. Apart from this, it is probable that the main contributing factor comes from the urbanised regions upstream of the river which releases urban effluent (wastewater) and toxic chemical discharges from different types of land uses. Other possible sources might include untreated or semi-treated sewage from residential and poultry farms; uncontrolled landfill leachate and land developments; riverine squatters, septic tanks, factories; storm water and washouts as a result of rainfall; oil, sullage (wastewater) from kitchens and bathrooms and sewage treatment plant effluent. The high TSS concentration may be the result of intensive land clearing for development and farming activities.

4. Conclusion

In conclusion, the quality of water for four (4) sampling stations along Sg. Perak can be concluded as CLASS III with only one station concluded as CLASS IV according to the National Water Quality Standards (NWQS). For Sg. Kinta, four sampling stations were classified as CLASS III whereas only one station falls into CLASS II. The diversity of water quality in river should be monitored closely as it often fluctuates easily. Close monitoring of river water quality would allow possible sources of pollution to be identified. It is therefore recommended that the pollution is identified and minimised at point source. Currently, there is no standard for ammoniacal nitrogen discharge and all existing sewage treatment plants in Malaysia are not designed for the removal of ammoniacal nitrogen [9]. Therefore, it is suggested that authorities and environmental protection agencies monitor discharges into the rivers and facilitates the removal of ammoniacal-nitrogen at point source. It is also important to consider other parameters, which are not physico-chemical based, such as coli form based indicators – as it is affects relevant to skin contact (recreation) and even for potable water supply. Other parameters not considered in the WQI, which is potentially carcinogenic such as certain heavy metals, should also be monitored.

5. References

[1] River water in Perak among best in the country: LAP, The Sun Daily, (2017). Available from: <https://www.thesundaily.my/archive/river-water-perak-among-best-country-lap FTARCH494486>. [19 October 2017].

[2] Manjit K. (2018), Efforts to improve river's water quality pay off, The Star Online, 1 November. Available from <https://www.thestar.com.my/metro/metro-news/2018/11/01/efforts-to-improve-rivers-water-quality-pay-off/>. [1 November 2018]

[3] Baharuddin, N., Nor'ashikin, S.A.I.M. and Zain, S.M. (2014), Characterization of spatial patterns in river water quality using chemometric techniques. Sains Malaysiana, Vol. 43, No. 9, pp.1355-1362.

[4] Shafii, H., Miskam, N., Yassin, A.M. and Musa, S.M.S. (2018), Status Kualiti Air Sungai Di Beberapa Kawasan Luar Bandar, Di Negeri Johor Dan Kesannya Kepada Kehidupan (The Status Of River Water Quality In Some Rural Areas, In State Of Johor And Its Effects To Life). Asian Journal of Environment, History And Heritage, Vol. 2, No.1.

[5] Hashim D. Water Quality (River) Monitoring System/ Programme and Pollution Control. Department of Environment (DOE). Ministry of Natural Resources & Environment.

[6] Singh, G. and Kamal, R.K. (2014). Application of water quality index for assessment of surface water quality status in Goa. Current World Environment, Vol. 9, No.3, p.994.

[7] National Water Quality Standards For Malaysia (NWQS): 2018.

[8] Malaysia Environmental Quality Annual Report (DOE) (2016), Chapter 2: River Water Quality.

[9] Indah Water (2019): Sewerage Facts – Ammonia. Available from: <https://www.iwk.com.my/do-you-know/ammonia>.

Acknowledgments
The authors would like to thank the School of Civil Engineering, Universiti Sains Malaysia (USM) and acknowledge USM for the financial support given to this research through [Short Term Grant no: 304.PAWAM.60313050] and Malaysia Research Universiti Network (MRUN) Grant [Grant no:203/PAWAM/67216002].

IOP Conf. Series: Materials Science and Engineering **920** (2020) 012014 doi:10.1088/1757-899X/920/1/012014

Assessment of the Effect of High Tide and Low Tide Condition on Stream Flow Velocity at Sungai Rompin's Mouth

M N Akhtar[1], M T Anees[2] and E A Bakar[1*]

[1]School of Aerospace Engineering, Universiti Sains Malaysia, 14300 Nibong Tebal, Penang, Malaysia

[2]Department of Geology, Faculty of Science, University of Malaya, Kuala Lumpur 50603, Malaysia

*Corresponding author: meelmi@usm.my

Abstract. The variation in stream flow velocities depends on the riverbed slopes. In upstream areas, where slope is steep, the magnitude of velocity is high, whereas in downstream areas, where slope is gentle, the magnitude of velocity is slow. The focus of this study is in the downstream area, where variations in the magnitude of velocities are due to high tide and low tide situations. Due to sea level rise, flooding occurs with high tides in several coastal regions which causes road closures due to submerged in water and public inconveniences. Velocities in high tide conditions need to be examine for safety and health risk of people. Therefore, the main objective of this study to assess the variation in velocity values of stream flow at both high tide and low tide condition near river mouth. One dimensional Hydrological Engineering Centre-River Analysis System (HEC RAS) simulation was conducted 9.4 km along reach of Sungai Rompin. A total of 188 river cross sections were used at an interval of 50 m. Other river geometries such as river central line, riverbanks and flow paths were constructed in RAS mapper of HEC RAS. The unsteady flow was run in both high tide and low tide conditions. The outputs of the simulation were compared and analysed. The results revealed that the stream flow velocities reduced up to 60% in high tide condition when compared to the low tide condition near the river mouth.

1. Introduction

The rise and fall of sea level, which is also called as tides, is a natural phenomenon. The rising and lowering down of sea level is called high tide and low tide conditions respectively. The stream flow velocity reduces its power when it encounters with gentle slope areas. The river mouth lies in gentle slope area of any watershed but here, variation in velocities are due to high and low tide conditions. The high tide and low tide condition are also controlled by the monsoon and temperature variation of a region [1]. In high tide condition, water from sea starts moving backward towards upstream side which causes flooding near the river mouth [2].

With the advancement of computation technology, several hydrological models were developed to simulate a real stream flow and runoff situation. These models have been used for accurate flood vulnerability assessment either at upstream or at downstream [3-5]. HEC RAS hydrological model is publicly available software generally use for the assessment of flood vulnerability. However, several ground-based or surveyed data also need to run HEC RAS model [6].

The focus of this study was to conduct one dimensional HEC RAS simulation at river mouth of Sungai Rompin to analyse the effect of stream flow velocity variation in downstream area. Therefore, the objectives of this study were (i) to conduct 1D HEC RAS simulation at 9.5km reach of Sungai Rompin near the river mouth and (ii) to assess the stream flow velocity variation due to high tide and low tide condition in the study area. The information and assessment from study will be helpful for policy makers to mitigate coastal flood risk and vulnerability.

2. Materials and methods

2.1 Study area

The study area is located in Pahang State, eastern part of Peninsular Malaysia that lies between Latitudes 2° 48'N and 2°49' N, longitudes 103° 27'E and 103° 29'E. The coastal area is bounded by South China Sea as shown in Figure 1. The elevation ranges from 0 to 25m above mean sea level with gentle slope in the study area. The climate is tropical, humid with average temperature ranges from 20 °C to 30 °C. The period from November to January receives maximum rainfall, while between June and July, the season remains dry. The average annual rainfall of the area is approximately 3000 mm, while the average daily wind speed is 1.50 m/s.

2.2 Data

The bathymetry data was obtained from the field survey in study area using ODOM MKIII Echo Sounder mounted on a boat along with other sensors such as Trimble SPS 461 Receiver Differential Global Positioning System (DGPS), HydroPRO 2.4v Nav Software for Data Acquisition and TOPCON GNSS RTK Systems GR-5 Receiver. DGPS was used to navigate and positioning for survey vessel and sensors with great data accuracy, HydroPRO was used for navigation of integrated survey vessel and survey sensor to data recording & tide processing and TOPCON GNSS RTK was used to Measure ground level for topographic XYZ data using GPS base station reference. The data obtained from different sensors were processed through the aforementioned software. The Digital Elevation Model (DEM) of 25 m resolution was developed by using the surveyed data in the GIS environment. The DEM was used as a reference terrain during hydrological simulation. Because of unavailability of stream flow data for the study area, the data of Kelantan was used. Stream flow data of December 2010 were used in this study.

2.3 Methodology

For stream flow simulation, HEC RAS 5.0.7 was used. The project was defined, and the projection of the project was set as WGS_1984_UTM_zone_48N in RAS Mapper. The terrain was created by using the DEM in RAS Mapper. A geometry file was created for the project and river central line, bank line, flow path, river cross sections were digitized in the RAS Mapper. The digitized cross sections read automatically the elevation points and saved it in the geometric file. The spacing of the cross sections were at 50 m interval. A total of 188 river cross sections were digitized along the 9.4 km reach. The Manning's n value for the left and right riverbanks and for central line were set as 0.5 and 0.025 respectively. For unsteady flow, the upstream boundary condition was used as hydrological flow in which the streamflow data was input, while for downstream boundary condition, stage option was chosen. The slope for normal depth was used as 0.0003195 which was calculated from DEM. In high tide condition, the average height of sea level was 4.49 m while the average height of sea level at low tide was 1.32 m [7].

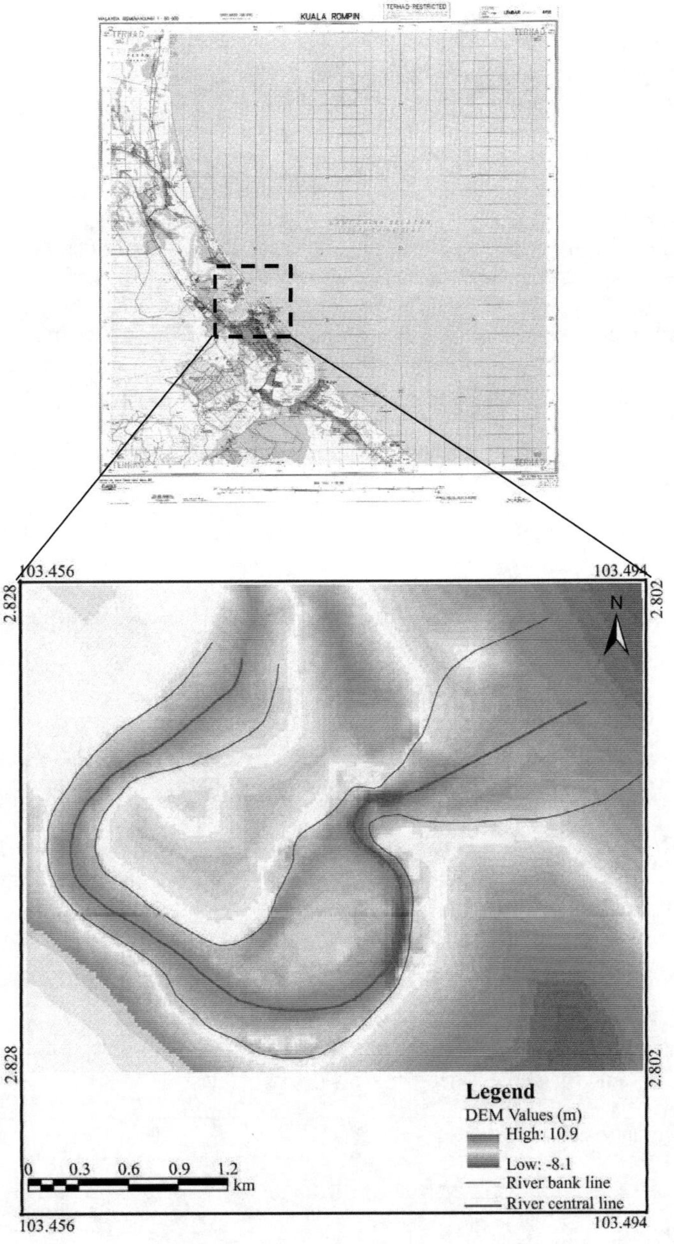

Figure 1. Location of study area

3. Results and discussions

The width of in situ river cross sections ranges from 720 m to 1310 m. The maximum depth of -7.7332 (at downstream boundary of the reach) while minimum depth of -2.4322 (upstream boundary of the reach) were reported from in situ cross sections. The streamflow values range from 29.4 m^3/s to 1113.4 m^3/s with an average of 87.4 m^3/s.

The average velocity in the study area during high tide was observed as 0.3 m/s while during low tide it was 0.7 m/s. During high tide, minimum velocity which is almost zero was found at the mouth of the river while maximum velocity (0.9 m/s) was found at upstream boundary of the reach. During low tide, minimum velocity (0.44 m/s) was found at the mouth of the river while maximum velocity (1.22 m/s) was found at midstream of the reach. The average minimum velocity, during high tide in the month of December was found to be 0.04 m/s at the mouth of the reach, while the average minimum velocity in case of low tide was 0.6 m/s. Average maximum velocity during high tide in the month of December was found to be 0.42 m/s at the mouth of the reach, while the average maximum velocity in case of low tide was 0.8 m/s at the river mouth of the reach as shown in Table 1. The results indicate that due to low velocity during high tide makes the water stagnant which increases water level towards upstream side and hence, flooding occurs. However, due to high velocity during low tide makes the way of river water to flow frequently and does not make it stagnant at the upstream side. The extent and depth of flooding depends on the magnitude of velocities during high tides. This type of information can be used to demarcate flooding area during high tide, restrict any settlement around the demarcated area and avoid any construction in this area.

Table 1. Velocity values (m/s) at 5 different places along the reach

Places	Min		Max		Average	
	High tide	Low tide	High tide	Low tide	High tide	Low tide
1	0.00	0.44	0.26	0.89	0.04	0.58
2	0.06	0.64	0.61	1.22	0.22	0.84
3	0.11	0.55	0.73	1.08	0.29	0.73
4	0.17	0.51	0.81	1.04	0.35	0.70
5	0.23	0.44	0.89	0.99	0.42	0.63

The maximum difference of stream flow velocities during high tide and low tide condition was observed at the river mouth which indicates the maximum daily fluctuation occurring at river mouth. On the other hand, minimum difference of stream flow velocities during high tide and low tide condition was observed at the upstream boundary which indicates the minimum daily fluctuation occurring at the upstream boundary of the reach (Table 2).

Table 2. Difference in the velocities of high and low tides at 5 difference locations and the percent decrease in velocity values

Difference between the tides	% decrease
0.535 m/s	92.6
0.622 m/s	74.1
0.440 m/s	60.0
0.344 m/s	49.2
0.209 m/s	33.1

These results indicate that the fluctuation in the stream flow velocity decreases towards the upstream side of the river from the river mouth. The results also indicate that the effect of high tide and low tide was observed up to 9.4 km towards upstream boundary from the river mouth. The fluctuation at the downstream reflects the social and economic impact as number of fishermen lives and do their daily businesses and such environmental impacts can destroy coastal habitats. It also effects natural and man-made barriers due to swell condition during high tide. Moreover, waves generated from the fluctuation can be vulnerable to jetties situated in nearby area. Therefore, this information can be used by policy makers for engineered and natural defences for the safety of people and coastal infrastructure.

4. Conclusions

The study was conducted to know the effect of high tide and low tide conditions on stream flow velocities at downstream areas. The results showed that stream flow velocity becomes zero at the river mouth during high tide condition while it becomes normal during low tide condition which causes rise and fall in water level due to high and low tide conditions. The fluctuation in the stream flow velocity was found maximum at river mouth while minimum at the upstream boundary of the reach. The fluctuation can generate waves which will be vulnerable to the people and coastal infrastructure. Overall, almost 60% of the stream flow velocity decreased during the high tide condition. It can be concluded that due to 60% reduction in velocity near river mouth can cause increment in water level and hence coastal flooding. In addition to this, with respect to future work perspective, a deep learning model of the proposed study can also be developed, whereby parallel image processing using map-reduce [8] algorithm can be used to correlate the experimental results.

5. References

[1] Akhir M, Fadzil M, Zakaria N Z, and Tangang F 2014 Intermonsoon variation of physical characteristics and current circulation along the east coast of Peninsular Malaysia International *Journal of Oceanography* **2014** 1-9 doi: 10.1155/2014/527587

[2] Hinkel J, Lincke D, Vafeidis A T, Perrette M, Nicholls R J, Tol R S and Levermann A 2014 Coastal flood damage and adaptation costs under 21st century sea-level rise *Proc. of the National Academy of Sciences* **111**(9) 3292-3297 doi: 10.1073/pnas.1222469111

[3] Masood M, and Takeuchi K 2012 Assessment of flood hazard, vulnerability and risk of mid-eastern Dhaka using DEM and 1D hydrodynamic model *Natural hazards* **61**(2) 757- 770 doi: 10.1007/s11069-011-0060-x

[4] Ouma Y, and Tateishi R 2014 Urban flood vulnerability and risk mapping using integrated multi-parametric AHP and GIS: methodological overview and case study assessment *Water* **6**(6) 1515-1545 doi: 10.3390/w6061515

[5] Khattak M S, Anwar F, Saeed T U, Sharif M, Sheraz K, and Ahmed A 2016 Floodplain mapping using HEC-RAS and ArcGIS: a case study of Kabul River *Arabian* Journal for Science and Engineering **41**(4) 1375-1390 doi: 10.1007/s13369-015-1915-3

[6] Anees M T, Abdullah K, Nawawi M N M, Ab Rahman N N N, Piah A R M, Zakaria N A and Omar A M 2016 Numerical modeling techniques for flood analysis *Journal of African Earth Sciences* **124** 478-486 doi: 10.1016/j.jafrearsci.2016.10.001

[7] Zakariya, R., Ahmad, Z., Saad, S., & Yaakop, R 2013 Modelling Suspended Sediment Transport in Monsoon Season: A Case Study of Pahang River Estuary, Pahang, Malaysia. In EGU General Assembly Conference Abstracts (Vol. 15) https://ui.adsabs.harvard.edu/abs/2013EGUGA..15.2592Z

[8] Akhtar M N, Saleh JM, Awais H and Bakar E A 2020 Map-Reduce based tipping point scheduler for parallel image processing *Expert Systems With Applications* **139** 112848 doi:10.1016/j. eswa2019.112848

Acknowledgments
The authors would like to acknowledge School of Aerospace Engineering, Universiti Sains Malaysia and Department of Geology, University of Malaya, Kuala Lumpur for providing research facilities to conduct this work. The authors would also like to acknowledge the RUI grant 1001/PAERO/8014035 and RU-Top-Down grant 1001/PAERO/87052 provided by RCMO, Universiti Sains Malaysia to support this research.

IOP Conf. Series: Materials Science and Engineering **920** (2020) 012015 doi:10.1088/1757-899X/920/1/012015

A Review on metal 3D printing; 3D welding

Z K Wani and A B Abdullah*

School of Mechanical Engineering, Universiti Sains Malaysia, 14300, Nibong Tebal, Pulau Pinang

*mebaha@usm.my

Abstract. 3D printing is a well-known method to produce a 3D object and widely used in many types of industries including, mechanical, civil, medical, etc. The technology has been applied in industry for more than 20 years ago, but it is strictly limited to the slow production of small plastic objects, and some special fabric only. As its advancements, the technology of 3D printing becomes broad interest recently. In present industrial, 3D printers are able to produce high precision, quality products of different sizes fast at comparatively low cost. The aim of this paper is to highlight the importance of 3D printing in meeting low and medium production volumes besides having a review on metal 3D printing techniques available. The issue and challenges in producing metal 3D printing also will be discussed extensively in this paper. Suggestions on the subtractive method to complete the 3D printing process according to research interest also explained in this paper.

1. Introduction

3D printing is additive manufacturing (AM) formerly known as a method of reshaping product development in industries. This 3D printing is resulting in the rapid growth in showing the usefulness in design, small-batch production, and potentially distributed manufacturing [1-6]. In general, an AM is a technology that enables the fabrication of 3D parts from 3D data by depositing a thin layer of material layer-by-layer until a semi-finish part is produced. A few decades ago, 3D printing is commonly applied for rapid prototyping only but lately, it is used for actual manufacturing. Frazier [7] and Das et al [8] also addressed that AM has been around over two decades, but it has only recently begun to appear as a significant marketable manufacturing technology. According to Wong and Hernandez [9], the early version of 3D printing is to produce a fast prototype by speeding up the process in model development and shortening lead time between product development and market placement. The requirement to capture the market placement is significant to facilitate the product for not being outdated. In a fraction of time, the company able to produce prototype parts faster compared to conventional manufacturing methods, such as molding, forging, and milling [10]. Few researchers believed that the product produce by AM is unique as well it can be produced in a short time which is surely able to mass customization [11]. This is also considered as the growth of 3D printing where the changes in the uses of 3D printing in presenting the trend of new business model. In the meantime, 3D printing acknowledged as one of the fastest-growing fields in AM.

As a small and medium-sized production company, there are certainly a number of factors to consider in production costs. The total cost of production must include the cost of labor, the process, and the manufacture of a product which is of utmost importance. The importance of 3D printing in this kind of industry is to help in reducing waste. Since AM have a great technology to reduce energy and material cost, it could be an indicator for the industry in innovative development. The production steps are reduced when the assembly requirement is eliminated. Hence, by having a 3DD printing, practically small and medium companies can reduce production waste [3] as well as the number of workers. The AM process also able to shorter time to market the product with the benefit of lowering down the overall production cost [12]. AM necessarily correspond to the practices in small and medium companies. Few researchers carry out the studies on the importance of 3D printing in various fields such as in design, engineering, analysis and planning, pre-production parts, manufacturing and tooling, aerospace [13], automotive and biomedical [14]. For instance, a brief explanation of the importance and uses of 3D printing in mechanical engineering is done by Madhav et al. [15]. The growing trend in demand for metal-based parts for low and medium production is evident as various companies from the medical and aerospace industries continue to demand the advancement of this technology in meeting their needs [16]. It is vital for a company to comply with the demand to meet the customer's expectations. Further explanation of metal 3D printing or known as 3D welding, and its challenge, and issues will be discussed in sections 2 and 3 respectively. Suggestions on the subtractive process are discussed in section 4

2. 3D welding

Welding is a process of linking parts by melting the work piece and adding filler material. It is also being defined as a consistent joining technique to assemble metal parts with the heat application [17-18]. 3D welding is described as a constructing up metal beads layered by layer to produce a 3D metal object [20]. The differences between common welding and 3D welding are illustrated in Figures 1 and 2. The cost and productivity are increased to produce, metal volume objects by using the welding method [21]. Not only to produce a new 3D object, but research by Ren et al. [22] has also been done about the 3D repairing technology where surface patching has greatly improved the accuracy, efficiency, and reliability of part repairing. The repair shows it would save more cost than producing a new part to replace the broken part.

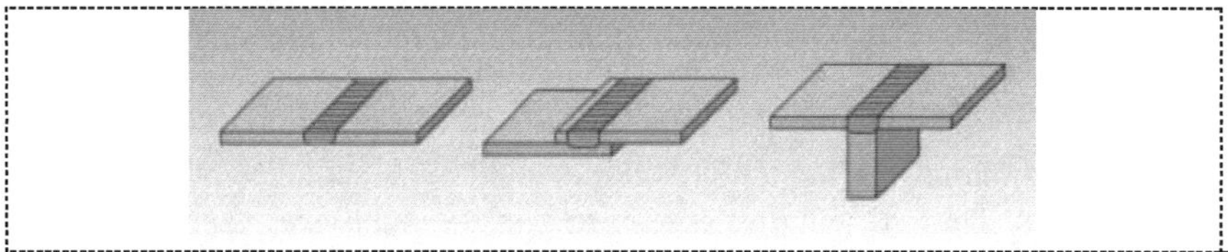

Figure 1. Example of common welding (joint) types: square butt, lap, and T-joint. [19]

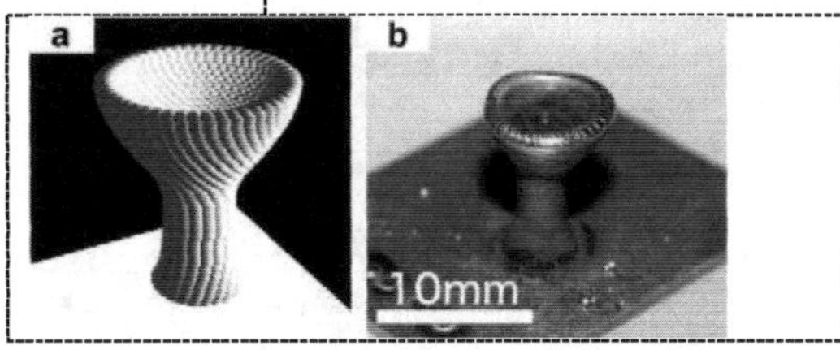

Figure 2. Example of 3D printing: (a) 3D computer graphic image of a cup object and (b) formed object by 3D Micro Welding. [20]

Additive welding technology is one of the processes to manufacture 3D parts for metal. The chronology (Figure 3) of development 3D welding for the fabrication of 3D structure has been discussed by Colegrove and Williams [23]. The most common technique for 3D printing consists of Powder Bed Fusion (PBF), Material Extrusion, Sheet Lamination, Directed Energy Deposition (DED), Material Jetting, and Binder Jetting. Material extrusion and Binder jetting methods used to be applied to plastics and ceramics only in the past 20 years, but now being used in metal 3D printing as well due to advancement in additive manufacturing technique. The material used for the metal 3D printing technique is most likely powder-based and wire-based [24]. A recent article written by Lewandowski and Seifi published about the mechanical properties of materials processed by various AM techniques, including powder bed fusion and directed energy deposition technologies [25].

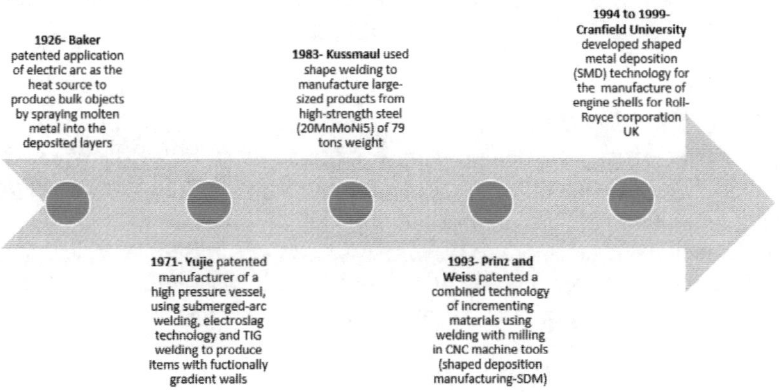

Figure 3. Chronology of the attempt to apply welding technologies in 3D structure of a complex shape.

Three universities named University of Nottingham, Wollongong University and Southern Methodist University in Dallas has conducted research work related to 3D arc welding [26]. According to Korzhik et al. [27], the development of plasma arc melting and welding technologies, as well as the other welding technologies of 3D printing of metal products is being researched by advanced research institutes and industrial corporations of economically developed countries at present. Metal 3D printing can be roughly categorizing into two main groups; Powder Bed Fusion (PBF) and Directed Energy Deposition (DED) based on technologies used. Table 1 shows the list of technology for both groups. Article from Bhavar et al. in 2014 [28] has discussed about this two type of technologies based on their energy source used. However, there are other techniques that have been recently developed, such as binder jetting [14] [29], cold spraying [30], friction stir welding [31], direct metal writing [32] and diode-based processes [33]. Besides Bhavar et al., Some other researchers [34-35] in their review about metal AM also highlight some of the differences between the various processes. A summary of the several AM technologies and the dictate AM equipment manufacturers is presented by Hederick in his review [36].

Table 1. List of Powder Based Fusion and Direct Energy Deposition technologies [28]

Powder Based Fusion (PBF)	Direct Energy Diffusion (DED)
Selective Laser Sintering (SLS)	Laser Engineered Net Shaping (LENS)
Selective Laser Melting (SLM)	Direct Metal Deposition (DMD)
Direct Metal Laser Sintering (DMLS)	Electron Beam Free Form Fabrication (EBFFF)
Electron Beam Melting (EBM)	Arc Based Additive Manufacturing

PBF is the leading technology in the industry which shows it is relevant to metal objects [37]. PBF technologies able to manufacture products with good mechanical properties and complex shapes with high accuracy (±0.02mm) [38]. According to Berger, DMLS is the most preferred PBF technology for welding 3D printing with short manufacturing time, cost-effective assembly, and wide variety of metal parts. DMLS is s laser-based rapid prototyping and tooling process by means of which net shape parts are fabricated in a single process. SLM, SLS, and DMLS are always being described as similar process however, the nature of the powder is the main difference. Research on the 3D welding machine is not new and various successful stories were reported from all over the world and had been reviewed by Korzhik et al. [27]. In terms of welding technique, there are various techniques employed e.g. arc-weld [39], laser [40], MIG [41], TIG [42-43], plasma-arc [44] and electron beam [45]. Table 2 shows the list of techniques for 3D welding focusing on metal that has been discussed by Wang et al [46].

Wire-Arc Additive Manufacturing (WAAM) is an arc-based technology considered by high productivity, high energy efficiency, and low raw material cost [47-48]. WAAM shows the high saving in the uses of material where buy to fly ratio compared to traditional welding method is 4.9 including the finishing machining [49]. In producing welded joint and metal layers a few technologies for welding methods such as MX3D, GMAW, GTAW being discussed by Peleshenko et al. [50] as a new level of 3D printing. However, among all additive welding technologies, plasma arc welding is the most promising with the high accuracy of manufacturing and the quality of the surface compared to other arc welding method [50].

Table 2. List of welding technology for metal 3D printing with the advantages [46]

Technology	Advantages	Material
TIG	High quality	-Low-alloy structural steel -High-strength steel -Stainless steel -Nickel heat-resistant alloys -Aluminum alloys -Titan, titanium alloys -Copper and copper alloy
High frequency TIG	High precision and quality	
Pulse MIG	Simplicity and economic efficiency	
Cold Metal Transfer (CMT)	Low heat input, High Stability of the process	
Tandem Pulsed MIG	High building-up speed	
Plasma PTA	High quality, High speed, Possibility to adjust the width of the build-up layer	

3. Issues and challenges on 3D welding

Significant research and further understanding are required in aspects of machine design and process integration, optimization, and level of automation including for process planning to meet future demands. One of the challenges in producing metal 3D printing is the selection of the material. For example, titanium and its alloys are commonly used in various industries due to their high performance [51-52] but it is the long lead time and high machining cost if being performed with conventional. The common steel used in 3D welding are tool steel [53], austenitic stainless steel [54], precipitation hardenable stainless steel [55], and maraging steel [56]. Due to the different properties of each metal, the correct material needs to be chosen in order to enlighten the purpose of AM which is reducing cost, waste, and energy. Besides the differences in properties, other things to be considered are such as the price, method, or technique and the strengthens [24] of the metal itself. Further research about each metal itself will help in selecting the appropriate material. Secondly, the defects after the 3D welding process also are the drawback of this process. The void formation, anisotropic microstructure, and mechanical properties, divergent from design to execution and layer by layer appearance are highlighted [24] as the challenges that are attributed to the nature of AM. High porosity created in product lead to the reduction of interfacial bonding between the printed layer which also will reduce its mechanical performance [57]. However, it is depending on the method and the type of material used.

Other issues regarding the metal 3D printing are poor surface finish, inaccurate dimensional and the requirement of post-processing (machining, heat treatment or chemical etching) to complete a part [24]. However, the combination of 3D-printing technology with equivalent manufacturing and subtractive manufacturing will create additional benefits for advanced manufacturing. The effectiveness of combination both process welding and milling was proven by Alhuzaim [58] in his research work. The welding process combined with milling is the conceptual idea presented by researcher from the India Institute of Technology and Automation [27]. The combination with the milling process gave great manufacturing flexibility besides increase surface quality and dimensional accuracy [59]. The residual stress and strain is highly dependent on the strategy of depositing the material which also leads to the constant height that can be measured layer by layer. Most of the available 3D printing machine available in the market is powder-based material. As it is a powder, the strength of the produced parts results in some issues. The post-welding process should not only be focused on the accuracy and the surface finish itself but also need to take into account about the product's strength.

Welding machine design also becomes one of the issues to perform welding because it is due to the complexity of the parts. Most of the existing machine available is a hybrid machine, where two or more processes were integrated on the same machine. A machine usually will consist main body, tools and mechanism involve to perform tasks (for this case welding), control system, and accessories. The body shape and size may affect the mechanism and arrangement of machine parts. Thus, at design stages, the complex design of the machine should be avoided to elude difficulty during the welding process. Figure 4 shows a few existing machine designs presented in previous researchers.

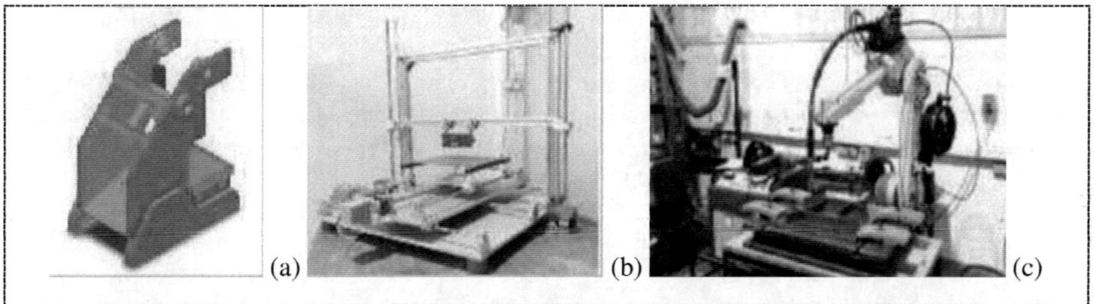

Figure 4. Solid casted iron body (a), structure wireframe (b) by Rosli et al., (41) and without frame as robotic arm is utilized by Colgrove and William (22) as shown in (c)

For welding operation, two important tools required are the platform and the welding nozzle. The nozzle uses to perform welding need to mount or clamp on a bracket or holder, while the platform is needed to prepare a space where the weld bead is produced. Two options for welding mechanisms to be described, it is either to have a fixed nozzle and moving platform or vice versa. However, it is to be highlighted that the efficient method where a stationed nozzle could reduce machine complexity, as the cable and bulky size will limit the movement of the nozzle. In addition, the moving platform is much easier to control.

4. Suggestions

Traditional manufacturing technologies [60], such as casting, forging, machining, and injection molding are portrayed as a process to make a 3D part in previous years. However, time-consuming for the preparations and waste in the material makes the process is not good compared to additive manufacturing. However, it cannot be simply compared to comparable manufacturing technologies such as casting, forging, and machining with subtractive manufacturing processes (milling, turning, and grinding); relatively, different manufacturing technologies should be applied where they can be of best use. 3D printing must provide end-use products by fabricating more than just simple structures with sufficient mechanical strength to retain shape in order to become more widely adopted in mainstream manufacturing technology. Through the powder-based technique, the necessary strength is not able to achieve. In future research, the author would like to suggest forging as the subtractive process of 3D welding. Forging, known as the metal forming process, that involves large deformation and can improve part's strength. In addition, time taken is shorter compared to machining and suitable to use due to the suitability for mass production.

Through 3D welding, the pre-form shape can be prepared, and this will reduce waste and can shorten the process steps. The critical issue to be addressed is on part complexity due to forging is not convenient with the harder materials because it is difficult in filling a complex cavity. Limitations of the process include the need to build larger volumes. Further research will be carried out accordingly in order to find the solution to the limitation of the forging process to become a subtractive process in producing 3D objects.

5. Conclusion

This paper presented the surface of 3D printing with the justification of 3D welding technology and technique available. Their relationship was explained based on the review the literatures in this research area. Many publications had explored the importance and techniques of 3D printing including metal 3D printing. However, there is limited research to discuss about strength of parts after welding process.

References

[1] Bourell D, Stucker B, Crane N B, Tuckerman J, and Nielson G N 2011 Self-assembly in additive manufacturing: opportunities and obstacles *Rapid Prototyping Journal*

[2] Lass N, Tropmann A, Ernst A, Zengerle R, and Koltay P 2011 June Rapid prototyping of 3d microstructures by direct printing of liquid metal at temperatures up to 500 C using the starjet technology *In 2011 16th International Solid-State Sensors, Actuators and Microsystems Conference* IEEE pp 1452-145

[3] Petrovic V, Vicente Haro Gonzalez J, Jordá Ferrando O, Delgado Gordillo J, Ramón Blasco Puchades J, and Portolés Griñan L 2011 Additive layered manufacturing: sectors of industrial application shown through case studies *International Journal of Production Research* **49(4)** pp 1061-1079

[4] Upcraft S, and Fletcher R 2003 The rapid prototyping technologies *Assembly Automation*

[5] Wittbrodt B T, Glover A G, Laureto J, Anzalone G C, Oppliger D, Irwin J L, and Pearce J M 2013 Life-cycle economic analysis of distributed manufacturing with open-source 3-D printers *Mechatronics* **23(6)** pp 713-726

[6] Lipson H, and Kurman M 2013 Fabricated: The new world of 3D printing *John Wiley & Sons*

[7] Frazier W E 2014 Metal additive manufacturing: a review *Journal of Materials Engineering and performance* **23(6)** pp 1917-1928

[8] Das S, Bourell D L, and Babu S S 2016 Metallic materials for 3D printing *Mrs Bulletin* **41(10)** pp 729-741

[9] Wong K V, and Hernandez A 2012 A review of additive manufacturing *ISRN Mech. Eng* **1** pp 1-10

[10] Banks J 2013 Adding value in additive manufacturing: researchers in the United Kingdom and Europe look to 3D printing for customization *IEEE pulse* **4(6)** pp 22–26

[11] Duda T, and Raghavan L V 2016 3D metal printing technology *IFAC-PapersOnLine* **49(29)** pp 103-110

[12] Berman B 2013 3D printing: the new industrial revolution *IEEE Engineering Management Review* **41(4)** pp 72-80

[13] Council N R 2004 Accelerating technology transition: bridging the valley of death for materials and processes in defense systems *National Academies Press*
[14] Bai Y, and Williams C B 2015 An exploration of binder jetting of copper *Rapid Prototyp J* **21(2)** pp 77–85
[15] Madhav C V, Kesav R S N H, and Narayan Y S 2016 Importance and utilization of 3D printing in various applications *International Journal of Modern Engineering Research (IJMER)* pp 24-29
[16] Castonguay K C 2018 Additive manufacture of propulsion systems in low earth orbit *Air Command and Staff College* Air University Maxwell AFB United States
[17] Hindsén M, and Bruze M 2020 Welding *Kanerva's Occupational Dermatology* pp 2351-2353
[18] Mat M F, Manurung Y H, Muhammad N, Ditler A, Abd Ghani M S, and Leitner M 2020 April Grain Growth Prediction of Bead-on-Plate with Filler Wire SS316L using FEM *In IOP Conference Series: Materials Science and Engineering* 834(1) pp 012009
[19] Gibson B T, Lammlein D H, Prater T J, Longhurst W R, Cox C D, Ballun M C, ... and Strauss A M 2014 Friction stir welding: Process, automation, and control *Journal of Manufacturing Processes* **16(1)** pp 56-73
[20] Horii T, Kirihara S, and Miyamoto Y. 2009 Freeform fabrication of superalloy objects by 3D micro welding *Materials & Design* **30(4)** pp 1093-1097
[21] Korzhyk V, Khaskin V, Voitenko O, Sydorets V N, and Dolianovskaia O 2017 Welding technology in additive manufacturing processes of 3D objects *In Materials Science Forum* **906** pp 121-130
[22] Ren L, Padathu A P, Ruan J, Sparks T, and Liou F W 2006 August Three dimensional die repair using a hybrid manufacturing system *In Proceedings of the 17th Solid Freeform Fabrication Symposium, Austin, TX, USA* pp 14-16
[23] Colegrove P, and Williams S 2013 High deposition rate high quality metal additive manufacture using wire+ arc technology *Cranfield University*
[24] Ngo T D, Kashani A, Imbalzano G, Nguyen K T, and Hui D 2018 Additive manufacturing (3D printing): A review of materials, methods, applications and challenges *Composites Part B: Engineering* **143** 172-196
[25] Lewandowski J J, and Seifi M 2016 Metal additive manufacturing: a review of mechanical properties *Annu Rev Mater Res* **46** pp 151–186
[26] Kovacevic, R. Development of machine for rapid manufacturing/repair. in The 2003 CTMA Symp. (2003)
[27] Korzhik V N, Khaskin V Y, Grinyuk A A, Tkachuk V I, Peleshenko S I, Korotenko V V, and Babich A A 2016 3D-printing of metallic volumetric parts of complex shape based on welding plasma-arc technologies *The Paton Welding Journal* **5(6)** pp 117-129
[28] Bhavar V, Kattire P, Patil V, Khot S, Gujar K, and Singh R 2014 September A review on powder bed fusion technology of metal additive manufacturing *In 4th International conference and exhibition on Additive Manufacturing Technologies-AM-2014* pp 1-2
[29] Bogue R 2013 3D printing: the dawn of a new era in manufacturing? *Assembly Automation*
[30] Sova A, Grigoriev S, Okunkova A, and Smurov I 2013 Potential of cold gas dynamic spray as additive manufacturing technology *The International Journal of Advanced Manufacturing Technology* **69(9-12)** pp 2269-2278
[31] Sharma A, Bandari V, Ito K, Kohama K, Ramji M, and BV H S 2017 A new process for design and manufacture of tailor-made functionally graded composites through friction stir additive manufacturing *Journal of Manufacturing Processes* **26** pp 122-130
[32] Chen W, Thornley L, Coe H G, Tonneslan S J, Vericella J J, Zhu C, ... and Pascall A J 2017 Direct metal writing: Controlling the rheology through microstructure *Applied Physics Letters* **110(9)** p 094104
[33] Matthews M J, Guss G, Drachenberg D R, Demuth J A, Heebner J E, Duoss E B, ... and Spadaccini C M 2017 Diode-based additive manufacturing of metals using an optically-

addressable light valve *Optics Express* **25(10)** pp 11788-11800

[34] Frazier W E 2010 August Direct digital manufacturing of metallic components: vision and roadmap *In 21st Annual International Solid Freeform Fabrication Symposium, Austin, TX, Aug* pp 9-11

[35] Froes F H, and Dutta B 2014 The additive manufacturing (AM) of titanium alloys *In Advanced Materials Research* **1019** pp 19-25

[36] Herderick E 2011 Additive manufacturing of metals: A review *Materials science and technology* p 1413

[37] Berger R 2013 Additive manufacturing: A game changer for the manufacturing industry? *Munich*

[38] Selective Laser Melting Machine SLM 500 | SLM Solutions. https://slm-solutions.com/products/machines/selective-laser-melting-machine-slm-500. 23/10/2017

[39] Pan Z, Ding D, Wu B, Cuiuri D, Li H, and Norrish J 2018 Arc welding processes for additive manufacturing: a review *In Transactions on intelligent welding manufacturing* Springer Singapore pp 3-24

[40] Brandl E, Michailov V, Viehweger B, and Leyens C 2011 Deposition of Ti–6Al–4V using laser and wire, part I: Microstructural properties of single beads *Surface and Coatings Technology* **206(6)** pp 1120-1129

[41] Rosli N A, Alkahari M R, Ramli F R, Mat S, and Yusof A A 2019 Influence of process parameters on dimensional accuracy in GMAW based additive manufacturing *Proceedings of Mechanical Engineering Research Day 2019* pp 7-9

[42] Wang F, Williams S, and Rush M 2011 Morphology investigation on direct current pulsed gas tungsten arc welded additive layer manufactured Ti6Al4V alloy *The international journal of advanced manufacturing technology* **57(5)** pp 597-603

[43] Ding D, Pan Z, Van Duin S, Li H, and Shen C 2016 Fabricating superior NiAl bronze components through wire arc additive manufacturing *Materials* **9(8)** pp 652

[44] Aiyiti W, Zhao W, Lu B, and Tang Y 2006 Investigation of the overlapping parameters of MPAW-based rapid prototyping *Rapid Prototyping Journal*

[45] Murr L E, Gaytan S M, Ramirez D A, Martinez E, Hernandez J, Amato K N, ... and Wicker R B 2012 Metal fabrication by additive manufacturing using laser and electron beam melting technologies *Journal of Materials Science & Technology* **28(1)** pp 1-14

[46] Wang F, Williams S, Colegrove P, and Antonysamy A A 2013 Microstructure and mechanical properties of wire and arc additive manufactured Ti-6Al-4V *Metallurgical and materials transactions A* **44(2)** pp 968-977

[47] Venturini G, Montevecchi F, Scippa A, and Campatelli G 2016 Optimization of WAAM deposition patterns for T-crossing features *Procedia CIRP* **55** pp 95-100

[48] Bikas H, Stavridis J, Stavropoulos P, and Chryssolouris G 2015 Design and topology optimization for additively manufactured structural parts: a formula student case study *In 6th BETA CAE Int Conf*

[49] Colegrove P A, Martina F, Roy M J, Szost B A, Terzi S, Williams S W, ... and Jarvis D 2014 High pressure interpass rolling of wire+ arc additively manufactured titanium components *In Advanced Materials Research* **996** pp 694-700

[50] Peleshenko S, Korzhyk V, Voitenko O, Khaskin V, and Tkachuk V 2017 Analysis of the current state of additive welding technologies for manufacturing volume metallic products *Восточно-Европейский журнал передовых технологий* **3 (1)** pp 42-52

[51] Nie B, Yang L, Huang H, Bai S, Wan P, and Liu J 2015 Femtosecond laser additive manufacturing of iron and tungsten parts *Applied Physics A* **119(3)** pp 1075-1080

[52] Sheydaeian E, and Toyserkani E 2018 A new approach for fabrication of titanium-titanium boride periodic composite via additive manufacturing and pressure-less sintering *Composites Part B: Engineering* **138** pp 140-148

[53] Mazumder, J., Choi, J., Nagarathnam, K., Koch, J., & Hetzner, D. (1997). The direct metal

deposition of H13 tool steel for 3-D components. Jom, 49(5), 55-60.

[54] Carlton, H. D., Haboub, A., Gallegos, G. F., Parkinson, D. Y., & MacDowell, A. A. (2016). Damage evolution and failure mechanisms in additively manufactured stainless steel. Materials Science and Engineering: A, 651, 406-414.

[55] Murr, L. E., Martinez, E., Amato, K. N., Gaytan, S. M., Hernandez, J., Ramirez, D. A., ... & Wicker, R. B. (2012). Fabrication of metal and alloy components by additive manufacturing: examples of 3D materials science. Journal of Materials Research and technology, 1(1), 42-54.

[56] Casalino, G., Campanelli, S. L., Contuzzi, N., & Ludovico, A. D. (2015). Experimental investigation and statistical optimisation of the selective laser melting process of a maraging steel. Optics & Laser Technology, 65, 151-158.

[57] Wang, X., Jiang, M., Zhou, Z., Gou, J., & Hui, D. (2017). 3D printing of polymer matrix composites: A review and prospective. Composites Part B: Engineering, 110, 442-458.

[58] Alhuzaim, A. F. (2014). Investigation in the use of plasma arc welding and alternative feedstock delivery method in additive manufacture. Montana Tech of The University of Montana.

[59] Song, Y. A., Park, S., Choi, D., & Jee, H. (2005). 3D welding and milling: Part I–a direct approach for freeform fabrication of metallic prototypes. International Journal of Machine Tools and Manufacture, 45(9), 1057-1062.

[60] Fousová, M., Vojtěch, D., Kubásek, J., Dvorský, D., & Machová, M. (2015). 3D printing as an alternative to casting, forging and machining technologies?. Manufacturing Technology, 15(5), 809-814.

Breath sound analysis of sleep apnea syndrome using cluster model

M Shirasuna

Toyohashi University of Technology, 1-1 Hibarigaoka, Tempaku-cho, Toyohashi, Aichi, 4441-8580, Japan

Abstract. Sleep apnea syndrome (SAS) is a contemporary disease, with few subjective symptoms. For this reason, many people with SAS have not received a medical examination. In this paper, the possibility of a person with a high probability of having SAS in the near future (Pre-SAS) is determined from only the sound of the subject's breathing during sleep. The sound is extracted and the breathing pattern and its tendency analysed using the cluster method. Compared to the conventional SAS process for diagnosis in medical institutions, the method proposed in this paper is very simple and easy.

1. Introduction

Sleep apnea syndrome (SAS) is one of the modern illnesses, and not only is the affected patient treated, but many people are suspected of having SAS, but they often do not feel the need for treatment [1]. In addition, it takes time and effort for the subject to diagnose SAS, and the current SAS diagnosis has to diagnose data other than breath sounds during sleep for 6 to 8 hours. For this reason, many researchers are studying speech analysis in SAS. However, breathing sounds during sleep are very long and uneven. For this reason, there are a method in which a person cuts out the SAS sound analysis part in advance and analyses it in detail, and a method in which trends are analysed from the whole by machine learning.

In this paper, we efficiently cut out the part to be analysed from the entire long-term breathing sound recorded by the IC recorder. Then, a simple diagnosis of SAS was performed using cluster analysis. This proposal was a very simple method and the time required for analysis was very short. The SAS discrimination rate based on this proposal was 100%. This proposal would be useful for SAS medical support tools.

2. SAS diagnosis and problems in medical institutions

2.1. Current SAS diagnosis method

Current SAS diagnosis method take time and effort for both subjects and analysts to receive a SAS diagnosis at a medical institution [2]. First, subjects suspected of having SAS go to a medical institution to undergo prior measurements. There are two methods. One is for the patient to rent a simple portable device from the hospital, take it home overnight, return it, and the data is analysed [3]. The patient then visits the medical institution again to receive the diagnosis. Second, the subject stays overnight in a medical facility and undergoes a polygraph examination overnight [3]. In this case as well, measuring devices are installed on the body. Like the first method, it is necessary to come to the hospital for the results at a later date. At the medical institution, specialists comprehensively check the data obtained from subjects. They require about 6 hours per subject for this task. The doctor then diagnoses SAS based on the results submitted by the specialist. Fig. 1 shows the current SAS diagnosis method. As shown in this figure, there are so many steps.

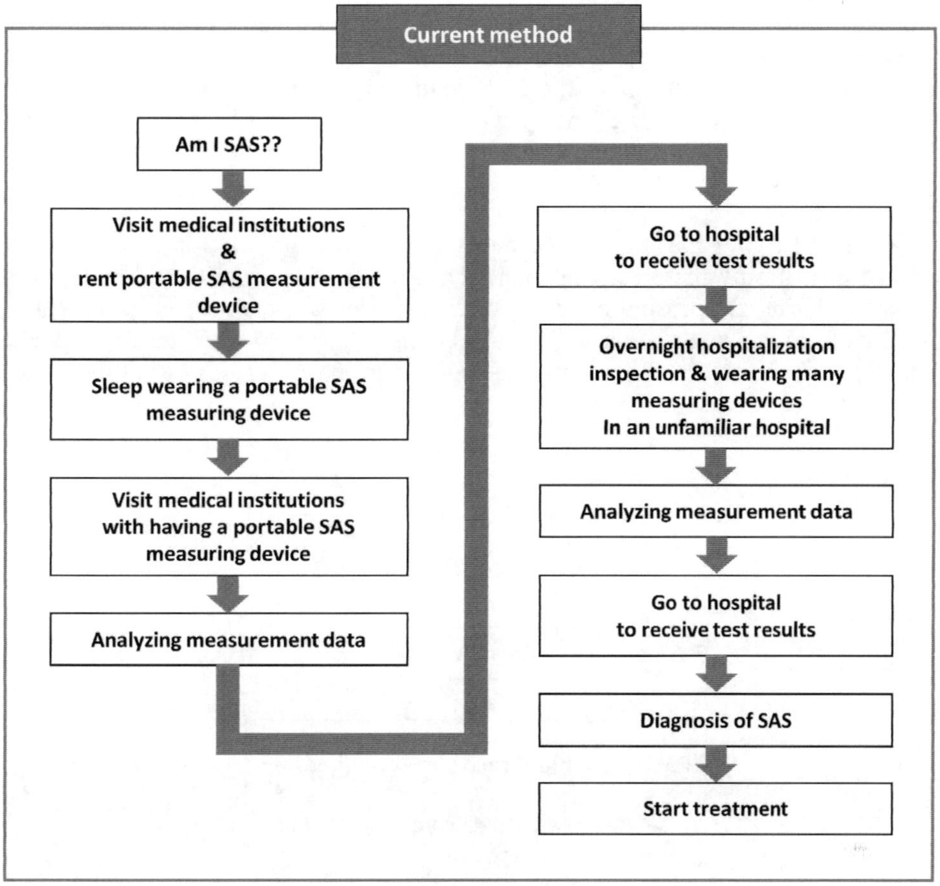

Figure.1 Current SAS diagnosis steps

2.2. The problems in medical institutions of SAS diagnosis method
In general, people are not able to sleep well when they are different from sleep conditions, and breathing sounds during sleep are also different [4]. For example, hospitals and unfamiliar hotel futons and pillows. The sounds and lights of the doors that nurses open when they visit hospital rooms. And sounds from outside the building and from the next room. Wearing a measurement device for SAS diagnosis is no exception. The current SAS diagnosis is a simple test: renting a measurement device from a hospital and recording measurement data, and a main test: an inspection that is hospitalized overnight and recording measurement data. For the reasons described above, these methods have a problem that accurate data cannot be measured from the subject.

2.3. Measurement data for SAS diagnosis
The measured data by these diagnoses measured from the subject at the medical institution does not include breathing sounds during sleep. There are several possible reasons, but the main reason is that it is difficult to analyse long-term sounds. However, there are many cases where it was noticed that a person was suffering from SAS by listening to the breathing sound during sleep.

2.4. SAS sound and its researches
However, SAS sounds during sleep are characteristic, such as abnormal sleep sounds during sleep: snoring and repeated apnea [4]. Many SAS people have started to receive treatment because their family members and friends pointed out these abnormal sleep sounds. here have been many previous studies

on SAS sounds. Many of them analysed the sound waveform narrowly, using Fourier transform, wavelet transform [5], or analysed the tendency of SAS sound widely using machine learning [6].

There are studies to analyse SAS sound waveforms, both in a narrow area and in wide area. The reason is that sleep time is as long as 6 to 8 hours, and the sound is uneven, which makes analysis difficult. In narrow-area analysis, a characteristic SAS sound portion is artificially cut out and analysed. Wide-area analysis studies the characteristic SAS sound waveform as a result of analysis. With the widespread use of smartphones, simple SAS diagnostic apps have been released, but these measure the state in which there is no breathing sound, which might indicate the person has stopped breathing.

2.5. Our proposed method

We propose a method for diagnosing the possibility of SAS and pre-SAS by a simple method using multivariate analysis of only the breathing sound during sleep. In our proposed method, a breathing sound is recorded during sleep using a simple voice recorder and the subject can perform a simple diagnosis of SAS from the recorded data. The subject has no stress to wear the device and can take a normal sleep. Fig.2 shows the SAS diagnostic method we propose. As shown in this figure, it is very simple compared to Fig.1.

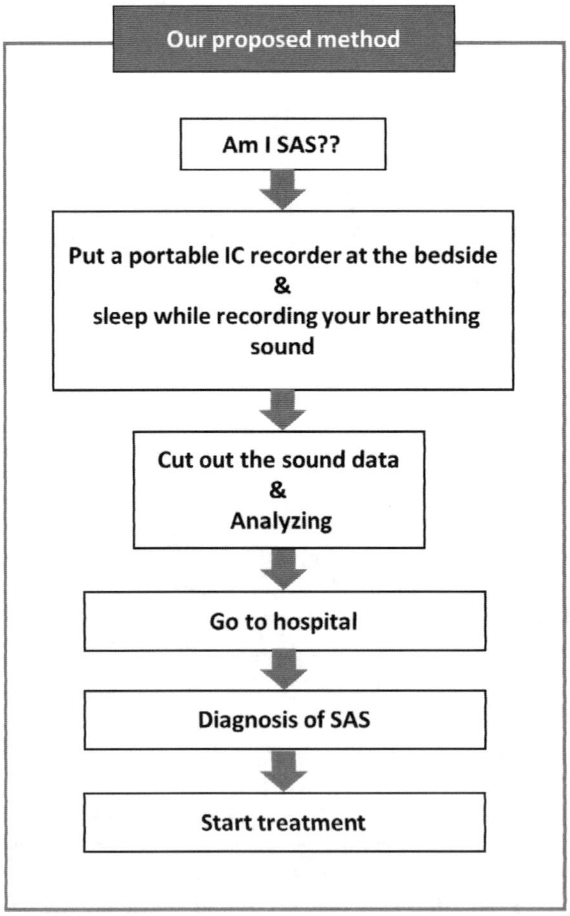

Figure 2. Subject details and points for analysis

3. Breathing sound during sleep used for analysis

3.1. Collect analysis data
In order to analyse the characteristic sounds of SAS, we collected respiratory sounds during sleep of SAS patients, Pre SAS people, and Non SAS people. In this paper, SAS, Pre SAS, and Non SAS are defined as follows. SAS is a person who currently suffers from SAS and is being treated at a medical institution. Pre SAS is a person who is not currently affected by SAS but is likely to become SAS in the near future. And Non SAS is not currently suffering from SAS, is unlikely to become SAS in the near future, and has no problems with health examinations. Regarding data collection, people with SAS have their data provided by a sleep medical institution in Toyohashi City, but Pre SAS and Non SAS are difficult to collect from medical institutions. I collected this data separately and had a specialist make a voice diagnosis.

3.2. Subject details
Table.1 shows the composition of the number of subjects we analysed. AHI in the table is the Apnea Hypopnea Index determined by the World Health Organization, and is an index for judging the severity of SAS symptoms [1]. There are three types of SAS, and all the subjects in this paper are obstructive SAS (OSAS): the main cause is that the tongue base and soft palate fall due to muscle relaxation during sleep and obstruct the airway.

Table 1. A Variable table of subjects.

---	AHI	No. of people
Non SAS	(5<AHI)	4
Mild	5≦AHI<15	5
Moderate	5≦AHI<30	11
Severe	30≦AHI	21
---	Total	42

3.3. Points for analysis
The sound analysis will be described in the next chapter. Before that, we will define the sound used in this paper. What we analyse is the breathing sound of the subject during sleep, commonly referred to as snoring. Snoring is a collective term for abnormal breathing sounds during sleep and is medically distinct from SAS and snoring. However, the purpose of this study is not to distinguish between SAS and snoring, so in this paper we define the breathing sound during sleep obtained from these subjects as the breathing sound during sleep or breathing sound.

4. Subject details and points for analysis

4.1. SAS sound features
When comparing breath sounds during sleep between SAS and Non SAS, there are some notable differences. In this paper, we focus on two of them and analyse them.

- Features1. The sound pressure of SAS is larger than Non SAS.
- Features2. The respiratory rhythm of SAS is less constant than Non SAS.

An example of feature 1 is shown in FIG. 3. FIG.3 (A) shows example of SAS waveform sound, which is a 3-second breath sound during sleep. (B) shows the power spectrum of (A). The vertical axis is the energy dB of the audio component of each frequency, and the horizontal axis is the normalized frequency. It can be seen that the area enclosed by the horizontal and vertical axes of SAS is larger than Non SAS. The features 2, Non SAS repeat the breath to be constant during sleep in the same manner as when awake, but, SAS is the cessation of breathing during sleep, and the breathing rhythm is not constant.

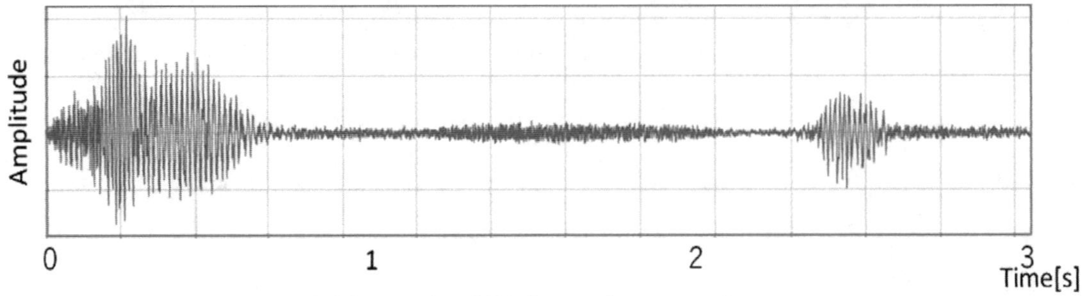

(A) Example of SAS waveform sound.

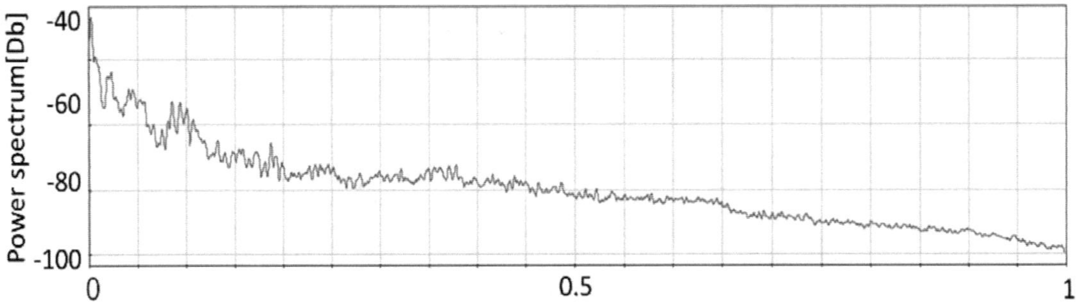

(B) Example of SAS waveform sound.

Figure 3. Example of SAS sound spectrum.

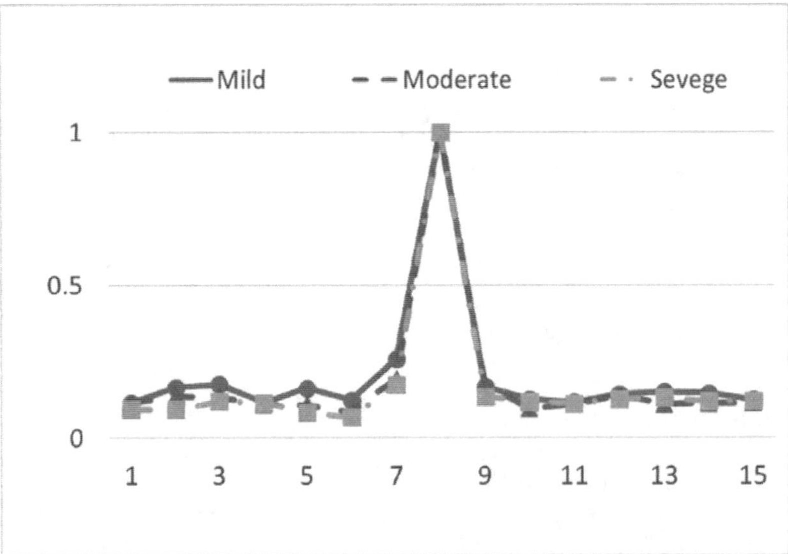

Figure4. Example of Cluster analysis of SAS breath sounds.

4.2. Analysis method

From the above features 1 and 2, we thought that there would be a difference between SAS and Non SAS at the point where the sound pressure of breathing sound during sleep was high. Therefore, from the recorded breathing sounds during sleep, the top five points with high sound pressure for each subject were cut out. The sound to be cut out was 7 seconds before and after the high sound pressure. Then, each data was normalized based on the centre, and all data were clustered into 6 by k-means method [8] - [9].

4.3. Results

The results were good, and classified into 6 clusters of characteristic breathing patterns. Fig. 2 shows the most common pattern of patients with SAS. The horizontal axis is elapsed seconds, and the vertical axis is normalized sound pressure. In this pattern, the patient takes a large breath while stopping breathing and then stops breathing again. In the other five patterns, breathing patterns different from those of Non SAS were obtained. In addition, Pre SAS had the same result as SAS, and it was possible to distinguish between Non SAS and Pre SAS. Also, the analysis time was 10 to 15 minutes per person because the recorded part of the SAS sound was cut out semi-automatically rather than analysing all the recorded data of the subject. It was more efficient than the time required for conventional SAS diagnosis.

5. Conclusion

We cut out the breath sounds of SAS during sleep and performed cluster analysis based on them. The result was 100%. This proposed method has been shown to be useful for simple diagnosis of SAS. In the future, I would like to increase the number of subjects, examine the cluster method, and improve the accuracy.

6. References

[1] Takahashi H 1993 *Sleep Apnea Syndorome - Its diagnosis and treatment* (Kinpodo) pp 65-92 (in Japanese)

[2] Uchida N 2013 *Sleeping medicine that you like 2nd* (Kodansya) pp 79-83 (in Japanese)

[3] Rachim V.P Li G and Chung W.Y 2014 Sleep apnea classification using ECG-signal wavelet - PCA features (Biomed Mater Eng Vol 24 Issue 6) pp 2875-2882

[4] Tomita Y and Narui K 2016 *Sleep Apna Syndorome - Inspection of sleep apnea syndrom and PCA features* (Biomed Mater Eng Vol 24 Issue 6) pp 2875-2882

[5] Qian K Janott C Pandit V Zhang Z Heiser C Hohenhorst W Herzog M Hemmert W and Schuller,B 2016 Classification of the Excitation Location of Snore Sounds in the Upper Airway by Acoustic Multi-Feature Analysis (IEEE Transactions on Biomedical Engineering) Vol 64 Issue 8 pp 1731-1741 DOI:10.1109/TBME.2016.2619675

[6] Jin H Lee L.A. Song L Li Y Peng J Zhong N Li H.Y and Zhang X 2015 *Acoustic Analysis of Snoring in he Diagnosis of Obstructive Sleep Apnea Syndrome : A Call for More Rigorous Studies* (Journal of Clinical Sleep Medicine) Vol 11 No 7 pp 765-771

[7] Ide T 2015 *Anomaly detection by introductory machine learning - practical guide by R* (Corona) pp 84-93 (in Japanese)

[8] Munechika M 2001 *Introduction to multivariate analysis method* (Library New mathematicians major) cience pp 174-184 (in Japanese)

IOP Conf. Series: Materials Science and Engineering **920** (2020) 012017 doi:10.1088/1757-899X/920/1/012017

Oscillations of a flexible beam in cavity flow with tonal sound

H Yokoyama*, K Hibino, M Nishikawara and H Yanada

Department of Mechanical Engineering, Toyohashi University of Technology, 1-1, Hibarigaoka, Tempaku, Toyohashi, Aichi, Japan

*Corresponding author: h-yokoyama@me.tut.ac.jp

Abstract. To establish the methodologies for utilization of oscillation energy in a cavity flow, the self-sustained oscillations in a flow over a cavity with and without a flexible beam are investigated, where stainless steel and aluminium beams are utilized. The coupled simulations for aeroacoustic phenomena and bending vibration of the beam are conducted along with wind tunnel experiments. The measurement of sound pressure level with changing the freestream Mach number shows that intense oscillations occur for the second acoustic mode for both the cavities with a stainless and an aluminium beam along the upstream edge of the cavity while the oscillations with the first mode occur for the cavity without a beam. The computational results show that the beam is vibrated at the natural frequency of each beam, where this frequency can be different from the fundamental frequency of the radiating sound. Also, it is shown that the vibrations become more intense for the cavity flow with the beam along the downstream edge, where the acoustic radiation also becomes more intense.

1. Introduction

It is well known that the acoustic radiation occurs from the self-sustained oscillations in the cavity flow as shown in figure 1. The cavity flow exists in various industrial products such as the car gap of a high-speed train, sunroof of an automobile, and pipe arrangement in a plant. Under the current circumstances, the oscillation energy in these cavity flows is wasted without utilization. The final objective of this research is to establish the methodologies of utilization of the oscillation energy in the cavity flow.

Many researchers over the past 50 years have investigated the mechanism of self-sustained oscillations with acoustic radiation (cavity tone) in cavity flows. Rossiter [1] described an oscillation mechanism similar to that presented for edge tones by Powell [2]. In this mechanism, the interactions of vortices with the downstream edge of the cavity radiate acoustic waves, which cause the formation of new vortices at the upstream edge.

Rockwell *et al.* classified the oscillations into three categories: fluid-dynamic oscillations, fluid-resonance oscillations and fluid-elastic oscillations [3]. The fluid-dynamic oscillations occur mainly due to the unsteadiness of the shear layer of the cavity flow with acoustic feedback [4], the fluid-resonance oscillations occur due to the coupling of this unsteadiness and acoustic resonance [5], and the fluid-elastic oscillations occur with vibrating components. The oscillations due to the coupling of all these oscillations (fluid-resonance-elastic oscillations) can occur intensely in a deep cavity with a vibrating beam as shown in figure 1. By installing a piezoelectric material as the vibrating beam, the oscillation energy can be converted to electric energy. To establish the methodology for the utilization of the oscillation energy, the flow field, acoustic radiation, beam vibrations in the fluid-resonance-elastic oscillations are necessary to be clarified.

The objective of this paper is to clarify the conditions of intense self-sustained oscillations in the flow over a cavity with a flexible beam. Particularly, the effects of the material of the beam and installation position on the oscillations are focused on. The direct aeroacoustic simulations coupled with the computation of the beam vibrations are performed along with wind tunnel experiments.

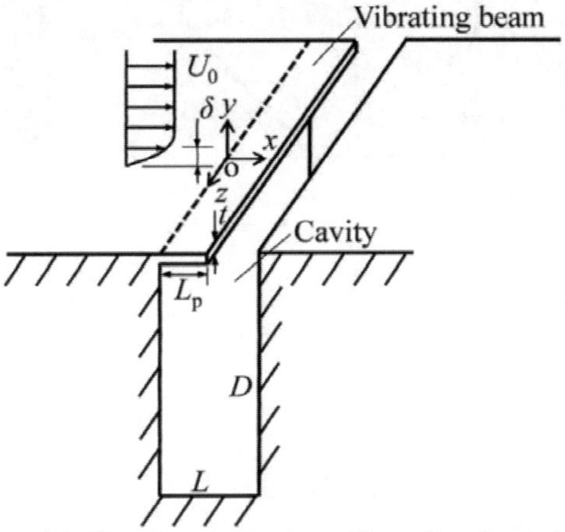

Figure 1. Flow over a cavity with a vibrating beam.

2. Methodologies

2.1. Flow configurations

The experimental setup is shown in figure 2. The coordinate origin is set at the spanwise center of the upstream edge of the cavity as shown in the figure. The cavity length is $L = 20$ mm, and the depth-to-length ratio is $D/L = 2.5$.

For the above-mentioned cavity geometry, the acoustic resonance for the first mode with one-quarter wavelength mode and the second mode with three-quarter wavelength mode were estimated to be 1300 Hz and 3900 Hz, respectively, with reference to the literature [5]. A stainless steel beam with the thickness of $t = 0.2$ mm and an aluminium beam of $t = 0.5$ mm were selected as a flexible beam to fit the natural frequency of the plate estimated by equation (1) with the frequencies of first and second acoustic modes, respectively.

$$f_c = \frac{1}{2\pi}\frac{\lambda^2}{L_p^{\ 2}}\left(\frac{EI}{\rho_p A}\right)^{\frac{1}{2}}, \tag{1}$$

where $\lambda = 1.875$ is a constant for the first vibration mode, ρ_p is the density of the material, E is the modulus of elasticity, and I is the second moment of area. The beam length was set to be $L_p/L = 0.5$. The beam was attached along the upstream edge of the cavity in the experiment. To clarify the effects of the installation position of the beam, the computation was also performed for the installation of the beam along the downstream edge.

The freestream Mach number based on the freestream velocity, M, was changed from 0.03 to 0.14 in the experiment, while the computations were performed at $M = 0.13$ The incoming boundary layer is laminar, and the boundary layer thickness was $\delta/L = 0.053$ at $M = 0.13$ as shown in figure 3.

The distance of the upstream edge of the cavity from the imaginary origin of the boundary layer was estimated to be 4.1L.

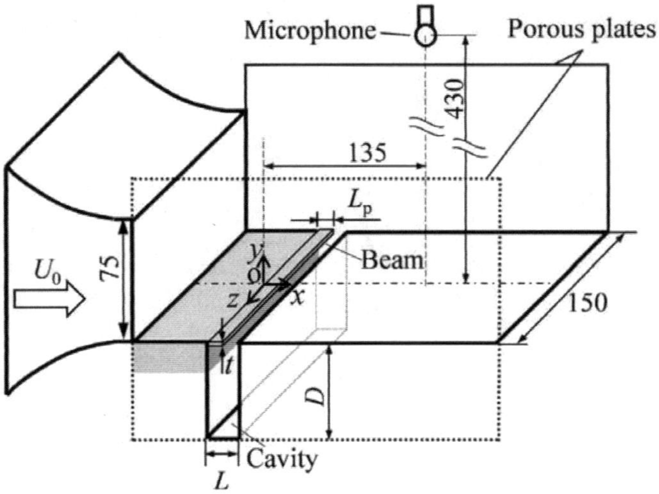

Figure 2. Experimental setup.

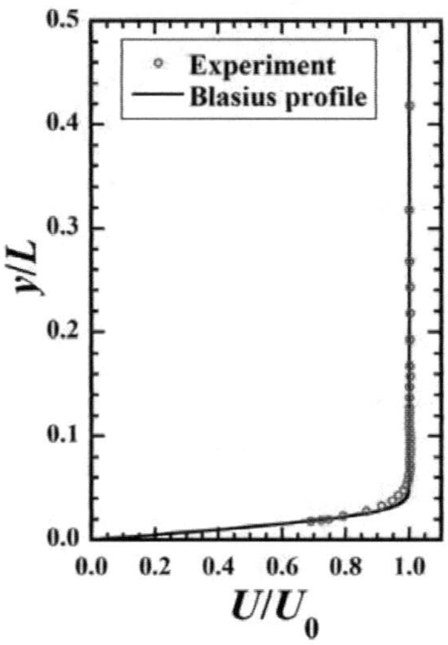

Figure 3. Measured boundary layer profile with Blasius profile.

2.2. Experimental methodologies

The experiments were carried out using a suction-type, low-noise wind tunnel with a rectangular test section having a cross-section with dimensions of 150 mm × 75 mm. The intensity of freestream turbulence was less than 0.5%, and the background-noise level was 55.2 dB (A) at $M = 0.087$. The test section of the cavity was terminated in the spanwise direction by end walls composed of porous plates, which suppressed acoustic resonance in the spanwise direction. The width of the test section was $W/L = 7.5$. The beam and test section in the upstream of the cavity were carved out from one block so that the beam can be vibrated as a cantilever.

The magnitude of the acoustic pressure in the far field ($x/L = 6.75$, $y/L = 21.5$) was measured with a non-directional 1/2-inch microphone (UC-53A, RION, Tokyo, Japan) and a precision sound-level meter (NL-52, RION, Tokyo, Japan). The measurement time required for data acquisition was 30 s, and the sampling frequency was 80 kHz.

2.3. Numerical simulations

In order to simulate the interactions between flow and acoustic fields, the three-dimensional compressible Navier-Stokes equations with mass and energy conservation laws were directly solved using a finite difference scheme. The spatial derivatives were evaluated by the six-order-accurate compact finite difference scheme (the forth-order-accurate at the boundary) [6]. The time integration was performed using the third-order-accurate Runge–Kutta method. The detailed numerical methodologies such as computational domain and boundary conditions for the simulations of the cavity flow are described in the previous paper [7].

To reproduce the vibrating beam on rectangular grids, the volume penalization (VP) method [8] was utilized. The VP methods utilized in this paper are the same as those described in the previous papers [9–11]. Also, while the beam vibrations are given as a sinusoidal wave in the previous paper, the computation of the beam vibration was also coupled with the direct aeroacoustic simulations in the present paper. The following governing equations of the beam vibration was calculated at each time step of the aeroacoustic simulation.

$$\frac{\partial^2}{\partial x_p^{\,2}}\left[EI(x_p)\left(1+\eta\frac{\partial}{\partial t}\right)\frac{\partial^2 y_p}{\partial x_p^{\,2}}(x_p,t)\right]+\rho_p S(x_p)\left[\frac{\partial^2 y_p}{\partial t^2}(x_p,t)+\gamma_B\frac{\partial y_p}{\partial t}(x_p,t)\right]=F(x_p,t), \qquad (2)$$

where t is time and x_p is the horizontal position along the beam length, $y_p(x_p, t)$ is the vertical position of the top-surface of the beam, and $S(x_p)$ denotes the cross-sectional area. The coefficient η represents the magnitude of the internal viscoelastic losses, and γ_B accounts for damping of the surrounding fluid. In this paper, the values were set to be $\eta = 6\times10^{-7}$ s and $\gamma_B = 100$ s^{-1} referring to the literature [12]. $F(x_p, t)$ is a driving force per unit length, which was calculated from the fluid pressure around the beam at each time step. Equation (2) was solved by using the implicit-θ-scheme [12], which is a specific class of implicit schemes known as the "variable-weighted implicit approximation method" [13]. Finally, clamped-free boundary conditions lead to the constraints.

$$y_p(0,t)=\frac{\partial y_p}{\partial x_p}(0,t)=0, \qquad (3)$$

$$\frac{\partial^2 y_p}{\partial x_p^{\,2}}(L_p,t)=\frac{\partial^3 y_p}{\partial x_p^{\,3}}(L_p,t)=0 \qquad (4)$$

3. Numerical validation

Figure 4 shows the predicted and measured sound pressure spectra for the cavity flow with the aluminium beam along the upstream edge of the cavity at M = 0.13. The frequency resolution is 150 Hz, and the spectra are averaged 15 times and 9400 times in computation and experiment, respectively. As shown in the figure, the intense tonal noise occurs at the fundamental frequency (4000 Hz), and higher harmonic frequencies. The predicted levels and frequencies for the tones are in good agreement with those of the measured results. This indicates that the present computations can correctly capture the acoustic radiation from the cavity flow with a plate.

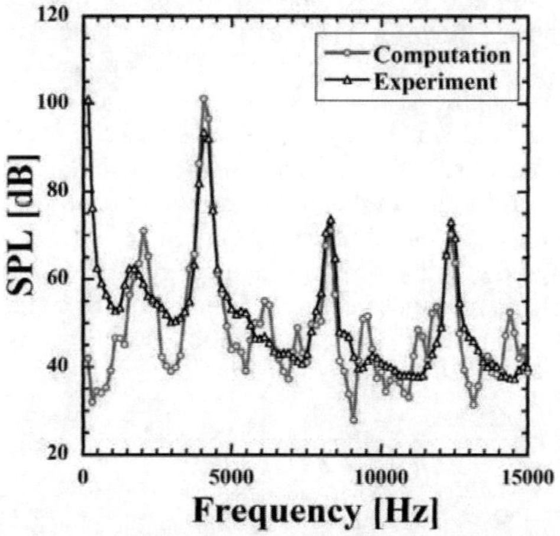

Figure 4. Predicted and measured sound pressure spectra for the cavity with the aluminium beam along the upstream edge of the cavity at $M = 0.13$ ($x/L = 6.75$, $y/L = 21.5$).

4. Results and discussions

4.1. Effects of installation of beam on acoustic radiation

Figure 5 shows the measured sound pressure spectra for the cavity flows with and without a beam at M = 0.13, where the frequency resolution is 5 Hz. The tonal sound at the fundamental frequency around

4000 Hz becomes more intense by attaching the beam. While the tonal sound at a lower frequency around 2000 Hz becomes weaker, the tones at higher harmonic frequencies become more intense.

Figure 6 shows the variation of the sound pressure level at the fundamental frequency with the freestream Mach number. While a peak appears at $M = 0.10$ in the case without a beam, two peaks appear around $M = 0.06$–0.08 and $M = 0.12$–0.14 in the case with a beam. The Mach number for the maximum peak is $M = 0.134$ and 0.128 for the stainless steel and aluminium beams, respectively.

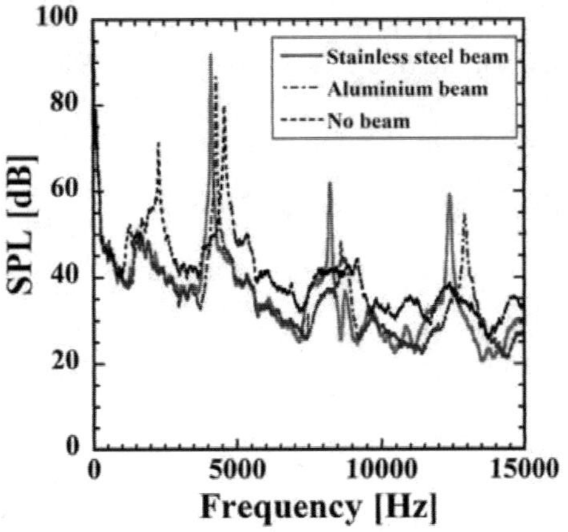

Figure 5. Measured sound pressure spectra with and without a beam at $M = 0.134$.

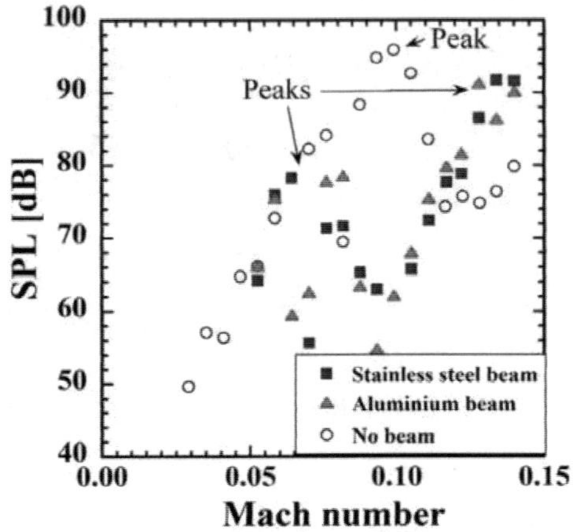

Figure 6. Variation of sound pressure level at the fundamental frequency with freestream Mach number.

Figure 7 shows that variation of the non-dimensionalized fundamental frequency (Strouhal number) with the Mach number, where the Strouhal number is based on the freestream velocity and the opening length of the cavity, $L - L_p$ and L, for the cavity flow with and without a beam. Also, the curves of the frequencies for the first and second acoustic modes predicted by the semi-empirical equation in the literature [5] are shown. Figure 7(a) shows that the measured fundamental frequency for $M = 0.05$–0.12 is close to that for the first acoustic mode. This indicates that the tonal sound occurs due to the acoustic resonance of the first mode in the cavity at the above-mentioned peak at $M = 0.10$. Also, figure 7(b) shows that the measured fundamental frequency at $M = 0.11$–0.14 is close to that for the second mode in the cavity flow with a beam. This means that the above-mentioned peak around $M = 0.13$ occurs due to the acoustic resonance of the second acoustic mode. These results show that the installation of the beam can vary the dominant acoustic mode in the cavity.

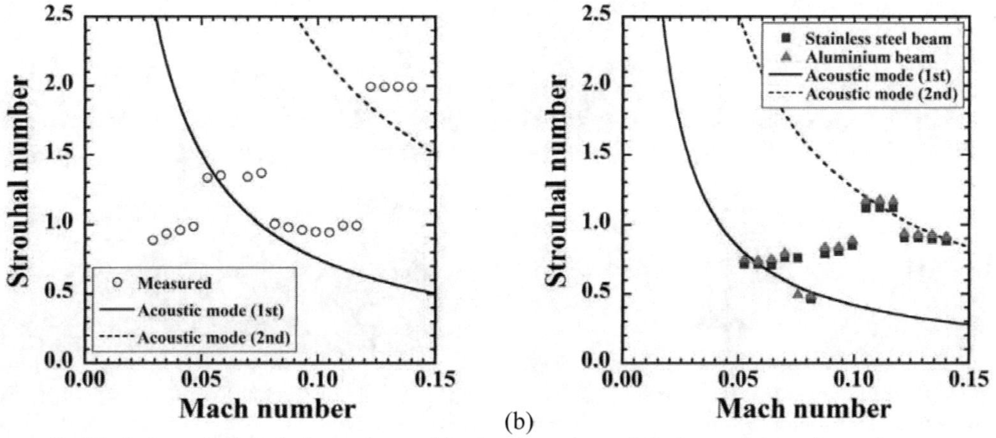

(a) (b)

Figure 7. Variation of Strouhal number with the freestream Mach number, where the curves of acosutic modes are also shown. (a) Case for the cavity flow without a beam. (b) Case for the cavity flow with a beam.

4.2. Predicted flow and acoustic fields

Figure 8 shows predicted vortices and acoustic fields by the iso-surfaces of second invariant of velocity gradient tensor and contours of fluctuation pressure for the cavity flow with the aluminium beam at M = 0.13. The time of t = 0 corresponds to the time when the vortex is collided with the downstream edge of the cavity. As shown in the figure, the apparent two vortices are convected in the free shear layer, which indicates the occurrence of the second Rossiter mode [1]. In the cavity, the pressure fluctuations near the opening of the cavity and those near the bottom show anti-phase behaviour. This supports the above-mentioned occurrence of the acoustic resonance with the second (three-quarter wavelength) mode.

Figure 8(a) shows that the expansion wave is radiated into the cavity when the vortex is collided with the downstream edge. After the half fundamental period, the compression wave is radiated into the cavity as shown in figure 8(b). The acoustic radiation sustains the acoustic resonance in the cavity.

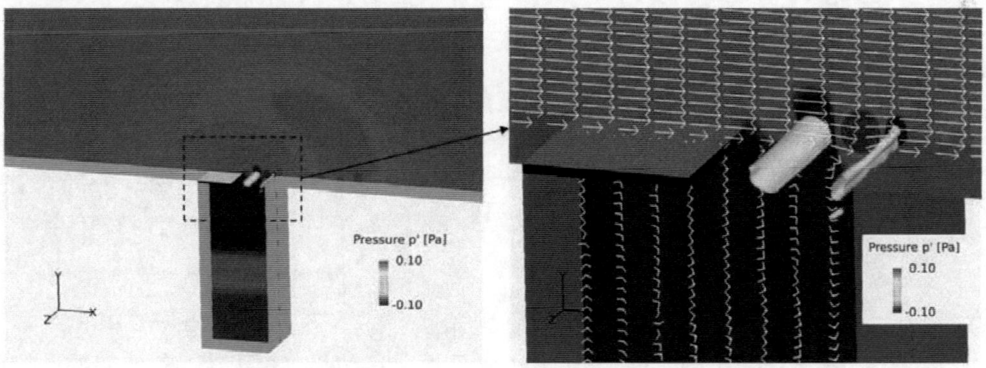

(a)

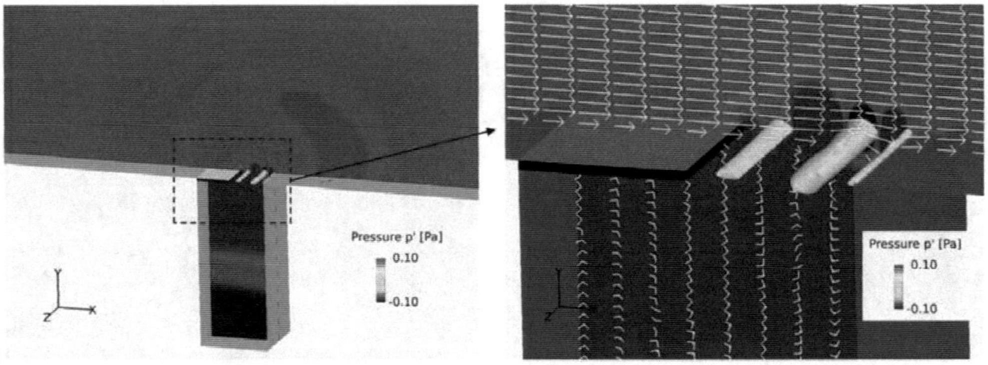

(b)

Figure 8. Predicted iso-surfaces of second invariant and contours of fluctuation pressure for the cavity flow with an aluminium beam along the upstream edge of the cavity. (a) $t = 0$. (b) $t/T = 0.5$.

4.3. Vibrations of beam

Figure 9(a) shows the time variation of the beam displacement at the tip ($M = 0.13$). The peak-to-peak amplitude is 13 μm and 11 μm for the stainless and aluminium beams, respectively. The power spectra of the displacement fluctuations are shown in figure 9(b). The stainless and aluminium beams were found to be vibrated at the 1400 and 4100 Hz, which approximately agree with the natural frequencies of these beams. As shown in figure 5, for both the cases, the acoustic radiation occurs around 4000 Hz, which is corresponding to the frequency of the second acoustic mode. The result for the cavity flow with the stainless steel beam present that the beam can be vibrated at the natural frequency even if the frequency is different from the fundamental frequency. This is possibly because the pressure fluctuations occur not only at the fundamental frequency but in the broadband frequency range, which causes the source for the beam vibrations at each natural frequency.

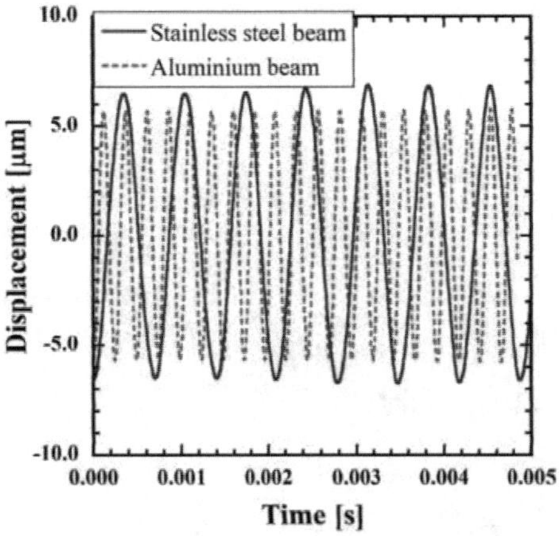

(a)

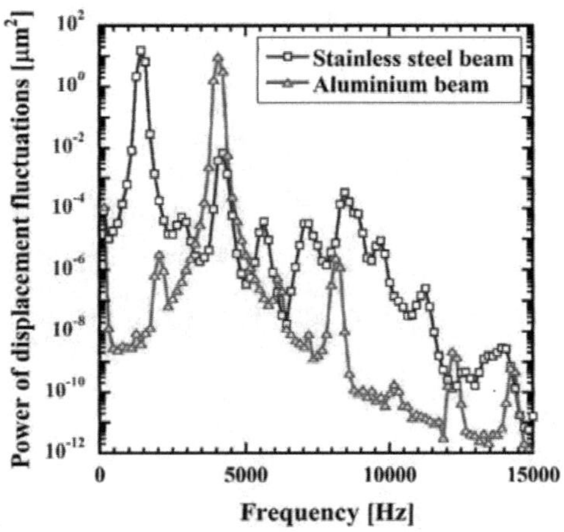

(b)

Figure 9. Predicted displacement of beam. (a) Time variation. (b) Power spectra.

4.4. Effects of installation position

The effects of the installation position of the beam on the oscillations are discussed. Figure 10 shows the predicted vortices and acoustic fields for the cavity flow with the aluminium beam along the downstream edge of the cavity. As shown in the figure, the acoustic radiation into the outward of the cavity becomes more intense compared with the case with the beam along the upstream edge of the cavity (figure 8). Also, the vortices are found to be deformed by the collision with the tip of the beam and convected along the beam. The deformation of the vortices can lead to the intensification of the acoustic radiation [12].

Figure 11 shows the time variation of the beam displacement at the tip. The peak-to-peak amplitude for the downstream installation is 32 μm, which is threefold larger compared with that for the upstream installation. This is because the deformed vortices and intense acoustic resonance cause the pressure fluctuations around the beam.

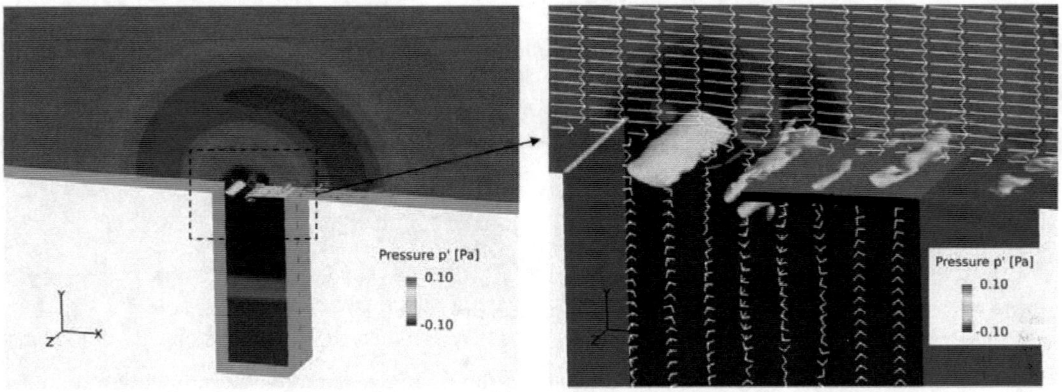

Figure 10. Predicted iso-surfaces of the second invariant and contours of the fluctuation pressure for the cavity flow with the aluminium beam along the downstream edge of the cavity ($t = 0$).

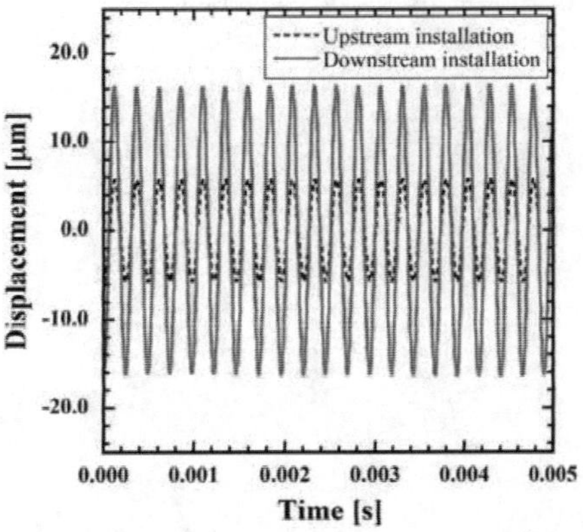

Figure 11. Predicted displacement of the beam tip for the cavity flow with the aluminium beam along the downstream edge of the cavity compared with those along the upstream edge.

5. Conclusions

In order to clarify the conditions of intense fluid-resonance-elastic oscillations in the flow over a cavity with a flexible metal beam, wind tunnel experiments and the direct aeroacoustic simulations coupled with the computations of beam vibrations were conducted.

The measured variation of sound pressure level of cavity tone with the freestream Mach number shows that intense oscillations occur with the second acoustic mode for the cavity flows with a beam along the upstream edge of the cavity while the oscillations occur with the first acoustic mode for the cavity without a beam. The predicted acoustic field confirms the occurrence of the second acoustic mode in the cavity with the beam.

IOP Conf. Series: Materials Science and Engineering **920** (2020) 012017 doi:10.1088/1757-899X/920/1/012017

The predicted power spectra of the beam displacement show that the beam can be vibrated at the natural frequency of each beam independently of the fundamental frequency of the radiating sound. Also, the amplitude of the beam displacement become threefold larger in the cavity flow with the beam along the downstream edge compared with that along the upstream edge.

Acknowledgments
This work was supported by JSPS KAKENHI Grant Number 17K06153.

References
[1] Rossiter J E 1964 *Aeronautical Research Council Report* 3438
[2] Powell A 1961 *J. Acoust. Soc. Am.* **33** pp 395–409
[3] Rockwell D and Naudascher E 1978 *Trans. ASME J. Fluids Eng.* **100** pp 152–65
[4] Rowley C W, Colonius T and Basu A J 2002 *J. Fluid Mech.* **455** pp 315–46
[5] East L F 1966 *J. Sound Vib.* **3**, pp 277–87
[6] Lele S K 1992 *J Comp. Phys.* **103** pp 16–42
[7] Yokoyama H, Tanimoto I and Iida A 2017 *Appl. Sci.* **7**, 790
[8] Liu Q, Vasilyev O V 2007 *J Comp Phys.* **227** pp 946–66
[9] Tanaka Y, Yokoyama H and Iida A 2018 *J. Sound Vib.* **431** pp 248–64
[10] Sato S, Yokoyama H and Iida A 2019 Appl. Sci. **9**(4) 428109
[11] Kobayashi M., Yokoyama H., Iida A. 2017 *Proc. Inter-Noise 2017* (Hong Kong, 27–30 August 2017) pp 6707–18
[12] Avanzini F and van Walstijn M 2004 Acustica **90** pp 537–47
[13] Lapidus L and Pinder G 1999 *Numerical solution of partial differential equations in science and engineering* (New York, John Wiley & Sons) pp 281–2
[14] Yokoyama H and Kato C 2009 *Phys. Fluids* **21** 105103

IOP Conf. Series: Materials Science and Engineering **920** (2020) 012018 doi:10.1088/1757-899X/920/1/012018

Developing the Regression Relationships for the Critical Period of the Reservoir Systems at Sungai Bekok and Sungai Kesang Catchments in Malaysia

I S Oskoui and R Abdullah*

School of Civil Engineering, Universiti Sains Malaysia, 14300 Nibong Tebal, Penang, Malaysia

*Corresponding author: cerozi@usm.my

Abstract. The knowledge of critical period (CR) for determining the behavior type of reservoir systems is very beneficial in the planning process for distinguishing the period of the required streamflow data in the simulation procedure. Generally, for within-year systems (CR < twelve months) utilizing the critical twelve months of the flow data record is required, whereas for the carryover systems (CR > twelve months) employing the annual streamflow data will suffice. Currently, there is no relationship in the literature that can be used for predicting the behavior type of reservoir systems employing reliability and vulnerability performance indices in Sungai Bekok and Sungai Kesang catchments. Hence, the objective of this study is to develop a novel equation involving performance indices for estimating the critical period of the aforementioned reservoir systems. The reservoirs were analyzed employing a Monte Carlo technique by improved sequent peak algorithm. Afterward, new regression equations for the critical period were developed based on simulation results and then verified by comparing the equations' outcomes with simulation results. Finally, it was found that the performance of the new equations is excellent in reproducing the critical period.

1. Introduction

Surface water reservoir systems are built on rivers to regulate the streamflow with saving water for fulfilling the demand at the specified performance level for domestic, industrial and agricultural purposes during shortage periods. Reservoir capacity planning process is usually based on a single historical data record; however, the historical data record is only a sample of the various data that the reservoir might experience in future [1-3]. Hence, each estimate of the reservoir capacity can be considered as a random variable [4-7]. In this study 1000 replicates of synthetic data having similar statistical characteristics to the historical data were used for the reservoir simulation.

The capability of a reservoir to operate satisfactorily under a wide range of probable hydrologic conditions and demands is an essential system characteristic [8-10]. The likely performance of reservoir systems is usually measured by performance indices that are defined based on some particular aspects in the failure periods [11-13]. The failure period is the time during drought periods that the reservoir is not able to provide the target demand completely [14]. The performance indices are very useful in selecting operational policies, storage capacities, system configurations and targets [9, 15-17]. Hashimoto et al. (1982) presented the performance indices to measure the efficiency of reservoir systems. The performance measures were reliability (i.e. how often the system is reliable) and vulnerability (i.e. how acute the consequences of the failure could be) that were used in this study [9, 11]. For instance, the reliability of 95% means that the reservoir could provide the target demand in 95% of the time and the vulnerability of 15% means the reservoir might not be able to deliver up to 15% of the target demand during failure periods.

The storage-yield analysis of reservoir systems could be carried out by two general methods. The first method is the probability matrix procedure that is based on direct modelling of the reservoir volume [18]. The second method is the critical period procedure that could be carried out by either simulation or optimization processes [19]. Since among the aforementioned methods only simulation process of improved sequent peak algorithm was capable to include both reliability and vulnerability indices in the reservoir system analysis hence, improved sequent peak algorithm was employed in this study to analyze Sungai Bekok and Sungai Kesang reservoirs [16, 20, 21].

Critical Period (CP) is described as a time that a full reservoir goes empty without overflowing in this period [20, 22, 23]. CP is employed to distinguish two major behavior types of reservoir systems. The behavior that concerns with CP > twelve months is termed as carryover behavior whilst the behavior relates to CP < twelve months is called the within year behavior. These behaviors have an important influence on some aspects of reservoir systems such as sustainability, resiliency, reliability, vulnerability and storage capacity [24, 25]. Moreover, the within-year and carryover behaviors could be used to determine the interval of the streamflow data that is required for the reservoir simulation. For instance, if the reservoir shows within-year behavior utilizing twelve critical monthly flows is required for the reservoir analysis; however, when the reservoir has carryover behavior employing only annual streamflow data will suffice. In the case that reservoir shows both within-year and carryover behaviors employing the whole monthly flows will be necessary in the analysis process.

The only general method that is used in literature for distinguishing the behavior type of reservoir systems is the parameter m (i.e. the standardized demand) method as follows [26]:

$$m = \frac{1-D}{C_V} \tag{1}$$

In Equation (1), m is the standardized demand parameter, D is the demand as a ratio of the average yearly runoff (AYR), and C_V is the coefficient of variation of the annual flow. According to parameter m method if $m < 1$ the system's behavior will be carryover and if $m > 1$ the system's behavior will be within-year. Moreover, when $m = 1$ a transitional behavior between within-year and carryover happens [26-28].

As can be seen parameter m method determines the reservoir behavior applying only demand and coefficient of variation of annual flows; however, other parameters such as performance measures also play the significant role in the behavior type of reservoir systems [24]. Currently, there is not a relationship in the literature for determining the behavior type of reservoirs considering performance indices for Sungai Bekok and Sungai Kesang catchments as best of the authors knowledge. Hence, the objective of this study is to develop the regression equations for modeling the critical period of Sungai Bekok and Sungai Kesang catchments in Malaysia.

2. Materials and Methods

2.1. Catchments and Data

This study is performed on Sungai Bekok and Sungai Kesang catchments in Peninsular Malaysia. A regression equation employing performance indices was developed for each catchment separately to model the behavior type of reservoir systems. The catchments of Sungai Bekok and Sungai Kesang receive the average yearly rainfall of 1974 and 2688 mm, respectively. They generally have a continuous humid equatorial rainforest weather that is sunny and warm. The monthly flow records from the catchments are employed to simulate the critical period of the reservoir systems. The main specifications of the catchments that are required for reservoir designing are shown in Table 1. The variation coefficients of the annual flow of the study rivers are quite high (0.49 and 0.40) that justifies the need for designing reservoir systems to regulate streamflow in different months.

Table 1. Specifications of the study catchments

Site	River	Gauging station	Record length (years)	Catchment area (km^2)	Annual flow statistics				
					Mean (mm)	Mean ($\times 10^6 m^3$)	C_V	Skewness	ρ
1	Bekok	Jln. Y. Peng, Labis	39 (1974-2012)	350	707	247	0.49	0.54	0.28
2	Kesang	Chin Chin	53 (1960-2012)	161	456	73	0.40	1.64	-0.07

Note: ρ is serial correlation (lag-1);
mm is millimeter.

The streamflow data were examined using different statistical tests prior to Monte Carlo simulation. That includes homogeneity test that verifies data for consistency [21, 29]. The data were also examined applying trend test and run test to ensure that they are stationary and the result of a random procedure [30-32].

In orders to normalize the data for data generation there is need to find the most appropriate density function of the annual and monthly streamflow data. For finding the most appropriate probability distribution the standard error test was used [25, 33]. Since Log Pearson type III produced the minimum standard errors among other probability density functions, hence, Log Pearson type III was selected as the most appropriate density function for both monthly and annual flow data. The fitness of Log Pearson type III density function to the historical annual flow data of Sungai Bekok and Sungai Kesang Catchments is shown in Figure 1.

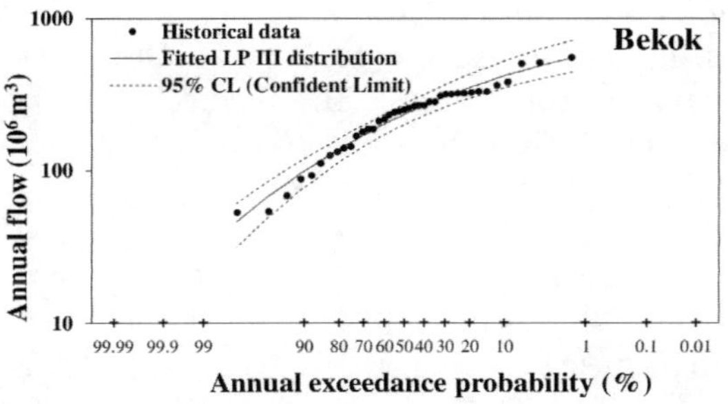

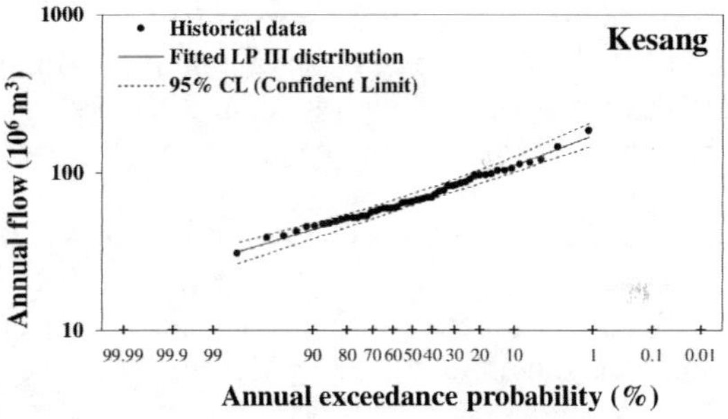

Figure 1. The fitness of Log Pearson type III distribution to the historical streamflow data of Sungai Bekok and Sungai Kesang catchments

2.2. Synthetic Streamflow Data Generation

Using synthetic data in reservoir system analysis enables the designer to test alternative choices against the range of the streamflow sequences that might occur in future [20, 34-36]. Hence, in this study instead of the traditional methods that use only available data for reservoir simulation, the stochastic procedure that is based on large number of synthetic flow sequences was employed [37-39]. Generally, streamflow data generation models should reproduce the important statistics of the annual and monthly flows such as mean, standard deviation, coefficient of variation, skewness and auto correlation lag one [4, 5, 40]. Since Auto regressive lag one (AR (1)) model for generating annual flows and Valencia and Schaake (V-S) model for disaggregating the generated annual flows to monthly flows could reproduce the important statistics of the streamflow data at both monthly and annual levels, hence, the pair of AR (1) and V-S models was used to generate the flow data for the reservoir analysis in this study [41-43]. Vogel and Stedinger (1987) showed that AR (1) model is more appropriate than other models to generate annual flows for reservoir system analysis. AR (1) model to generate annual flows is as follows [26]:

$$r_{j+1} = \mu\left(r_j - \bar{r}\right) + \bar{r} + v_j t\sqrt{(1-\mu^2)} \quad (2)$$

In Equation (2) r_{j+1} and r_j are the yearly runoff of the $(j+1)^{th}$ and j^{th} years, respectively; $\bar{r}$ is the mean annual runoff; t is the standard deviation of yearly runoff; v_j is the standard normal random variable and μ is the auto correlation lag one of the yearly runoff.

Valencia and Schaake (V-S) model was then employed to disaggregate the generated annual flows of the study systems to monthly flows. The pair of AR (1) and V-S models were used to generate 1000 replicates of monthly flow data that has the same length of historical data (i.e. 39 years for Sungai Bekok and 53 years for Sungai Kesang).

2.3. *Simulation of Reservoir Systems*

The simulation of reservoir systems was performed using improved sequent peak algorithm (SPA). The improved SPA can identify the critical period (CR) of the reservoir with involving time-based reliability and vulnerability (i.e. maximum amount of shortage during failure period [9, 17, 44, 45]. The time-based reliabilities of 1.00, 0.99, 0.98, 0.97, 0.96, 0.95, 0.93 and 0.90 were employed in the simulation. Moreover, the vulnerability indices of 0.00, 0.05, 0.10, 0.15, 0.20, 0.25 and 0.30 were applied in the improved SPA.

The annual demands of 0.20, 0.30, 0.40, 0.50, 0.60, 0.70 and 0.8 that are considered as a ratio of annual yearly runoff (AYR) were used in the analysis. For demands below 0.20 AYR there is usually no need to build a reservoir and the natural flow of the river can satisfy the demand and for demands above 0.80 AYR the improved SPA does not usually yield a unique critical period. Hence, the lower limit and upper limit of demand is considered as 0.20 and 0.80 AYR, respectively in this study. The monthly demands were also considered to be constant during different months of the year.

The simulation of the study reservoirs was carried out by improved SPA for the aforementioned time-based reliability, vulnerability and demands in 1000 sequences using 1000 replicates of the flow data that was generated in Section 2.2 and 1000 CRs were obtained for each case. The average of 1000 CRs for every case was considered as the most appropriate value of CR in this study.

2.4. *Regression Relationships for Critical Period*

There are variables that play a significant role in the critical period of reservoir systems. The most significant variable is parameter *m* (i.e. Equation 1). Hence, the relationship between CR and parameter *m* can be assumed as follows [24]:

$$\mathrm{CR} = \mathrm{P}\,.\,m^{Q} \quad (3)$$

In Equation (3), CR is the critical period and *m* is the standard demand parameter. P and Q are parameters that could be expressed as a function of other important variables in the critical period of reservoir systems. These variables involve time-based reliability and vulnerability indices. Therefore, the following relationship is proposed for P and Q in this study:

$$\phi = \gamma_0 + \gamma_1.\mathrm{R_L} + \gamma_2.\mathrm{V_L} \quad (4)$$

In Equation (4), ϕ is dependent variable (i.e. P or Q); R_L is time-based reliability; V_L is vulnerability; γ_0, γ_1 and γ_2 are the regression coefficients.

By combining Equation (3) and Equation (4) the regression relationship for critical period was developed. The performance of the obtained regression equation was verified by comparing the equation results with simulation outcomes.

3. Results and Discussion

3.1. Synthetic Replicates of the Storage Capacities

The box plot diagrams of the generated storage capacities of the reservoirs introduced by Tukey are shown in Figure 2 [24, 46]. The diagrams are presented for the demand of 70% average yearly runoff (AYR). To generalize the results the storage capacities were stated as a ratio of AYR. The box plots show maximum, minimum, 25, 50 and 75 percentiles of the generated storage capacities. The box plots show that reliability and vulnerability indices play a significant role in the storage capacity and by implication in critical period of reservoir systems. (i.e. by increasing reliability or decreasing vulnerability indices the storage capacity and critical period increase.)

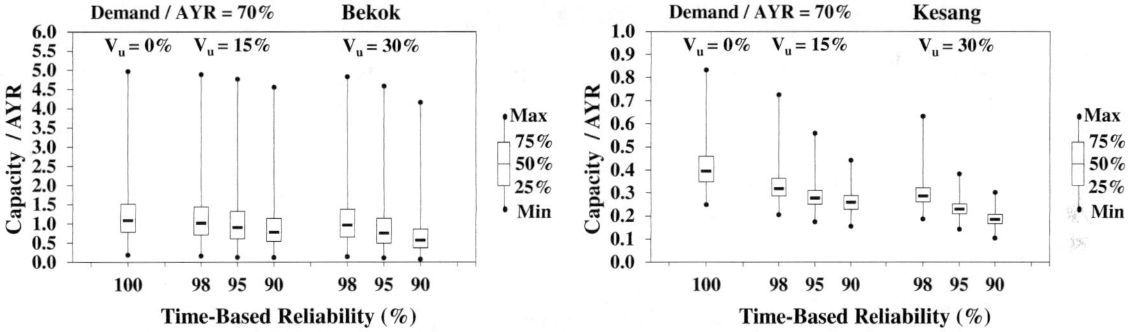

Figure 2. The box plots of storage capacities in demand of 70% AYR for different reliability and vulnerability indices in Sungai Bekok and Sungai Kesang catchments

3.2. Regression Relationship for Critical Period

3.2.1. Relationship Between Critical Period and Standard Demand Parameter

The relationship between critical period and standard demand parameter was presented in Section 2.4 (i.e. Equation (3)). In Equation (3), P and Q are parameters that could be found by runs of the improved SPA for each possible pair of reliability and vulnerability indices.

Equation (3) was calibrated for every possible pair of reliability and vulnerability indices applying 7 data points that relate to the demands of 20%, 30%, 40%, 50%, 60%, 70% and 80% of average yearly runoff (AYR). The fitness of Equation (3) to the simulation results in Sungai Bekok and Sungai Kesang systems is shown in Figure 3 as a sample. Although the fitted curve is for reliability of 0.95 and vulnerability of 0.15; however, the fits for all pairs of reliability and vulnerability indices are also very appropriate. The fits are generally excellent with R^2 varying between 0.9769-0.9984 for the study sites. Table 2 involves the sample estimates of P and Q along with R^2 for the applied pairs of reliability and vulnerability indices in this study.

3.2.2. Linear Regression Relationships for P and Q

Considering other important variables in the critical period of reservoir systems such as time-based reliability and vulnerability Equation (4) was proposed for P and Q in Section 2.4. The regression coefficients of Equation (4) were estimated for all available data points of each site separately (i.e. 48: 8 (time-based reliability indices) × 6 (vulnerability indices)) using the least square method.

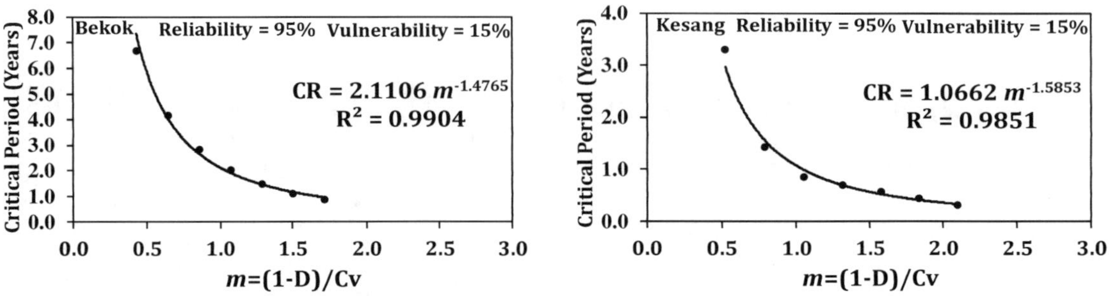

Figure 3. Fitness of the proposed relationship for critical period at reliability of 95% and vulnerability of 15% for Sungai Bekok and Sungai Kesang catchments

Table 2. Regression coefficients P and Q along with R^2 for different pairs of vulnerability and time-based reliability indices for Sungai Bekok and Sungai Kesang catchments

		Bekok			Kesang		
V_L	R_L	P	Q	R^2	P	Q	R^2
0.00	**1.00**	2.36	-1.49	0.9860	1.39	-1.82	0.9912
	0.99	2.30	-1.49	0.9869	1.32	-1.84	0.9933
	0.98	2.27	-1.49	0.9874	1.27	-1.80	0.9905
	0.97	2.24	-1.49	0.9869	1.23	-1.76	0.9878
0.10	**0.96**	2.21	-1.48	0.9879	1.19	-1.70	0.9859
	0.95	2.19	-1.48	0.9888	1.16	-1.66	0.9831
	0.93	2.15	-1.46	0.9906	1.11	-1.58	0.9812
	0.90	2.11	-1.43	0.9916	1.06	-1.49	0.9826
	0.99	2.24	-1.49	0.9868	1.25	-1.84	0.9948
	0.98	2.18	-1.50	0.9870	1.15	-1.77	0.9943
	0.97	2.12	-1.49	0.9877	1.09	-1.70	0.9902
0.20	**0.96**	2.06	-1.48	0.9905	1.04	-1.63	0.9886
	0.95	2.03	-1.47	0.9908	0.99	-1.54	0.9844
	0.93	1.95	-1.44	0.9941	0.92	-1.40	0.9806
	0.90	1.87	-1.37	0.9957	0.86	-1.25	0.9764
	0.99	2.19	-1.50	0.9873	1.19	-1.87	0.9954
	0.98	2.09	-1.50	0.9887	1.05	-1.79	0.9960
	0.97	2.00	-1.49	0.9898	0.97	-1.70	0.9961
0.30	**0.96**	1.93	-1.48	0.9926	0.91	-1.59	0.9938
	0.95	1.87	-1.46	0.9941	0.86	-1.48	0.9922
	0.93	1.76	-1.39	0.9965	0.79	-1.32	0.9875
	0.90	1.65	-1.29	0.9984	0.73	-1.15	0.9769

V_L = Vulnerability; R_L = Time-Based Reliability

The regression coefficients along with *t*-statistics and *p*-values for P and Q at the study sites are presented in Table 3. Moreover, Table 4 shows the main statistics of the linear regression equations such as R^2 (i.e. coefficient of determination), SEE (i.e. standard error of estimate, DF (i.e. degrees of freedom), *F*-stat (i.e. *F*- statistics), *p*-value (i.e. the probability that a high value of *F*-stat happened by chance). Since according to Table 4 *F*-stats are high and the values of R^2 are close enough to 1 hence, the developed regression equations for P and Q at the study sites are statistically sufficient. Additionally, in Table 3 the *p*-values corresponding to *t*-statistics of the regression coefficients are less than 0.05 that indicate both time-based reliability and vulnerability indices are important in the regression analysis for P and Q at the usual 5% probability level.

Table 3. Coefficient estimates, *t*-statistics and *p*-values for P and Q

Site		P			Q		
		Estimate	*t*-statistic	*p*-value	Estimate	*t*-statistic	*p*-value
Bekok	γ_0	-0.543	-2.816	0.004	-0.435	-3.418	0.001
	γ_1	2.978	9.720	0.000	-1.084	-8.305	0.000
	γ_2	-1.182	-12.351	0.000	0.058	2.414	0.016
Kesang	γ_0	-1.760	-6.866	0.000	2.968	7.308	0.000
	γ_1	3.203	12.187	0.000	-4.885	-11.729	0.000
	γ_2	-1.223	-14.896	0.000	0.441	3.387	0.001

Table 4. Main statistics of the linear regression relationships for P and Q

	Bekok		Kesang	
	P	Q	P	Q
R^2	0.88	0.65	0.92	0.81
SEE	0.06	0.03	0.05	0.09
DF	45.0	45.0	45.0	45.0
F-stat	172	42	259	94
p-value	0.00	0.00	0.00	0.00

3.2.3. The Performance of the Developed Regression Relationship in Critical Period Prediction
According to regression statistics in Sections 3.2.1 and 3.2.2, the developed regression relationships for critical period at Sungai Bekok and Sungai Kesang are statistically sufficient; however, the practical method to verify the performance of a regression equation is to see how precise it could reproduce the observed data. By uniting Equations (3) and (4) and applying statistics of Table 3 the final relationships for predicting the critical period of Sungai Bekok and Sungai Kesang could be presented as follow:

$$CR_{(Bek)} = (-0.54 + 2.98R_L - 1.18V_L)m^{(-0.44-1.08R_L+0.06V_L)} \quad (5)$$

$$CR_{(Kes)} = (-1.76 + 3.20R_L - 1.27V_L)m^{(2.97-4.88R_L+0.44V_L)} \quad (6)$$

In Equations (5) and (6) $CR_{(Bek)}$ and $CR_{(Kes)}$ are the critical periods of Sungai Bekok and Sungai Kesang systems, respectively in years; R_L is time-based reliability measure between 0.90 and 1.00; V_L is vulnerability measure between 0.00 and 0.30 and m is the standard demand parameter.
Equations (5) and (6) were applied to produce the critical period of Sungai Bekok and Sungai Kesang systems in Figure 4. 336 points (i.e. 7 (demands) × 6 (vulnerability indices) × 8 (time-based reliability indices)) were applied for each system. As can be seen in Figure 4 Equations (5) and (6) could reproduce the simulated critical periods efficiently specially at low values corresponding to within-year behavior. As the novelty, these equations are the first relationships that were developed for modeling the behavior of reservoir system employing performance indices at Sungai Bekok and Sungai Kesang systems as the best of the authors knowledge. The calculated R^2 between the predicted and simulated critical periods are 0.9827 and 0.9752 for Sungai Bekok and Sungai Kesang systems, respectively that also indicate the excellent performance of the developed regression equations in predicting the critical period.

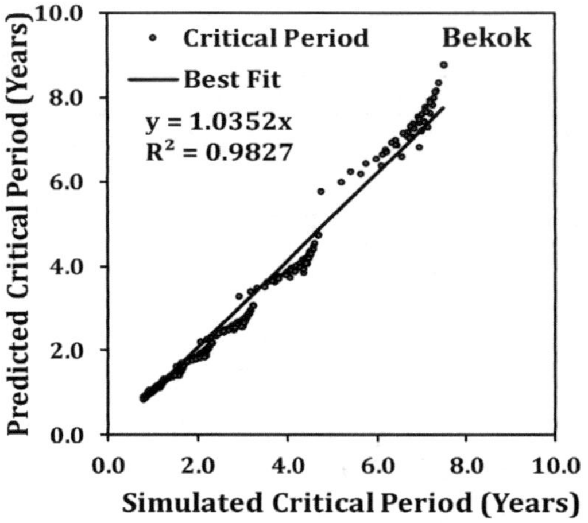

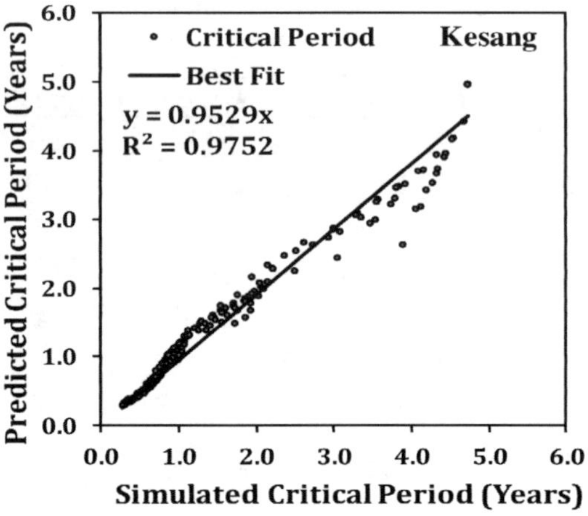

Figure 4. Comparison between simulated critical periods and predicted critical periods applying Equations (5) and (6) in Sungai Bekok and Kesang catchments, respectively

4. Conclusion

Multiple regression relationships were proposed for predicting the behavior type of reservoir systems at Sungai Bekok and Sungai Kesang systems. The presented equations could be used for the planning of reservoir systems in the study catchments to determine the interval of the required streamflow data for the simulation purpose. If the reservoir shows within-year behavior the critical twelve months of the flow records are applied in the analysis; however, when reservoir shows carryover behavior the annual flow data would be enough for the analysis. In the case that reservoir has both within-year and carryover behaviors the whole historical monthly flows is applied in the system analysis. The reservoir systems were analyzed by improved sequent peak algorithm applying 1000 replicates of synthetic streamflow data. The multiple regression relationships were developed applying stochastic analysis results for different reliability and vulnerability indices. Generally, the developed regression equations could reproduce the critical period efficiently. This is promising given that the developed equations are the first relationships that were developed for modeling the critical period of study systems applying the performance indices.

References

[1] Prairie J, Nowak K, Rajagopalan B, Lall U and Fulp T 2008 A stochastic nonparametric approach for streamflow generation combining observational and paleoreconstructed data *Water Resour. Res.* **44**

[2] Ahmed J A and Sarma A K 2007 Artificial neural network model for synthetic streamflow generation *Water Resour. Manag.* **21** 1015-29

[3] Lamontagne J R and Stedinger J R 2018 Generating Synthetic Streamflow Forecasts with Specified Precision *J. Water Resour. Plan. Manag.* **144**

[4] Ünal N E, Aksoy H and Akar T 2004 Annual and monthly rainfall data generation schemes *Stoch. Env. Res. Risk A.* **18** 245-57

[5] Srikanthan R and McMahon T A 2001 Stochastic generation of annual, monthly and daily climate data: A review *Hydrol. Earth Syst. Sci.* **5** 653-70

[6] Portela M M and Silva A T 2016 Disaggregation Modelling of Annual Flows into Daily Streamflows Using a New Approach of the Method of Fragments *Water Resour. Manag.* **30** 5589-607

[7] Silva A T and Portela M M 2013 Stochastic Assessment of Reservoir Storage-Yield Relationships in Portugal *J. Hydrol. Eng.* **18** 567-75

[8] Oskoui I S, Abdullah R and Montaseri M 2014 The Critical Period of Reservoir Systems Considering Performance Indices on Malaysia Rivers *Int. J. Eng. Technol.* **3** 166-73
[9] Hashimoto T, Stedinger J R and Loucks D P 1982 Reliability, resiliency, and vulnerability criteria for water resource system performance evaluation *Water Resour. Res.* **18** 14-20
[10] Vogel R M and McMahon T A 1996 Approximate reliability and resilience indices of over-year reservoirs fed by AR(1) Gamma and normal flows *Hydrolog. Sci. J.* **41** 75-96
[11] McMahon T A, Adeloye A J and Zhou S-L 2006 Understanding performance measures of reservoirs *J. Hydrol.* **324** 359-82
[12] Ahbari A, Stour L and Agoumi A 2019 Ability of the Performance Criteria to Assess and Compare Reservoir Management Approaches *Water Resour. Manag.* **33** 1541-55
[13] Chaves H M L and Lorena D R 2019 Assessing reservoir reliability using classical and long-memory statistics *J. Hydrol. Reg. Stud.* **26** 100641
[14] Vogel R M and Bolognese R A 1995 Storage-Reliability-Resilience-Yield Relations for Over-Year Water Supply Systems *Water Resour. Res.* **31** 645-54
[15] Sandoval-Solis S, McKinney D C and Loucks D P 2011 Sustainability Index for Water Resources Planning and Management *J. Water Resour. Plan. Manag.* **137** 381-90
[16] Adeloye A J, Montaseri M and Garmann C 2001 Curing the misbehavior of reservoir capacity statistics by controlling shortfall during failures using the modified Sequent Peak Algorithm *Water Resour. Res.* **37** 73-82
[17] Loucks D P 1997 Quantifying trends in system sustainability *Hydrolog. Sci. J.* **42** 513-30
[18] Wurbs Ralph A 1993 Reservoir-System Simulation and Optimization Models *J. Water Resour. Plan. Manag.* **119** 455-72
[19] Azari A, Hamzeh S and Naderi S 2018 Multi-Objective Optimization of the Reservoir System Operation by Using the Hedging Policy *Water Resour. Manag.* **32** 2061-78
[20] McMahon T A and Adeloye A J 2005 *Water Resources Yield* (Highlands Ranch, Colo.: Water Resources Publications)
[21] Oskoui I S, Abdullah R and Montaseri M 2015 Multiple Regression Model Using Performance Indices for Storage Capacity of a Reservoir System in Johor Catchment *Appl. Mech. Mater.* **802** 563-8
[22] You J Y and Cai X 2009 Reexamination of critical period for reservoir design and operation *J. Water Resour. Plan. Manag.* **135** 392-6
[23] Oskoui I S, Abdullah R and Montaseri M 2015 Modelling Critical Period of a Reservoir System Using Performance Indices in Johor Catchment *Appl. Mech. Mater.* **802** 569-74
[24] Oskoui I S, Abdullah R and Montaseri M 2015 Modeling Critical Period of Reservoir Systems Considering Performance Indices on Malaysia Rivers *Am. Eur. Asian J. Agric. Environ. Sci.* **15** 1809-17
[25] Oskoui I S, Abdullah R and Montaseri M 2015 Storage Capacity Modeling of Reservoir Systems Employing Performance Measures *Int. J. Eng. Technol.* **6** 2641-9
[26] Vogel R M and Stedinger J R 1987 Generalized storage-reliability-yield relationships *J. Hydrol.* **89** 303-27
[27] Srinivasan K and Philipose M C 1998 Effect of Hedging on Over-Year Reservoir Performance *Water Resour. Manag.* **12** 95-120
[28] Oskoui I S 2016 Stochastic and Modified Sequent Peak Algorithm for Reservoir Planning Analysis Considering Performance Indices. In: *School of Civil Engineering* (Pulau Pinang, Malaysia: Universiti Sains Malaysia)
[29] Bras R L 1990 *Hydrology : An Introduction to Hydrologic Science*: Addison-Wesley Publishing Company)
[30] McGhee J W 1985 *Introductory Statistics* (New York, USA: West Publishing Co.,)
[31] Kottegoda N T 1980 *Stochastic Water Resources Technology* (London: Palgrave Macmillan)
[32] Salas J D, Boes D C, Yevjevich V and Pegram G G S 1979 Hurst phenomenon as a pre-asymptotic behavior *J. Hydrol.* **44** 1-15

[33] Oskoui I S, Abdullah R and Montaseri M 2015 Storage-Yield-Performance Measure Relationship for Reservoir Planning in Malaysia *Adv. Environ. Biol.* **9** 151-9
[34] Loucks D P, Stedinger J R and Haith D A 1981 *Water Resource Systems Planning and Analysis* (Englewood Cliffs New Jersey USA: Prentice-Hill)
[35] Maheepala S and Perera B J C 1996 Monthly hydrologic data generation by disaggregation *J. Hydrol.* **178** 277-91
[36] Vogel R M 2017 Stochastic watershed models for hydrologic risk management *Water Secur.* **1**
[37] Kendall D R and Dracup J A 1991 A comparison of index-sequential and AR(1) generated hydrologic sequences *J. Hydrol.* **122** 335-52
[38] Silva A T and Portela M M 2012 Disaggregation modelling of monthly streamflows using a new approach of the method of fragments *Hydrolog. Sci. J.* **57** 942-55
[39] Ortiz-Partida J P, Kahil T, Ermolieva T, Ermoliev Y, Lane B, Sandoval-Solis S and Wada Y 2019 A Two-Stage Stochastic Optimization for Robust Operation of Multipurpose Reservoirs *Water Resour. Manag.* **33** 3815-30
[40] Medda S and Bhar K K 2019 Comparison of single-site and multi-site stochastic models for streamflow generation *Appl. Water Sci.* **9** 67
[41] Valencia R D and Schakke Jr J C 1973 Disaggregation processes in stochastic hydrology *Water Resour. Res.* **9** 580-5
[42] Harms A A and Campbell T H 1967 An extension to the Thomas-Fiering Model for the sequential generation of streamflow *Water Resour. Res.* **3** 653-61
[43] Oskoui I S and Abdullah R 2017 Modeling the Storage Capacity of a Reservoir System in Kelantan Catchment Employing Performance Indices *Ecol. Environ. Conserv.* **23** 16-20
[44] Adeloye A J, Soundharajan B-S, Musto J N and Chiamsathit C 2015 Stochastic assessment of Phien generalized reservoir storage–yield–probability models using global runoff data records *J. Hydrol.* **529** 1433-41
[45] Zongxue X, Jinno K, Kawamura A, Takesaki S and Ito K 1998 Performance Risk Analysis for Fukuoka Water Supply System *Water Resour. Manag.* **12** 13-30
[46] Rousseeuw P J, Ruts I and Tukey J W 1999 The Bagplot: A Bivariate Boxplot *Am. Stat.* **53** 382-7

Acknowledgment
The authors are grateful to the Department of Irrigation and Drainage Malaysia for providing the requested runoff records of Peninsular Malaysia for carrying out this research.

Exfoliation of charcoal by electrochemical method to synthesize few layers graphene sheets

K L Yap[1], W W Liu[2]

[1] School of Materials Engineering, Universiti Malaysia Perlis (UniMAP), 01000 Kangar, Perlis, Malaysia
[2] Institute of Nano Electronic Engineering (INEE), Universiti Malaysia Perlis (UniMAP), 01000 Kangar, Perlis, Malaysia

*Corresponding author: wwliu@unimap.edu.my

Abstract. Graphene is a new 2D carbon nanomaterial that has unique properties for several applications. However, the synthesis of large quantity of graphene is a key challenge for researchers. Previous conventional methods such as Chemical Vapour Deposition (CVD) and epitaxial growth resulted in low production rate and high cost. The research on the electrochemical exfoliation by using charcoal as an electrode is still remains unknown as most researchers employed graphite to produce graphene. The electrochemical exfoliation of charcoal and graphite rod is easy to operate and less time-consuming in the synthesis of few layers graphene sheets. An electrochemical cell consists of charcoal/graphite (as anode) and copper foil (as cathode) which immersed in several types of electrolytes was developed. The surfaces of exfoliated graphene sheets were observed as wrinkle, flaky and layered structures. Fourier transform infrared spectroscopy (FTIR) showed the presence of oxygen functional groups in the exfoliated graphene sheets while the characteristic peak of GO showed up at 2θ = 9-10° was not seen clearly in all XRD spectrums. The results proved that high amount of exfoliated graphene sheets was formed but less amount of graphene oxide (GO) was produced in the electrochemical exfoliation process.

1. Introduction

In 2004, Geim's group first mechanically exfoliated the monolayer graphene with one-atom-thick plate of carbon from graphite [1]. In 2010, Andre Geim and Konstantin Novoselov were awarded The Noble Prize in Physics for the project titled "ground-breaking experiments regarding the two-dimensional material graphene" [2]. Since the discovery of "scotch tape" method by Geim's group, numerous researches have been sparked in the worldwide due to its extraordinary properties and depletion of indium (In) [3]. The extraordinary properties such as excellent electrical conductivity and high charge carrier mobility are the some of the attractions and thus, the synthesis and application of graphene become the main research of the researchers [3]. Graphene is a two-dimensional (2D) carbon allotrope in the crystalline form that possesses single-layer of sp^2 hybridisation atoms and in a honeycomb lattice arrangement [3]. Graphene has different structures with other carbon allotropes such as diamonds and fullerenes but possesses similar chemical composition. Graphene has been widely used to make plenty of electronic devices. The applications that have been utilized the graphene sheets are fuel cells, capacitors and sensors [3].On the other hand, graphene oxide (GO) is a compound that mainly consists of carbon, hydrogen and oxygen functional groups. The chemical structure of the GO was same as graphite oxide which comprises many oxygen functionalities on its carbon basal plane and edges [3]. GO consists of monolayer or few layered stacked structure while graphite oxide consists of multilayer stacked structure like graphite [3].

Several methods have been established for producing graphene such as epitaxial growth [4], Chemical Vapour Deposition (CVD) [5] and electrochemical exfoliation of graphite [6]. However, synthesis of large quantities of graphene is a key challenge to researchers. The first method, "scotch tape" method has been developed to produce good quality of monolayer graphene but resulted in low production rate. Other techniques such as CVD and epitaxial growth require expensive cost production. Furthermore, chemical exfoliation method such as Hummers' method involves the use of hazardous chemicals such as $KMnO_4$ and thermal reduction needs a lot of electricity to reduce GO to graphene sheets which take long process. Thus, a more effective method such as electrochemical exfoliation of charcoal and graphite is preferable to synthesize few layers graphene sheets. The electrochemical techniques are more beneficial as compared to other techniques for the mass production of high-quality graphene [7]. It is easy to operate, cheap and less time-consuming. Therefore, electrochemical production of graphene in single stage or multiple steps and followed by electrochemical reduction of graphene oxide (GO) have gained extensive interest from researchers. Consequently, number of researches based on electrochemical methods have been significantly increased in the last decade [8].

In this paper, we demonstrated the synthesis of few layers graphene sheets by using electrochemical exfoliation method. The effect of types of electrolytes and anodes on the exfoliated graphene sheets were investigated using several characterization methods. The results obtained from the characterization methods such as Fourier transform infrared spectroscopy (FTIR), X-ray diffraction (XRD) and scanning electron microscopy (SEM) were discussed in this paper.

2. Methodology

In current research, the graphene sheets were prepared by using the electrochemical exfoliation technique. The materials, chemicals and method used in the research were described in detail. After the synthesis of the graphene sheets, the characterization on the exfoliated graphene sheets will be carried out.

2.1. Materials

In this research, the raw materials were charcoal (anode), graphite rod (anode) and copper foil (cathode). BBQ Grade Charcoal Coconut Shell Briquettes with purity 98.0% was ordered from The Clorox Company. Copper foil (alfa aesar, VWR Singapore) with purity 99.999 % and graphite rod (pencil core) was bought in stationery shop. The following chemicals were ordered from Sigma-Aldrich and used without further purification: copper (II) sulfate pentahydrate ($CuSO_4 \cdot 5H_2O$), sodium sulfate decahydrate ($Na_2SO_4 \cdot 10H_2O$, ACS Reagent, $\geq$ 99.0%), sulfuric acid (H_2SO_4, ACS Reagent, 95.0 - 98.0%) and hydrochloric acid (HCl, ACS reagent, 37%).

2.2. Methods

In this work, 125 mL of 0.2 M Na_2SO_4 electrolyte was prepared by mixing 64.438g of $Na_2SO_4 \cdot 10H_2O$ powder mixed with the deionized (DI) water and stirred for 10-15 mins. The electrochemical cell was prepared by using two electrodes where the copper foil was used as the cathode and a charcoal was used as the anode. Both electrodes were immersed into 125 mL of Na_2SO_4 electrolyte. Then, the electrochemical process was carried out by applying optimized DC voltages for 25 mins. After exfoliation of charcoal for 25 mins, the exfoliated graphene sheets were collected and filtered to remove electrolytes from graphene sheets by using filter paper with a diameter of 9 cm. After filtration, the filter paper with black samples were dried at room temperature. After dried, the graphene powder was extracted by using a spatula. The final product, graphene powders were put into a crucible for drying in an oven (Memmert) at 80 °C for 24 hours. The same procedures will be repeated by using graphite rod as an anode and different electrolytes such as Na_2SO_4, H_2SO_4 and $CuSO_4$ solution. The voltage was applied at 3.0 V, 6.0 V, 9.0 V, 12.0 V and 15.0 V, respectively for 25 mins. One sample only was synthesized using $CuSO_4$ solution at 15.0 V for 25 mins.

3. Results and discussions

3.1. Fourier transform infrared spectroscopy (FTIR)

All samples names: A, B, C, D and E are based on the descriptions as shown in the Table 1. From the FTIR analysis (Figure 1), the exfoliated graphene samples were mainly composed of hydroxyl groups, alkyne, alkene and sulfate ions. In the FTIR spectra, a broad and intense peak was noticed at the wavelength of 3624 cm^{-1} was assigned to the O-H stretching of the alcohol groups. Moreover, the O-H stretching of the carboxylic acids was presented at the wavelength of 2541 cm^{-1}, 2888 cm^{-1} and 2976 cm^{-1}. Besides, the characteristic peaks showed up at wavelength of 2177 cm^{-1}, 2000 cm^{-1} and 840 cm^{-1} were attributed to the presence of alkyne groups (C≡C), aromatic compound (C-H) and alkene groups (C=C), respectively. In addition, the presence of the free (nonbonded) sulfate ion, SO_4^{2-} at the wavelength of 624 cm^{-1} was observed because of the absorption of SO_4^{2-} ions in the electrolyte occurred on the surface of the exfoliated graphene sheets [9].

Table 1. Description of legends in the FTIR spectra and different types of anodes and electrolytes were used.

Samples	Types of Anodes	Types of Electrolytes
A	Charcoal	Na_2SO_4
B	Graphite Rod	$CuSO_4$
C	Graphite Rod	$CuSO_4$
D	Graphite Rod	H_2SO4
E	Graphite Rod	Na_2SO_4

Some functional groups presented in graphene sheets synthesized using charcoal do not presented in the graphite. For spectrum A (charcoal), there was no significant peak presented at the wavelength of 1594 cm^{-1} and 1250 cm^{-1}. In the spectrum A, the characteristic peaks showed up at the wavelength of 1749 cm^{-1} and 1465 cm^{-1} was assigned to the aliphatic C=O stretching of esters group and C-H bending of the alkane, respectively. On the other hand, in B to D spectrums, the peak presented at the wavelength of 1594 cm^{-1} and 1250 cm^{-1} was ascribed to the C=C stretching of the cyclic alkene and an aromatic C-O stretching of esters, respectively. H_2SO_4 is a strong acid and demonstrated high exfoliation rate than other electrolytes [1]. For $CuSO_4$, high voltage of 15.0 V was applied in the exfoliation process of the graphite rod. When a high voltage was applied, more positive charges were produced at the graphite rod (anode) as compared with low voltage. This large number of positive charges attracted the anions and encouraged more anions were intercalated into the graphene layer. Thus, resulted in high exfoliation rate and $CuSO_4$ was as effective as H_2SO_4. The graphene oxide (GO) sheets were successfully produced through the exfoliation of charcoal and graphite rod which was well-evidenced by the presence of O-H peaks.

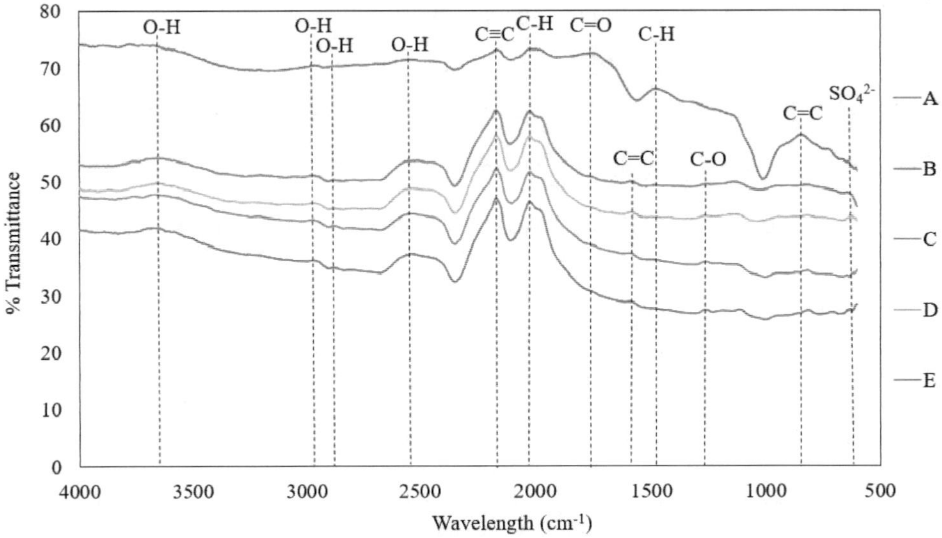

Figure 1. FTIR spectra of exfoliated graphene samples using different types of electrolytes.

3.2. X-Ray Diffraction (XRD)

For all the exfoliated graphene samples, a strong and sharp peak was observed at $2\theta = 26.2°$ or 26.5° along the (002) orientation (Figure 2). This (002) peak confirmed the presence of graphite structure with d-spacing of 0.3395 nm or 0.3356 nm between the different graphene layers [10], [11]. In XRD spectrum of A, lowest peak intensity can be seen at $2\theta = 26.2°$ with d-spacing of 0.3395 nm, along the (002) orientation with ICSD File 01-075-1621 (Reference pattern 98-001-7130). It is due to the incomplete exfoliation of charcoal as charcoal was brittle and easily broken down during the exfoliation process. Thus, the exfoliation duration of charcoal was shorter than graphite rod. On the other hand, the XRD spectrum of E has a higher peak intensity at $2\theta = 26.2°$ as compared with spectrum A even the same electrolyte of Na_2SO_4 was used in the exfoliation. Therefore, higher crystalline structure of graphite (spectrum E) was retained as compared with charcoal (spectrum A).

The XRD pattern of B and C are almost similar because of the same electrolyte of $CuSO_4$ was used in the exfoliation. A peak showed up at $2\theta = 36.20°$ with (111) orientation in samples B and C was attributed to the cuprite [12]. This indicates the presence of contaminant due to the used electrolyte. However, cuprite was not detected in FTIR results. Sulfuric acid is known as a strong oxidation acid but the oxidation process took place was mild and created less damage on the honeycomb lattice of graphene [9]. Thus, high quality of graphene can be obtained through the exfoliation method. The characteristic peak of GO appeared at $2\theta = 9\text{-}10°$ [10, 13] was not seen clearly in all XRD spectrums due to less amount of GO sheets were produced in the exfoliation process. However, the presence of GO with broad O-H peak was well evidenced by the FTIR result as shown in Figure 1. The degree of oxidation on the charcoal and graphite rod is significantly affected by the types of electrolytes and anodes.

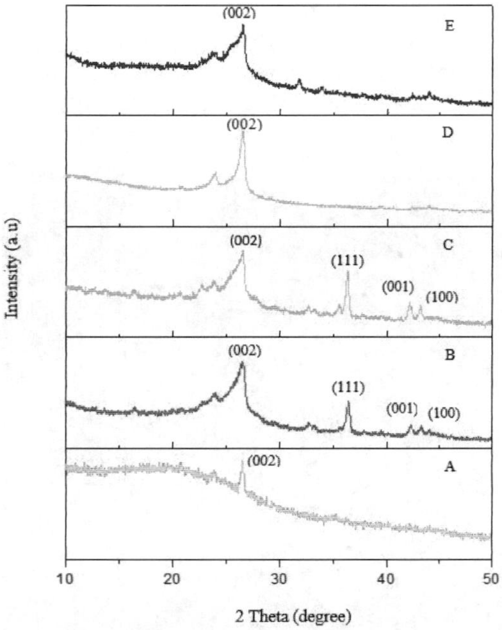

Figure 2. XRD patterns of exfoliated graphene samples.

3.3. Scanning Electron Microscopy (SEM)

Figure 3(a) and (b) reveal the SEM micrographs of exfoliated graphene sheets by using two different types of anodes with the same electrolyte of Na_2SO_4. Wrinkle and flaky structures can be observed after the exfoliation [14]. It was reported by other researchers that mild oxidation of graphite induced the formation of defects at the stacked graphene sheets, which appeared as wrinkles in the pile [3, 15]. In addition, few layers of irregular shape flakes were observed. This layered structure implies that the charcoal and graphite were exfoliated to form low amount of GO but high quantity of graphene sheets which correlated with above XRD results [16]. The surfaces of samples in Figure 3(a) and (b) were almost similar with flaky structure. This is because of charcoal and graphite rod have the same basic structure. However, charcoal composes slightly more disordered phase [17] and porous structure [18] than graphite. This caused the charcoal was easily exfoliated into irregular small and big particles during the exfoliation process, resulting incomplete and faster exfoliation rate than graphite rod. Thus, besides large particles were formed, very tiny pieces charcoal was noticed on the surface as marked in red circle in Figure 3(a). On the contrary, graphite rod has stacked and higher uniformity structure due to the strong interlayer covalent bonding as compared with charcoal [19]. Therefore, the exfoliation of graphite rod to form proper graphene sheets was happened in lower rate as compared with charcoal.

Figure 3(c) and (d) indicate the SEM micrographs of graphene sheets synthesized by using $CuSO_4$ at the low and high voltages respectively. Wrinkled and layered structure can be observed in Figure 3(c) whereas crumpled structure was seen in Figure 3(d) [20]. Furthermore, more exfoliated layers can be observed in Figure 3(d) than Figure 3(c). In this experiment, different voltages were applied to synthesize samples B and C by using the same electrolyte of $CuSO_4$. When a high voltage of 15.0 V was applied for longer time of 25 mins, more positive charges were supplied at the graphite rod (anode) as compared to low voltage. Consequently, higher amount of positive charges attracted higher quantity of anions to intercalate into graphite rod, resulting in higher exfoliation rate, thus more exfoliated layers was produced. The SEM micrograph of sample D consisted of multilayers of irregular shape flakes (Figure 3(e)). This is because of the mild oxidation of graphite dislocated the sp^2 carbon chain of the stacked graphene sheets and caused the defects. Thus, a small increase in distance between adjacent sheets is expected due to the oxygen functional groups were attached at the edge and basal plane of the graphene sheets [3, 15].

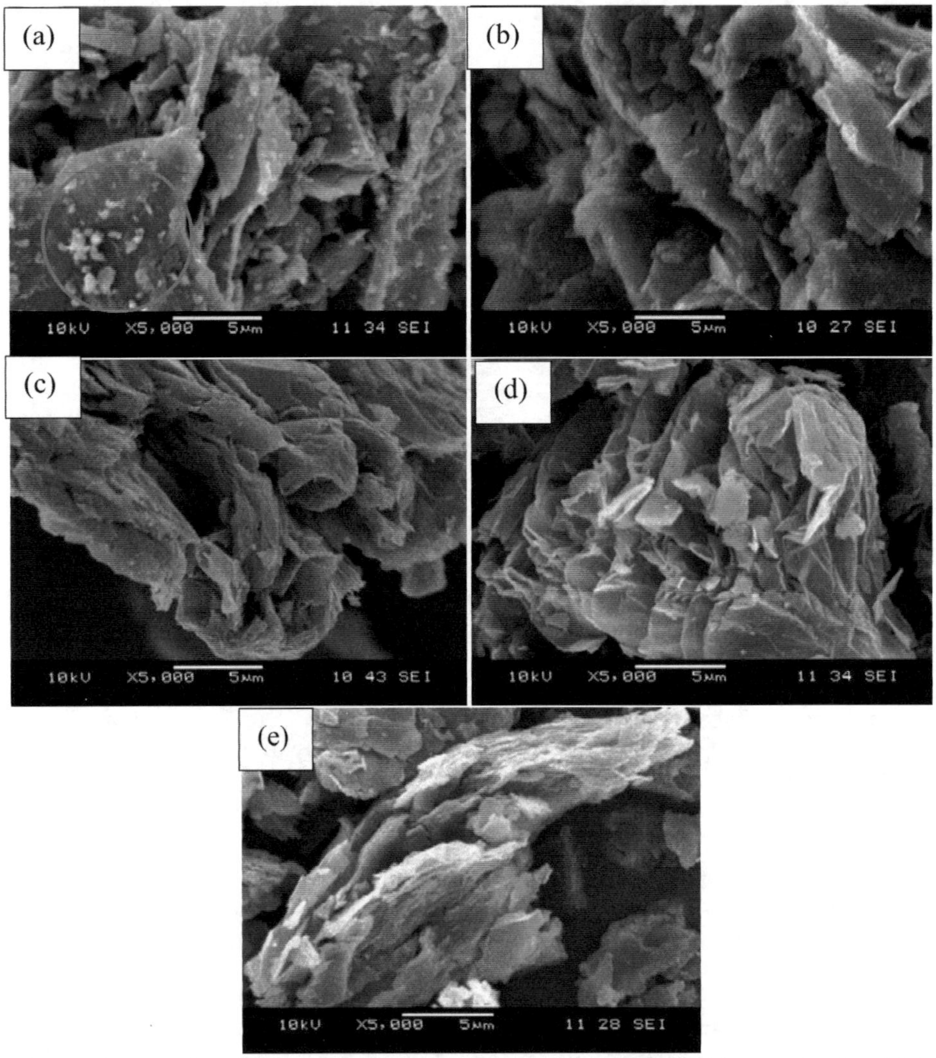

Figure 3. SEM micrographs of (a) sample A, (b) sample E, (c) sample B, (d) sample C and (e) sample D at magnification of 5KX respectively.

An EDX analysis was carried out to obtain the elemental composition in the graphene sample (Figure 4). Carbon and oxygen were the main elements traced in all samples. By using charcoal as an anode, the percentages of carbon and oxygen were 86.80 at% and 12.79 at%, respectively (Figure 4(a)). In addition, for graphite rod, the elemental compositions of carbon and oxygen were noticed at 71.87 at% and 27.75

at% respectively as shown in Figure 4(b). Besides, a small percentage of platinum were detected in all samples due to the coating of the platinum layer on the sample before SEM imaging. The percentage of oxygen as found in EDX is correlated with the amount of oxygen functional groups on GO and displayed by the characteristic GO peaks in FTIR [21, 22]. Since the sample E shows higher percentage of oxygen thus, more oxygen functional groups were believed attached to the basal and edge of the GO sheets as compared with the sample A. Therefore, sample A (synthesized by using charcoal as anode) contained a lower amount of oxygen functional groups as compared with other samples.

From the EDX analysis of sample B, the percentages of carbon (80.96 at%) and oxygen (18.65 at%) were detected whereas sample D shows the percentages of carbon and oxygen were 76.12 at% and 23.34 at%, respectively. Sample D (H_2SO_4) has a higher percentage of oxygen thus, more oxygen functional groups were attached onto the edge and basal plane of GO sheets as compared with sample B ($CuSO_4$). This is because of sulfuric acid, a strong oxidation acid caused more oxidation than that of $CuSO_4$ [9]. Thus, at high voltage, $CuSO_4$ was more efficient in exfoliation with much less oxidation on the graphene sheets than sulfuric acid.

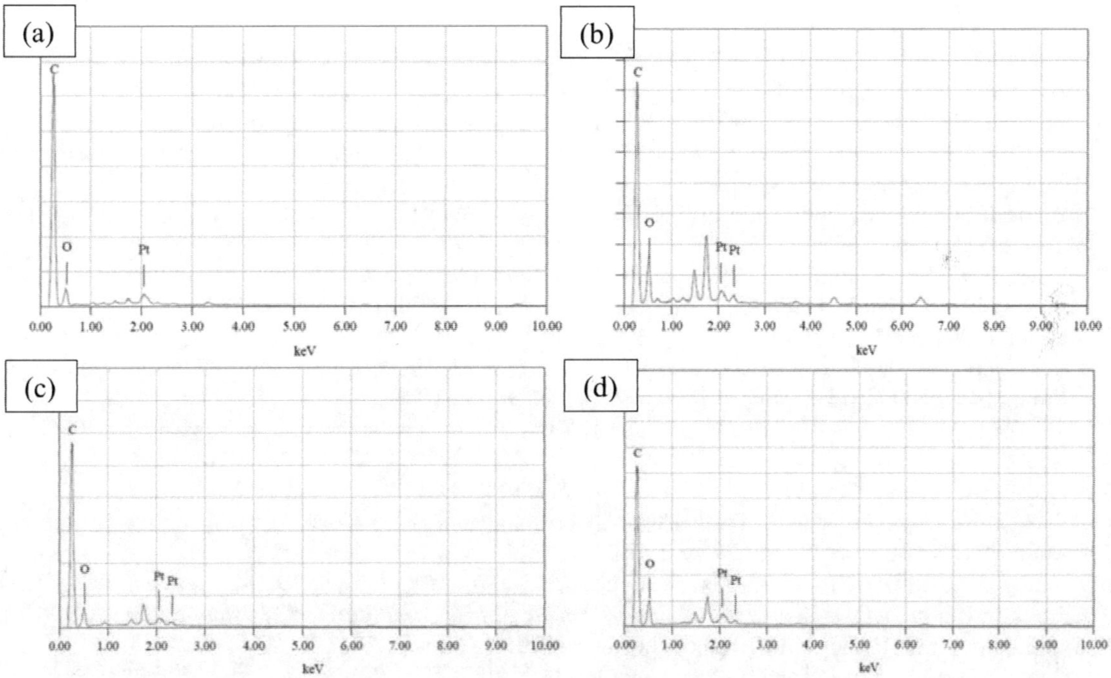

Figure 4. Elemental composition of (a) sample A (b) sample E, (c) sample B and (d) sample D.

4. Conclusion

The electrochemical exfoliation of charcoal and graphite by using different electrolytes was studied. Exfoliated graphene sheets were characterized by FTIR which showed the GO sheets were successfully produced in low amount in the exfoliation of charcoal and graphite rod. The XRD analysis confirmed the presences of graphite structure in all samples and less amount of graphene oxide (GO) was produced in the exfoliation process. For the SEM analysis, wrinkle and flaky structures can be observed due to the exfoliation. The layered structure as observed in the SEM micrographs indicates that low amount of GO but high quantities of graphene sheets were produced. The EDX spectra demonstrates carbon and oxygen were the main elements traced in all graphene samples. Overall, it was found that graphite rod is more efficient to be used to form proper graphene sheets than charcoal during the exfoliation process. All electrolytes showed near similar performance in exfoliation based on the obtained results.

Acknowledgement

I would like to thank Institute of Nano Electronic Engineering (INEE), School of Materials Engineering and Universiti Malaysia Perlis (UniMAP) for providing FTIR, SEM, XRD, and others characterization and laboratory facilities.

References

[1] Novoselov KS, Geim AK, Morozov SV, Jiang D, Zhang Y, Dubonos SV, Grigorieva IV and Firsov AA 2004 *Science* **306** 666.
[2] Low C, Walsh F, Chakrabarti M, Hashim M and Hussain M 2013 *Carbon* **54** 1.
[3] Toh SY, Loh KS, Kamarudin SK and Daud WRW 2014 Chem. Eng. J. **251** 422.
[4] Cambaz ZG, Yushin G, Osswald S, Mochalin V and Gogotsi Y 2008 *Carbon* **46** 841.
[5] Kim KS, Zhao Y, Jang H, Lee SY, Kim JM, Kim KS, Ahn JH, Kim P, Choi JY and Hong BH 2009 *Nature* **457** 706.
[6] Wang J, Manga KK, Bao Q and Loh KP 2011 *J. Am. Chem. Soc.* **133** 8888.
[7] Zhong YL, Tian Z, Simon GP and Li D 2015 *Mater. Today* **18** 73.
[8] Yu P, Lowe SE, Simon GP and Zhong YL 2015 *Curr. Opin. Colloid Interface Sci.* **20** 329.
[9] Su CY, Lu AY, Xu Y, Chen FR, Khlobystov AN and Li LJ 2011 *ACS nano* **5** 2332.
[10] Guo HL, Wang XF, Qian QY, Wang FB and Xia XH 2009 *ACS nano* **3** 2653.
[11] Schniepp HC, Li JL, McAllister MJ, Sai H, Herrera-Alonso M, Adamson DH, Prud'homme RK, Car R, Saville DA and Aksay IA 2006 *J. Phys. Chem. B* **110** 8535.
[12] Raffi M, Mehrwan S, Bhatti TM, Akhter JI, Hameed A, Yawar W and ul Hasan MM 2010 *Ann. Microbiol.* **60**, 75.
[13] Doğan HÖ, Ekinci D and Demir Ü 2013 *Surf. Sci.* **611** 54.
[14] Fu C, Zhao G, Zhang H and Li S 2013 *Int. J. Electrochem. Sci.* **8** 6269.
[15] Manoj B 2015 Russ. *J. Phys. Chem. A* **89** 2438.
[16] Dubey SP, Nguyen TT, Kwon YN and Lee C 2015 *J. Ind. Eng. Chem.* **29** 282.
[17] Franklin RE 1951 *Proc. Royal Soc. London. Series A. Mathematical and Physical Sciences* **209** 196.
[18] Isahak WNRW, Hisham MWM and Yarmo MA 2012 *J. Chem.* **2013** 1.
[19] Chen Z, Miao H, Wu J, Tang Y, Yang W, Hou L, Yang F, Tian X, Zhang L and Li Y 2017 *Ind. Eng. Chem. Res.* **56** 6939.
[20] Kashinath L, Kumar RS, Hayakawa Y and Ravi G 2013 *Int. J. Sci. Eng. Appl.* **10** pp. 3.
[21] Tai MJY, Liu WW, Khe CS, Hidayah NMS, Teoh YP, Voon CH, Lee HC and Adelyn PYP 2018 *AIP Conf. Proc.* **2045** 020032.
[22] Cheong MF, Liu WW, Khe CS, Hidayah NMS, Lee HC, Teoh YP, Foo KL, Voon CH, Zaaba NI and Adelyn PYP 2018 *AIP Conf. Proc.* **2045** 020031.

Study on fluttering characteristics of articulated flat plate flag

Masaki Yamagishi[1] and Hiroki Shida[2]

[1] Department of Mechanical Engineering, Toyohashi University of Technology, 1-1 Hibarigaoka, Tempaku, Toyohashi, 441-8580, JAPAN
[2] Electric & Mechanical System Engineering Advanced Course, National Institute of Technology, Nagaoka College, 888 Nishikatakai, Nagaoka, Niigata, 940-8532, JAPAN

yamagisi@nagaoka-ct.ac.jp / yamagisi@me.tut.ac.jp

Abstract. For wind/water power generation, we have been trying to harness flow-induced vibration, and we used a flag as the oscillating body. The flag was made of flat plates joined by articulations. To achieve a suitable design in engineering, it is important to clarify the fluttering characteristics of this articulated flat plate. The fluttering characteristics were therefore investigated experimentally in a wind tunnel. In the experiment, several rectangular plates were tested in several conditions of stream-wise length, span-wise length, thickness, and density. To investigate the fluttering characteristics, the fluttering frequency in a uniform flow velocity was focused on. The fluttering mode was observed simultaneously. It was found that an articulated flat plate flutters in a uniform flow and has a steady fluttering mode. The fluttering frequency increases with increasing uniform flow velocity for all conditions. The frequency was higher when the streamwise-length was short, thickness was thin, and density was small. It was found that there were two modes of fluttering: node-less mode and one-node mode. Furthermore, the dimensional analysis was performed, and the results show that the non-dimensional number consisting of the Strouhal number, Reynolds number, and mass ratio became constant.

1. Introduction

Oscillations induced by flow lead to many serious accidents in mechanical and structural engineering, such as the famous accident of Tachoma Bridge. Many researchers tried to reduce or control those oscillations. Oscillation power, however, could become the source of energy from the flow. Some examples were proposed to get the kinematic energy of flow by transforming to the oscillating energy [1-3]. These examples utilize the oscillation that is caused by the elasticity of the oscillating body as restoring forces. Thus, the oscillating body may be destroyed due to fatigue after repeated oscillation. Furthermore, from the viewpoint of the energy source, a controllable oscillation is required.

In this study, for the oscillating body, a flag-like body composed of several sized flat plates joined by articulations called multi-articulated flat plate was proposed [4], as shown in figure 1. The multi-articulated flat plate flutters in the uniform flow like a flexible sheet flag, and its fluttering motion is two-dimensional. Since the motion of the multi-articulated flat plate is more regular and periodic than that of a sheet flag, it is suitable for wind/water power generation by using fluttering. Generating electricity was attempted by attaching the piezo-generator on this multi-articulated flat plate. It was expected for an independent power source like as the water level sensor.

Many studies on sheet flags have been performed by researchers. For the pioneering study, Taneda investigated experimentally the motion and characteristics of sheet flags in several materials and showed

that a non-dimensional parameter consisting of Strouhal number, Reynolds number, and mass ratio is nearly constant independent of the Reynolds number [5]. Sato *et al.* investigated the motion of hard and soft flags [6]. They suggested that the fluttering frequency of the hard flag could be approximately predicted by using the kinematics of oscillation. In their study, the fluttering frequency of the hard flag is proportional to the flow velocity and inversely proportional to the square root of the mass per unit area and the length of flag. Eloy *et al.* studied experimentally and theoretically the flutter instability of cantilevered flexible plates in the uniform flow [7]. They clarified the relationship between the critical velocity, aspect ratio, and mass ratio. Zhang *et al.* carried out a two-dimensional experiment on the flexible filament in a soap film flow [8]. They indicated that there were two stable states in the waving motion of the filament: stretched-straight state, and flapping state. There exists a critical length of the filament in these two different stable states. On this hysteresis, Eloy *et al.* [9] and Nishiura *et al.* [10] also focused and studied experimentally and analytically. Nishiura et al. [10] and Huang and Sung [11] carried out numerical simulation on the flow around the flapping flag and showed the vortex structures and flapping mode of flag. Hiroaki *et al.* investigated on the flutter speed of the flexible sheet experimentally and analytically [12]. On a flexible cantilevered plate, Fujita and Matsumoto investigated on the relationship between the critical velocity and the mass ratio [13].
Although there are many studies on flexible flag motions in a flow as described above, a study on the oscillating characteristics of an object that is discontinuously deformed, such as articulated rigid plates, has not been reported so far. To use the oscillation for a wind/water power generator, it is important for design to clarify the fluttering characteristics of the multi-articulated flat plate. The aim of this paper is to clarify the frequency characteristics. In this work, as the first step, variations of the fluttering frequency against the flow velocity were investigated for several flat plate lengths and density. The characteristics of the fluttering frequency were estimated by the traditional non-dimensional parameters. The fluttering mode of the multi-articulated flat plate was also observed.

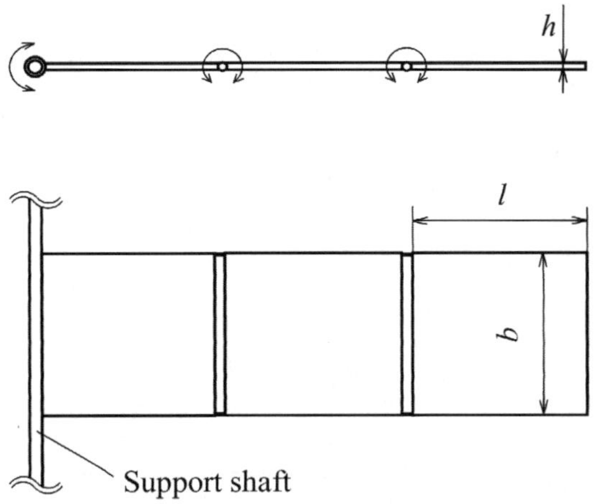

Figure 1. Schematic illustrations of the basic Multi-articulated flat plate.

2. Experimental apparatus and measurement method

In this study, all tested multi-articulated flat plates were composed of three plates of the same rectangular shape. They were made of polystyrene flat plates sandwiching a polyethylene sheet, as shown in figure 2. The gaps between the plates were 1 mm, and the naked polyethylene sheet at the gaps bended smoothly to provided articulations. Since the polyethylene sheet was very thin and flexible, the flexural rigidity and damping resistance of the articulations were neglected in this study. The plates are called the first plate, the second plate, and the third plate in order starting from upstream. The articulations are called the first articulation, the second articulation, and the third articulation. The first articulation corresponds to the support shaft. The standard shape of the unit plate had stream-wise length $l = 50$ mm, span-wise length $b = 50$ mm, and thickness $h = 1$ mm. The standard shape was decided based on the size of the piezo-generator [14][15]. The plate was made of polystyrene plate, which had a density $\rho_s = 1.056 \times 10^3$ kg/m^3. To investigate the effect of the size and material on the fluttering characteristics, different shapes and materials were tested. The tested conditions are shown in table 1.

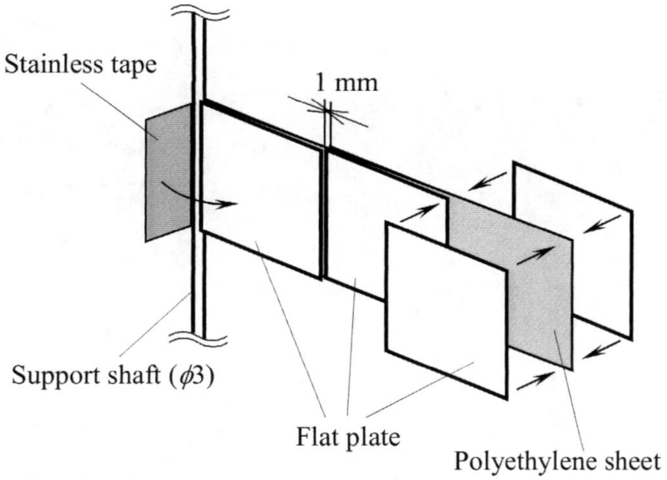

Figure 2. Making of Multi-articulated flat plate.

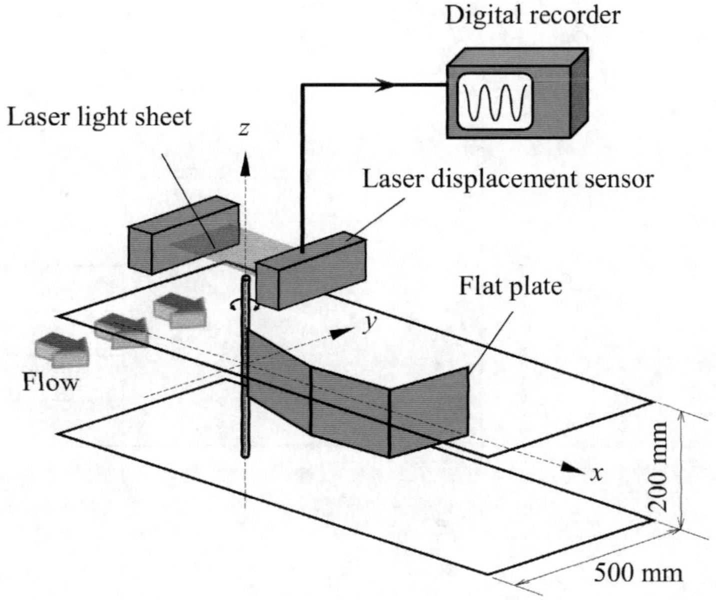

Figure 3. Schematic illustration of the experiment apparatus. Multi-articulated flat plate is settled horizontally in the flow.

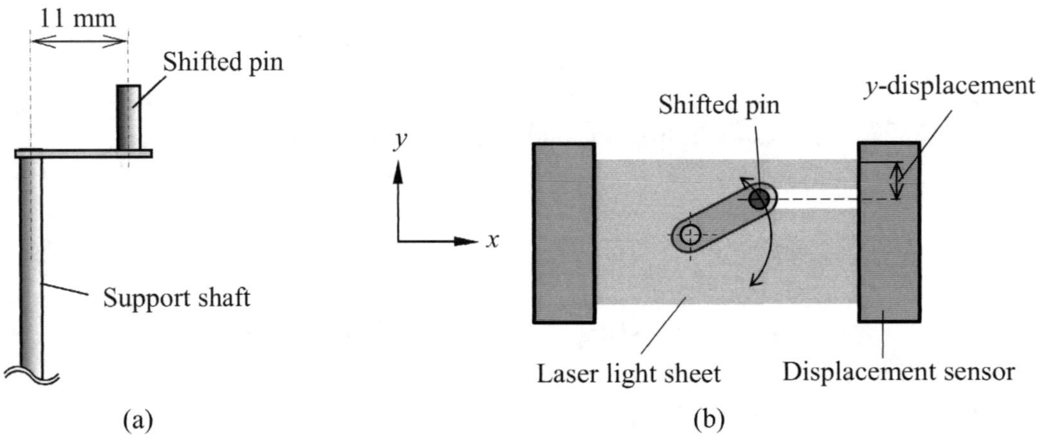

Figure 4. Shifted pin for acquisition of the displacement in the fluttering of multi-articulated flat plate. Laser displacement sensor acquired y-direction displacement of the shifted pin.

Table 1. Material and scale of shape of unit flat plate.

Material	Density ρ_s [$\times 10^3$ kg/m^3]	Length l [mm]	Width b [mm]	Thickness h [mm]	Mark
		30			▲
		40			▼
		50	50	1	●
		60			■
		70			◆
			30		△
			40		▽
Polystyrene	1.056	50	50	1	●
			60		□
			70		◇
				0.4	△ (with vertical line)
				1	●
		50	50	2	▽ (with vertical line)
				3	□ (with vertical line)
				4	◇ (with vertical line)
Magnesium	1.738				●
Aluminum	2.699				●
Zinc	7.130	50	50	1	●
Stainless steel	7.639				●
Brass	8.250				●
Copper	8.960				●

The arrangement of the wind tunnel experimental apparatus and the coordinate system are shown in figure 3. The wind tunnel was open-jet type with a nozzle of 400 mm × 600 mm square cross-section. The wind velocity could be varied to 7 m/s. The measuring section had upper and lower walls, while its sides were open. The distance between the walls was 200 mm. A multi-articulated flat plate was fixed on a 3 mm diameter shaft supported with bearings at the walls.

On one end of the support shaft, as shown in figure 4 (a), a small pin of diameter $\phi = 4$ mm attached with a 11 mm shift. The shifted pin will swing with the fluttering frequency of the articulated flat plate. The y-displacement of the shifted pin was acquired with a laser displacement sensor, as shown in figure 4 (b). The laser displacement sensor detects the distance of the center position of the object that obstructs the laser sheet from one side of the laser sheet. The displacement was recorded by the time-dependent signal, and the fluttering frequency f was obtained from the power spectrum of that signal.

The fluttering mode in the x-y plane was observed from the photographing by digital camera. For the fluttering mode, the amplitude distribution and the trajectory curve were observed. The amplitude distributions were obtained by synthesizing 60 pictures, which were photographed in sequence with 1/60 s intervals. For obtaining the trajectory curves, 4 s exposure photographing was conducted in a darkroom with a black light. The fluorescent paint was painted on the articulations and trailing edge of the multi-articulated flat plate. The trace lines of the shining fluorescent paint were recorded on the photograph as the trajectory curves.

3. Results

3.1. Relation between flow velocity and fluttering frequency

The multi-articulated flat plates fluttered in all cases. Figure 5 shows the fluttering frequency f against the uniform flow velocity U for the standard shape. The frequency was measured in increasing and decreasing uniform flow velocity. From the results, it was found that there is hysteresis in the fluttering motion of the multi-articulated flat plate. This hysteresis was also seen in the fluttering motion of a flag [6][7]. Zhang *et al*. expressed this hysteresis as bistability of the stretched-straight state and the flapping state [7]. The critical velocity at which the multi-articulated flat plate began to flutter depended largely on the small disturbances. In the low-turbulence wind tunnel, the disturbance was generated at the trailing edge of the flat plate oscillating due to the vortex street in the wake. There is another hysteresis in the higher velocity region. In higher velocity, the fluttering frequency discontinuously decreased because of the fluttering mode change, which is described in the next chapter.

3.2. Fluttering mode of articulated flat plate

Figures 6 - 8 show the fluttering mode of the standard shape articulated flat plate. In these figures, (a) is the amplitude distribution and (b) is the trajectory curves of the second articulation, the third articulation, and trailing edge of the multi-articulated flat plate. The amplitude increased with increasing uniform flow velocity. In higher velocity, it was seen that the trailing edge turned to the upstream, as shown in figure 7. This motion called whip was also seen in the fluttering motion of a sheet flag [5]. When the velocity exceeded some value, the fluttering mode changed. This was seen in the imperfect node on the second plate (the amplitude was not zero at the node). In this fluttering mode, the fluttering frequency increased. In this study, the node-less mode is called Mode I, and the mode with a node is called Mode II hereafter. On the trajectory curves, the second articulation naturally reproduces a circular arc. The third articulation and trailing edge traced a figure-of-eight trajectory. From these photographs, as the articulations and trailing edge traced the same trajectory curve at each cycle, it was found that the multi-articulated flat plate has a very steady fluttering mode.

IOP Conf. Series: Materials Science and Engineering **920** (2020) 012020 doi:10.1088/1757-899X/920/1/012020

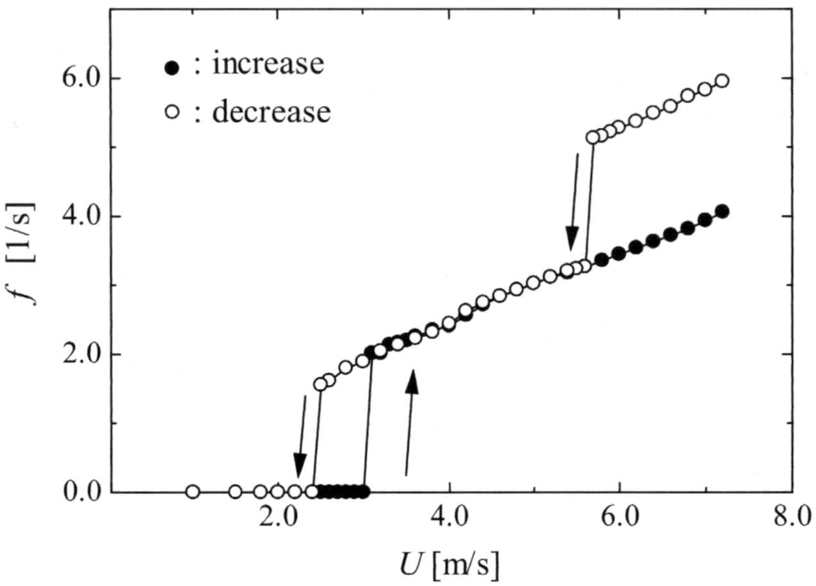

Figure 5. Relation between the velocity and frequency and amplitude.

3.3. Relation between fluttering frequency and plate conditions

The variations of fluttering frequency f against uniform flow velocity U in all conditions are shown in Figures 9 - 12. Because there was the hysteresis described above, the acquisitions were conducted while decreasing uniform flow velocity. In all conditions, the fluttering frequency increases with increasing the uniform flow velocity.

Figure 9 shows the results with different stream-wise length l. The frequency at the same velocity decreased with increasing stream-wise length. Except for the condition $l = 30$ mm, Mode II was appeared. The critical velocity at which the mode changed was decreased with increasing the stream-wise length.

From the results when using the different span-wise length b, as shown in Figure 10, it was found that the fluttering frequencies were almost the same when using the different span-wise length. Hence, the critical velocity at which Mode II appeared was different depending on the span-wise length. The articulated flat plates that had longer span-wise lengths over 60 mm did not show Mode II in the experiment velocity range.

The fluttering frequencies were decreased with increasing the flat plate thickness h as shown in figure 11. The articulated flat plate with thickness $h = 0.4$ mm fluttered at the highest frequency in all conditions. Mode II did not appear when the thickness was over 2.0 mm.

Figure 12 shows the variation of fluttering frequency f with several densities ρ_s. The values of the frequency at the same velocity decreased with increasing density. Mode II did not appear when the density was over 7.6×10^3 kg/m^3.

These results suggest that fluttering frequency f depends on stream-wise length l, thickness h, and density ρ_s. Interestingly, span-wise length b does not affect the frequency.

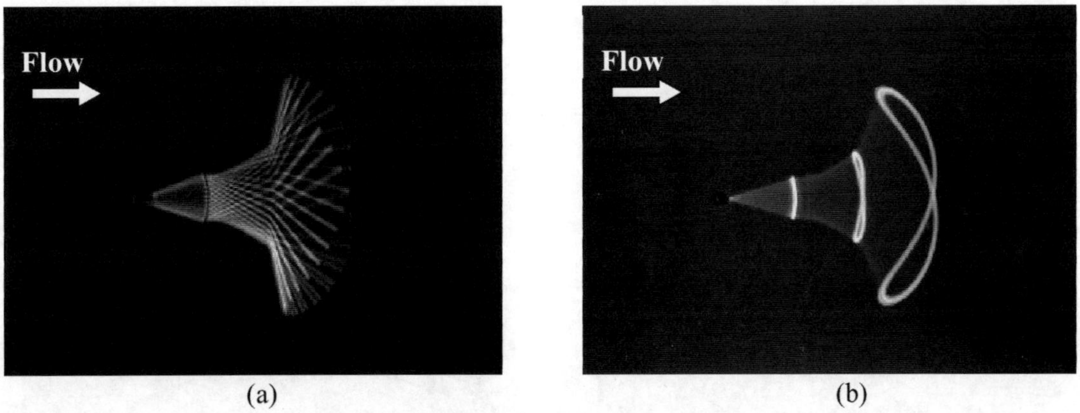

Figure 6. Amplitude distribution and trajectory curve of the fluttering mode (Mode I) at the mean flow velocity $U = 2.5$ m/s. It has node-less fluttering mode.

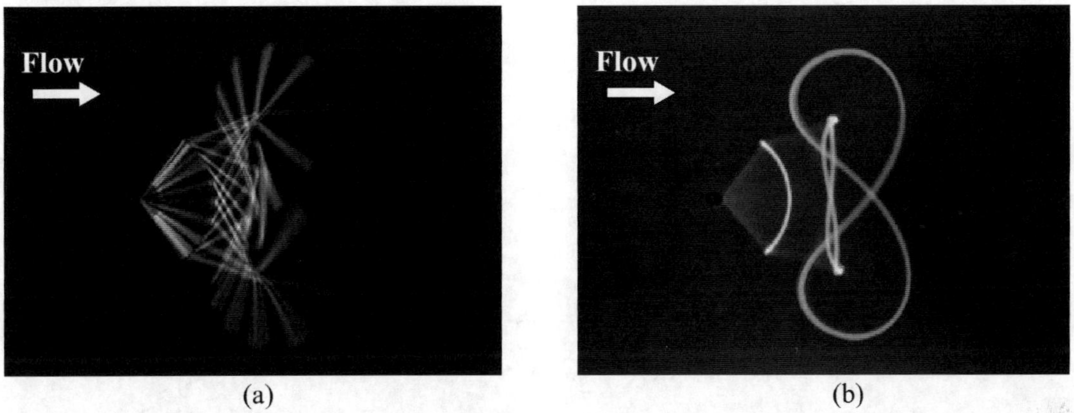

Figure 7. Amplitude distribution and trajectory curve of the fluttering mode (Mode I with whip) at the mean flow velocity $U = 5.0$ m/s. It has node-less fluttering mode and the trailing edge turns to the upstream.

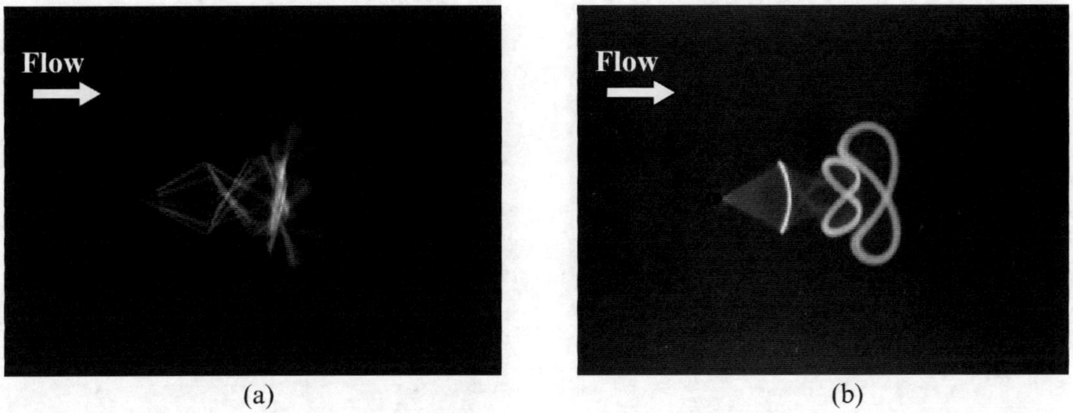

Figure 8. Amplitude distribution and trajectory curve of the fluttering mode (Mode II) at the mean flow velocity $U = 6.0$ m/s. It has imperfect node on the second plate and high fluttering frequency.

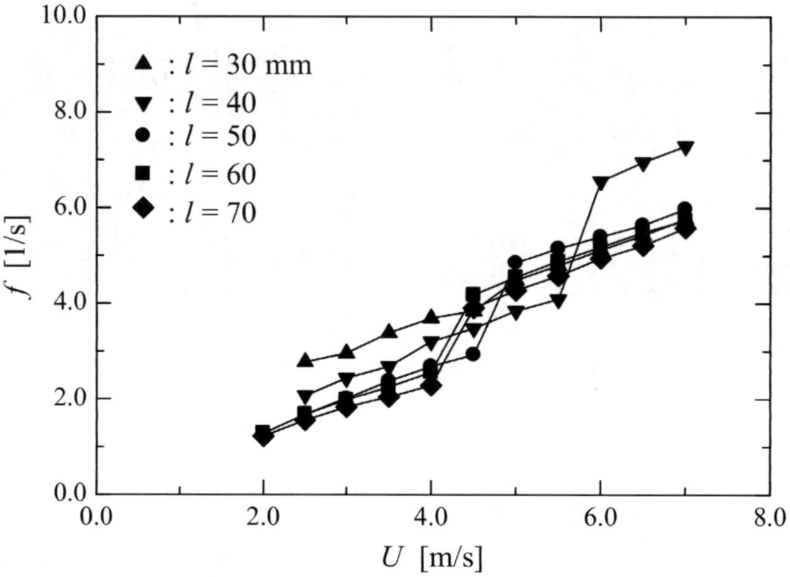

Figure 9. Relations between the velocity and frequency on the stream-wise length of the plate.

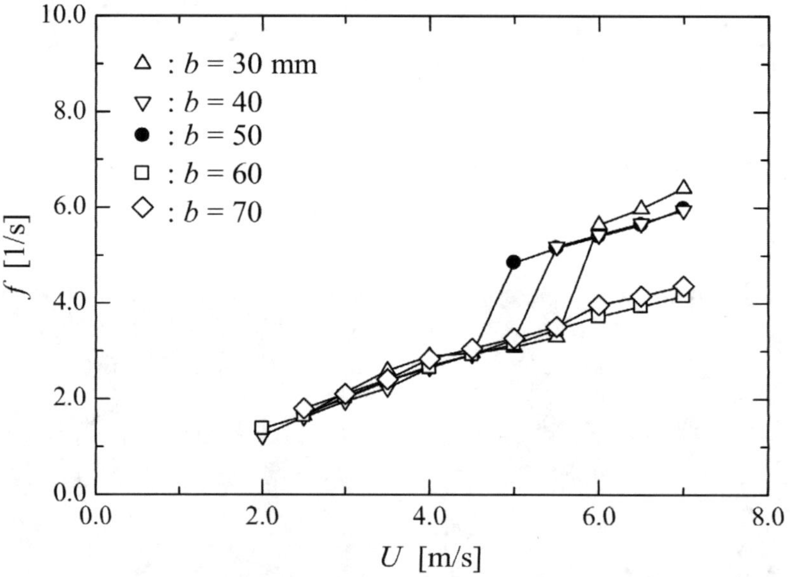

Figure 10. Relations between the velocity and frequency on the span-wise length of the plate.

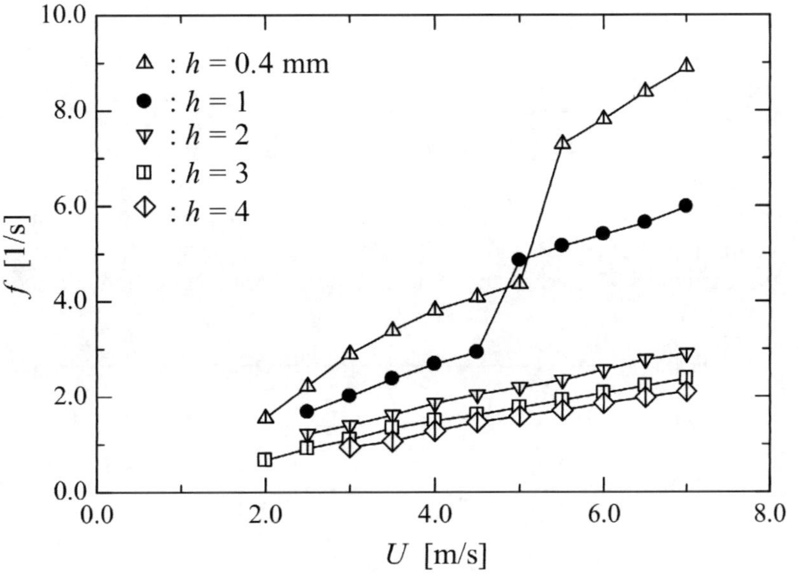

Figure 11. Relations between the velocity and frequency on the thickness of the plate.

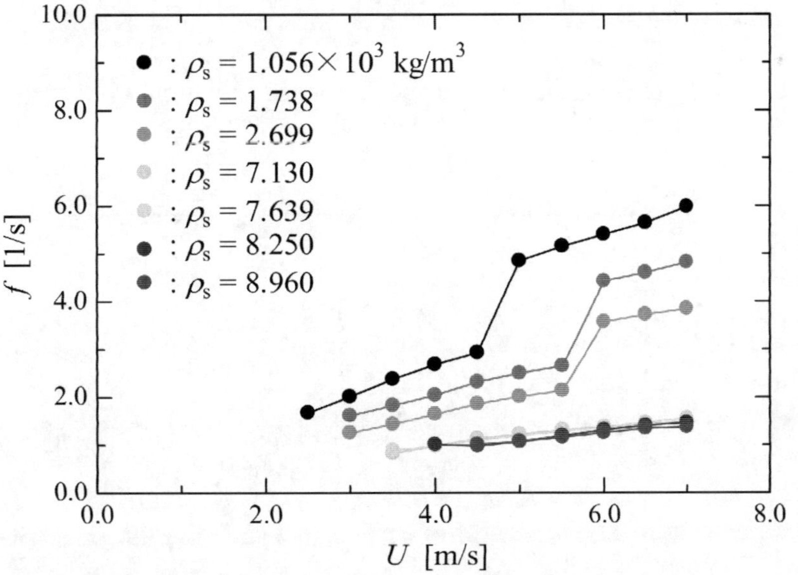

Figure 12. Relations between the velocity and frequency on the density of flat plate.

4. Dimensional analysis

The Strouhal number St, Reynolds number Re, and mass ratio Mr were defined as follows.

$$\mathrm{St} = \frac{fl}{U}\,, \quad \mathrm{Re} = \frac{Ul}{\nu}, \quad \mathrm{Mr} = \frac{\rho_s h}{\rho l} \qquad (1)$$

On the study of the flag, non-dimensional bending rigidity was also used. In this study, articulations were very smooth and freely bended. The rigidity of articulations, therefore, were neglected. From the dimensional analysis used by Taneda [5], it was found that the Strouhal number was proportional to -1/4 power of the Reynolds number and to -1/2 power of the mass ratio, as shown in figure 13 and 14. Furthermore, the values of $\mathrm{St}\cdot\mathrm{Re}^{1/4}\cdot\mathrm{Mr}^{1/2}$ became constant, as shown in Figure 15. In Mode I, the values were about 1.5, and in Mode II, they were about 2.3. The results of the non-dimensional analysis indicate that the fluttering frequency is defined by the flow and the flat plate conditions.

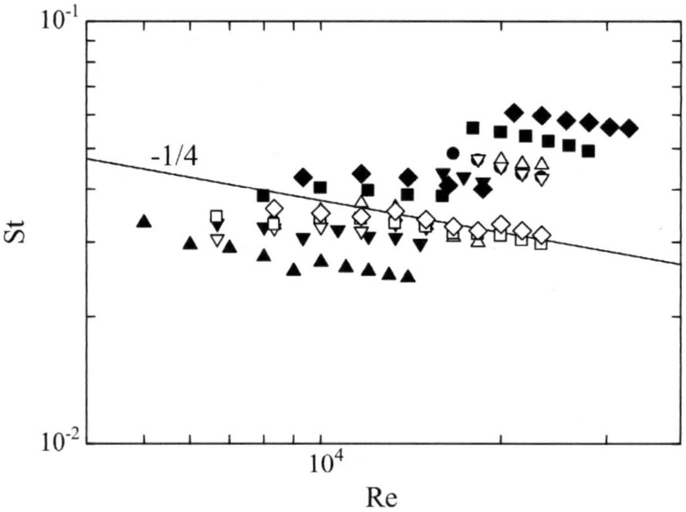

Figure 13. Relations between Reynolds number and Strouhal number on the stream-wise and the span-wise length of the plate.

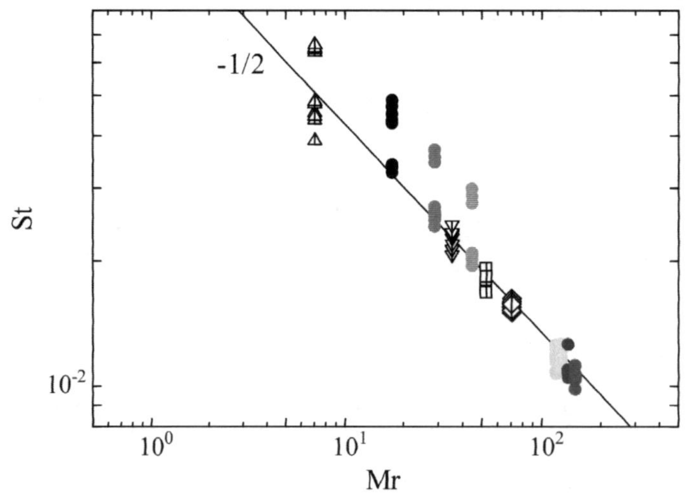

Figure 14. Relations between the mass ratio and Strouhal number on the thickness and the density of the plate.

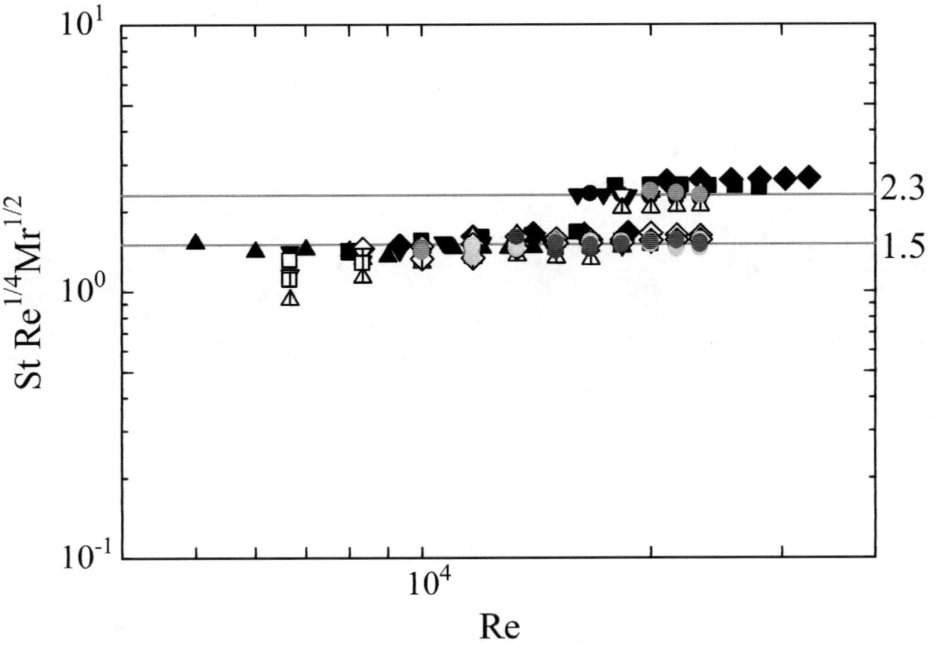

Figure 15. Relations between non-dimensional number St·Re·$^{1/4}$·Mr$^{1/2}$ and Reynolds number.

5. Conclusions

To develop a wind/water power generator by using flow-induced vibration, a multi-articulated flat plate was proposed, and its fluttering characteristics were investigated by wind tunnel experiments. The articulated flat plate flutters and has a very steady fluttering motion in the uniform flow. When the flow velocity increases or decreases, there is hysteresis in the fluttering motion. The fluttering frequency increases with increasing the uniform flow velocity. The fluttering frequency is higher when the plate is shorter, thinner, and lighter. An articulated flat plate has the node-less fluttering mode and the mode with the node. The non-dimensional number which consisted of the Strouhal number, Reynolds number, and mass ratio became constant.

6. References

[1] Masao H Patent Abstract of Japan, P2001－157433A
[2] Suzuki A, Kamiya K and Matsumoto T Patent Abstract of Japan, P2006-226221A
[3] Michelin S and Doaré D 2013 *J. Fluid Mec.*, Vol. 714, p 489
[4] Yamagishi M and Watanabe T 2009 *Proc. of CMMF'09*, Vol. 1, p 252
[5] Taneda S 1968 *J. Phys. Society of Japan*, Vol. 24, No. 2, p 392
[6] Sato H, Saito H and Nakamura H 2004 *JAXA Special Publication*, JAXA-SP-05-012, p 23 (in Japanese)
[7] Eloy C, Lagrange R, Souilliez C and Schouveiler L 2008 *J. Fluid Mech.*, Vol. 611, p 97
[8] Zhang J, Childress S, Libchaber A and Shelly M 2000 *Nature*, Vol. 408, p 835
[9] Eloy C, Kofman N and Schouveiler L 2012 *J. Fluid Mech.*, Vol. 691, p 583
[10] Nishiura K, Nishio Y, Izawa S and Fukunishi Y 2019 *Fluid Dyn. Res.*, Vol.51, No.2

[11] Huang W X and Sung H J 2010 *J. Fluid Mech.*, Vol. 653, p 301
[12] Hiroaki K, Kumakura C and Watanabe M 2016 *Proc. of ISFA2016*, JAXA-SP-16-008E, p 87, ISFA-1R06
[13] Fujita K and Matsumoto K 2016 *Proc. of ISFA2016*, JAXA-SP-16-008E, p 441, ISFA-4R17
[14] Umeda M, Nakamura K and Ueha S 1997 *Japan, J. Appl. Phys.*, Vol. 36, No. 5B, p 3146
[15] Umeda M, Sakai Y and Nakamura K 2003 *IEEJ Trans. SM*, Vol. 123, No. 12, p 534 (in Japanese)

Simulation on Air- Fly Ash Particles Using Discrete Phase Method

N.M. Zahari[1*], Mastika Falsha Harun[1], Mahyun Zainoodin[1], Daud Mohamad [1], Agusril Syamsir[1], M. Z. Ramli.[1], M. H. Zawawi[1], and Aizat Abas[2]

[1]Department of Civil Engineering, Universiti Tenaga Nasional (UNITEN), Kajang 43000, Selangor Darul Ehsan, Malaysia
[2]School of Mechanical Engineering, Universiti Sains Malaysia, Engineering Campus, 14300 Nibong Tebal, Pulau Pinang, Malaysia

[*]Email: mubinzahari@gmail.com

Abstract. In order to suggest for development of a better fluidized bed combustor designs, numerical simulations could be used to model the flow behavior inside the boilers. However, the complex nature of the multiphase flow of particles and combustion air makes the modelling very challenging. In this study the relatively recently introduced approaches for fly ash particles distribution model in CFD setting are studied. The applicability of the approaches is examined, and the computational requirements of the approaches are compared. Euler-Lagrange approach is used in order to track the particles in the simulation. The simulations results obtained are compared between different shape of simple models, square and cylindrical.

1. Introduction

Coal is widely available and less expensive compared to oil-based fuels, thus coal is broadly used in coal combustion by-products in power industry. Coal combustion also has become one of the prominent technologies in environment-friendly biomass combustion. In pulverized coal fired boiler the solid fuel particles are suspended on upward-blowing air resulting in a turbulent mixing of gas and solids. Rapid cooling in the post-combustion zone results in the formation of spherical, non-crystalline particles. The particles then may be collected in the flue gas particle equipment, or may be escaped and deposited to the environment, or also can be deposited on the furnace wall and gradually become sintered and harden. Therefore, manual cleaning and removal of the deposition need to be made and costly [1]. The fly ash particles in the flue gas stream experience forces from fluid drag and lift, gravity, and intermittent collisions with the duct walls and other particles. The higher density of fly ash particles approximately around 2500 kg/m^3 while the flue gas approximately around 1 kg/m^3. These differences densities cause the fly ash particles to diverge from the flue gas flow and separate from the flue gas at select locations depending on local flow conditions and, particle size and density [2].

Particles impacting an existing packed bed may transferred their kinetic energy to the fluid medium and be assimilated in the bed after. The fly ash particles typically have a range of sizes and densities and, a higher concentration of the heavier particles will almost always be seen near the bottom horizontal surfaces of the duct [3].

Recently, many studies have been done upon the fly ash utilization but still less in fly ash behavior in the combustion vessel in term of their characteristics. A study carried by Kaer et al developed a dynamic model to predict the ash deposition formation and investigated the ash deposition rate by different deposition mechanism [4]. García Pérez et al. modeled the deposit growth of fume particles based on the thermophoretic force, Brownian motion and inertial impaction [5]. However, the details on particle impaction and the movement patterns are not highlighted in the previous study.

Before that, there are many researchers using ANSYS-Fluent software to simulate few kinds of case study [6-9]. Thus, this paper will study on the fly ash transport by introducing particulates into the scale model in simulations by using discrete phase method in ANSYS-Fluent software. Two difference shape of scale model will be generated in SolidWorks and exported to the ANSYS for simulations. The simulations can give out varies result based on setting chosen in the processor such as the pressure, velocity, kinetic energy and others. Conclusion of the minimum flue gas velocity required for ensuring no drop out of transported particles and the minimum gas velocity required to re-entrain existing deposits of fly ash are made. Moreover, particle parameters and model scaling will need to be appropriate to ensure similar particle behavior.

2. Methodology

2.1. Discrete Phase Model

Simulation techniques commonly used in engineering analysis which account for coupling between the carrier gas phase and particulates, can be classified as Lagrangian dispersed phase simulation or Eulerian multiphase simulation. The Eulerian approach is currently commonly used in fluidized bed simulations and it is a relatively fast and usable approach. In the Eulerian-Lagrangian approach on the other hand, the particles are treated as separate, solid objects and their behavior is modeled using Newtonian equations of motion. Depending on the implementation, the Eulerian-Lagrangian methods can give very accurate and detailed results in wide range of different settings. However, the computational costs of these approaches are often very demanding especially in case of dense suspensions and their application is mainly limited to smaller scale computations and research use. In these approaches the fluid phase is still modeled with Navier-Stokes equations, but the particle phase is modeled using a large amount of individual particles obeying the Newtonian equations of motion [10].

The relative ability of the particles to follow the changes in the flue gas flow field can be determined by the particle Stokes number, St_p given Equation 1.

$$St_p = \frac{t_p}{t_f} \tag{1}$$

t_p = the ratio of the particle relaxation time scale
t = the applicable fluid dynamic time scale, expressed as Equation 2

$$t_p = \frac{\rho_{p} d_p^2}{18 \ _f} \tag{2}$$

μ_f = dynamic viscosity of the carrier fluid.

An ANSYS flow process is simplified in Figure 1,

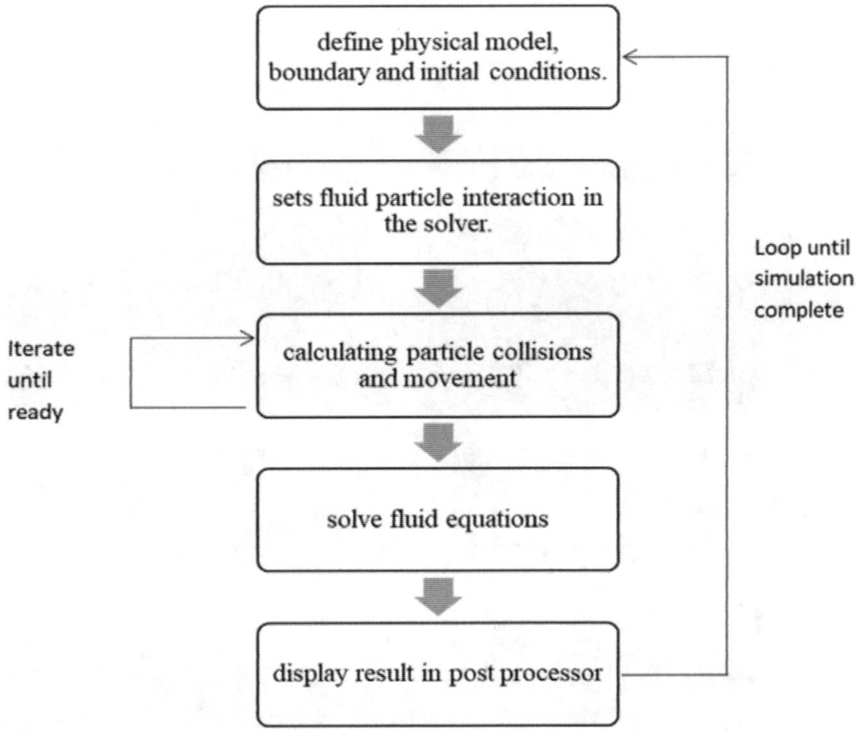

Figure 1. Simplified flow diagram.

2.2. Geometry

There are two different models in this study which are square and cylindrical in shape to be used in all simulations. Besides that, the input parameter for the simulations is the same in all simulations except for the inlet velocity. The models are displayed in Figure 2:

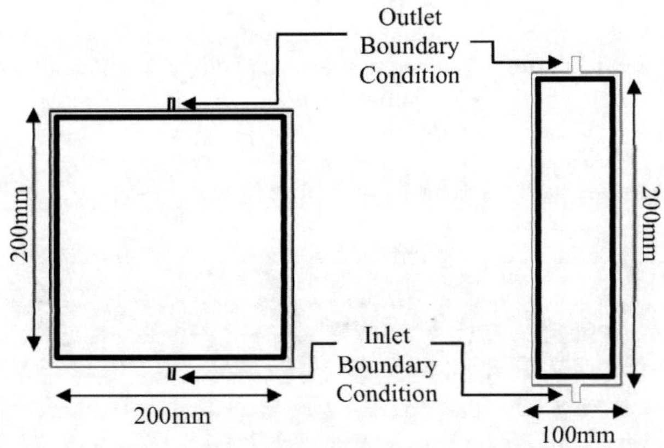

Figure 2. Front view of square (left side) and cylindrical (right side) shape of simple model for simulations

IOP Conf. Series: Materials Science and Engineering **920** (2020) 012021 doi:10.1088/1757-899X/920/1/012021

2.3. Meshing
Meshing generated for both models are shown as below:

Square **Cylindrical**

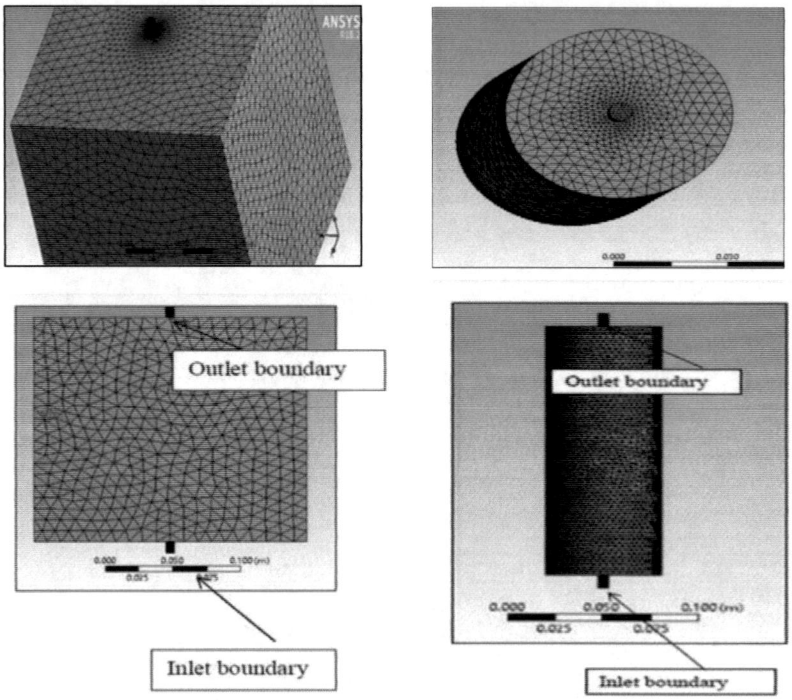

Sizing of mesh is set to fine mesh
Meshing pattern is in tetrahedral
Number of nodes: 18155
Number of elements: 94499

Sizing of mesh is set to fine mesh
Meshing pattern is in tetrahedral
Number of nodes: 7651
Number of elements: 38316

Figure 3. Meshing of square (left side) and cylindrical (right side) shape of simple model for simulations.

2.4. ANSYS-Fluent solver settings
The Ansys-Fluent solver settings are summarized in table below:

Table 1. Formatting sections, subsections and sub subsections.

Phase Properties	
Phase 1 (air)	
Density, kg/m^3	1.225
Viscosity, kg/m^3	1.7894 x 10^{05}
Phase 2	
Diameter, m	0.001
Density, kg/m^3	1000

Boundary Conditions	
Inlet	
Turbulent intensity,%	5
Hydraulic diameter, m	0.006
Phase 1 velocity, m/s	Case 1 : 1 m/s , Case 2: 10 m/s
Phase 2 velocity, m/s	-
Outlet	
Turbulent intensity,%	4.6
Hydraulic diameter, m	0.006
Gauge pressure, Pa	0

All simulations are run as a transient with a small step of 0.01 m/s and the initialization is done. Simulations are run until the initial transient had vanished and it reached a pseudo steady-state. Calculation of the time average will take place. The time averaging is mostly straightforward. During calculating time averages, to obtain more representative results it is assumed that a longer time is needed. Since the computations are very time consuming, the sampling time has to be limited.

3. Results and discussion

3.1. Particle trajectories between the models on different inlet velocity 1m/s and 10m/s

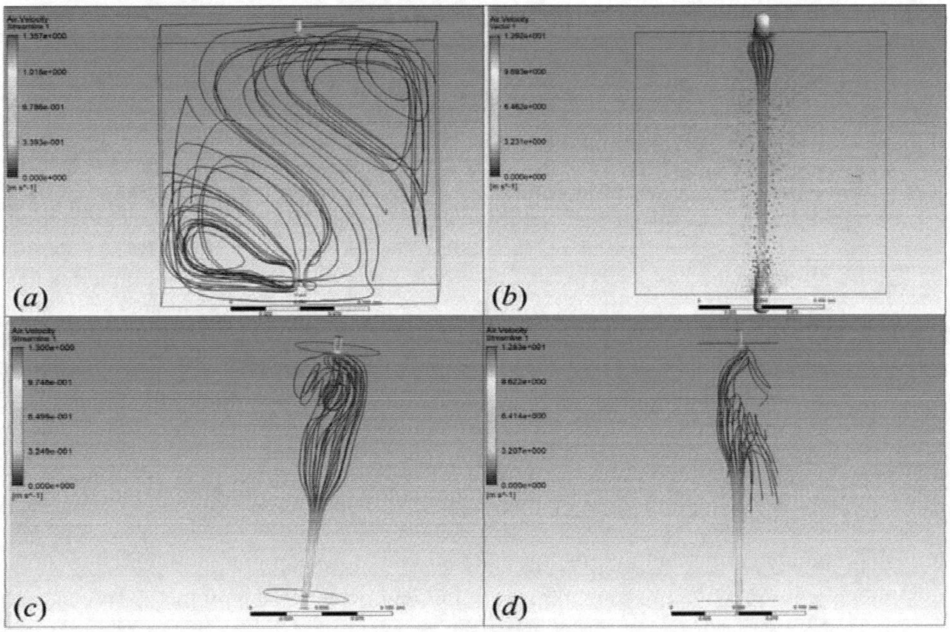

Figure 4. Particles impact at 1 m/s inlet velocity: (*a*) square model and (*c*) cylindrical model, Particle impact at 10 m/s: (*b*) square model and (*d*) cylindrical model.

The Particle trajectories that are formed show in Figure 4 to make better visualization for both models. It can find the differences in Particle trajectories to make comparison in different models (Square and Cylinder) and different velocity (1 m/s and 10 m/s).

3.2. Comparison of Pressure, Turbulence kinetic energy, and Turbulence eddy dissipation for both models.

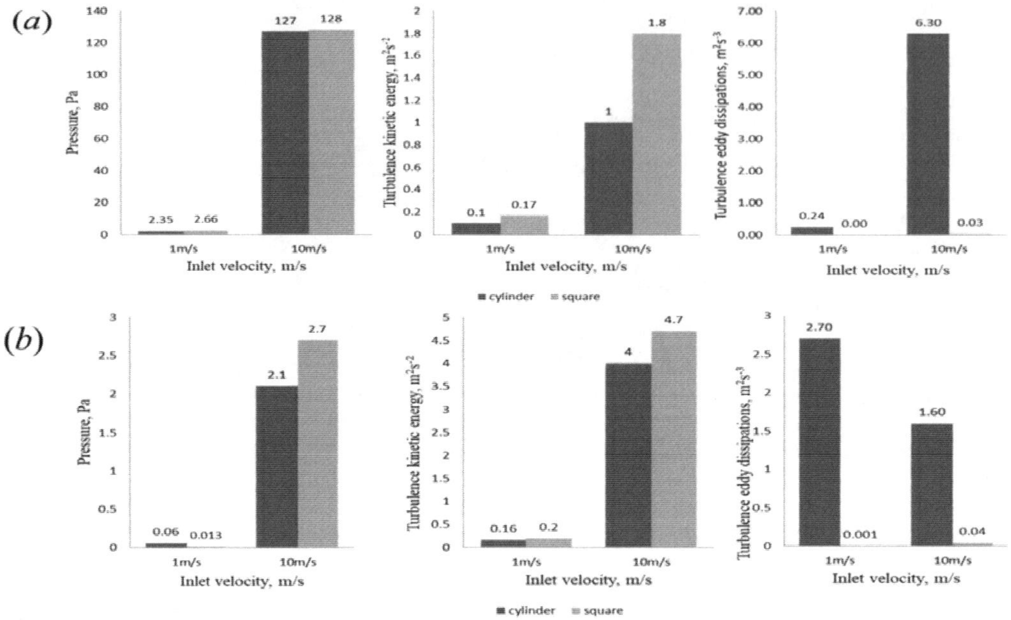

Figure 5. Comparison of Pressure, Turbulence kinetic energy, and Turbulence eddy dissipation for both models: (*a*) at inlet boundary, and (*b*) at outlet boundary.

Simulations results that are obtained graphs are formed to make a better evaluation for both models. The graph will be compared between the square and cylindrical models at a velocity of 1 m/s and 10 m/s. The maximum value of each result used to plot the graphs. The graph shows in Figure 5(a) the pressure of the square model at input 1 m/s velocity (2.66 Pa) is slightly higher (2.35 Pa) than the pressure in the cylindrical model. Inlet 10 m/s also shows that the pressure in the square (128 Pa) is slightly higher (127 Pa) than the cylindrical model. For the turbulence kinetic energy in the square model is (0.17 m^2s^{-2}) higher than the cylindrical model (0.1 m^2s^{-2}) for velocity input 1 m/s. Meanwhile, at input velocity 10 m/s the turbulence kinetic energy in the square model shows large different values compare to the cylindrical model. The difference value between the models at 10 m/s is 1.8 Pa while at 1m/s the pressure difference is 1 Pa. Referring to the turbulence eddy dissipation in the cylindrical model is greater than the square model at two inlet values, 1 m/s and 10 m/s. The turbulence eddy dissipation value difference between models shows a big difference which is (0.24 m^2s^{-3}) at input velocity 1 m/s, and (6.3 m^2s^{-3}) at input velocity 10 m/s.

This graph illustrates in Figure 5(b) the pressure of the square model at input 1 m/s velocity is 0.013 Pa less than the pressure in the cylindrical model (0.06). However, at inlet 10 m/s the pressure reading in the square is 2.7 Pa higher than the cylinder model which is 2.1 Pa. The graph shows that the turbulence kinetic energy in the square model is 0.2 m^2s^{-2} higher than the cylindrical model (0.16 $m^2s^{-2)}$ for velocity input 1 m/s. Meanwhile, at input velocity 10 m/s the turbulence kinetic energy in the square model shows 4.7 m^2s^{-2} different values compared to the cylindrical model is only 4 m^2s^{-2}. A bar graph shows that the turbulence eddy dissipation in the cylindrical model is greater than the square model at two inlet values, 1 m/s and 10 m/s. The turbulence eddy dissipation value difference between models shows big difference which is around (2.7 m^2s^{-3}) at input velocity 1 m/s, and (1.6 m^2s^{-3}) at input velocity 10 m/s

Results obtained showed in Figure 5 are comparison of pressure, turbulence kinetic energy, and Turbulence Eddy Dissipation. The cases were for both models (Cylinder and Square) at inlet boundary and at outlet boundary also compare with different velocity (1 m/s and 10 m/s). Clearly, it shows that between both models in the pressure and turbulence kinetic energy square model got high value to compare with the cylinder model either in the inlet boundary or in outlet boundary. It also compares with different velocity, where the value is higher in 10 m/s compared within 1 m/s. For turbulence eddy dissipation it shows that the cylinder model is higher than the square model and if comparing in different velocity, inlet boundary is higher in 10m/s and at the outlet, velocity is higher in 1m/s.

4. Conclusions

In the study presented in this paper single phase CFD flow been used for the analysis. All the phase has been tracking by Lagrange approach of the particulate optionally used for visualization purposes. The flow in these models depends strongly on particle–particle interaction and gas–particle interaction. For this reason, proper closure relations for these two interactions are vital for reliable predictions on the models. Gas–particle interaction can be studied with the use of the lattice Boltzmann model, while the particle–particle interaction can appropriately be studied with a discrete particle model. In this work it is shown that the discrete particle model has the capability to generate insight and finally closure relations in several processes. Finally, examples were shown demonstrating the capabilities of the discrete particle model in ANSYS-Fluent for the prediction of fly ash patterns in different models.

Acknowledgement

This research was supported by the internal grant under Universiti Tenaga Nasional (UNITEN). The authors acknowledge Civil Engineering Department, College of Engineering (UNITEN) for the facilities. Special thanks to those who contributed to this project directly or indirectly.

References

[1] M. U. Garba, D. B. Ingham, L. Ma, M. U. Degereji, M. Pourkashanian, and A. Williams, “Modelling of deposit formation and sintering for the co-combustion of coal with biomass,” Fuel, vol. 113, pp. 863–872, 2013.

[2] F. Goodarzi, “Characteristics and composition of fly ash from Canadian coal-fired power plants,” vol. 85, pp. 1418–1427, 2006.

[3] S. Thipse, M. Schoenitz, and E. Dreizin, “Morphology and composition of the fly ash particles produced in incineration of municipal solid waste,” Fuel Process. Technol., vol. 75, pp. 173–184, 2002.

[4] S. K. Kær, L. A. Rosendahl, and L. L. Baxter, “Towards a CFD-based mechanistic deposit formation model for straw-fired boilers,” Fuel, vol. 85, no. 5–6, pp. 833–848, 2006.

[5] M. García Pérez, E. Vakkilainen, and T. Hyppänen, “2D dynamic mesh model for deposit shape prediction in boiler banks of recovery boilers with different tube spacing arrangements,” Fuel, vol. 158, pp. 139–151, 2015.

[6] Zahari, N. M. et al. 2018. “Introduction of Discrete Phase Model (DPM) in Fluid Flow: A Review.” AIP Conference Proceedings, Vol. 2030, No. 1, p. 020234, 2018.

[7] Zawawi, M. H., A. Saleha, et al. 2018. “A Review: Fundamentals of Computational Fluid Dynamics (CFD).” AIP Conference Proceedings, Vol. 2030, No. 1, p. 020252, 2018.

[8] Zawawi, M. H., W. N. Yusairah, et al. 2018. “Computational Fluid Dynamic Analysis for Solar Powered Water Treatment Device.” AIP Conference Proceedings, Vol. 2030, No. 1, p. 020256, 2018.

[9] Zawawi, M. H., N. H. Hassan, et al. 2018. “Fluid-Structure Interactions Study on Hydraulic Structures: A Review.” AIP Conference Proceedings, Vol. 2030, No. 1, p. 020244, 2018.

[10] T. Niemi, “Particle Size Distribution in CFD Simulation of Gas-Particle Flows,” Diss, p. 79, 2012.

IOP Conf. Series: Materials Science and Engineering **920** (2020) 012022 doi:10.1088/1757-899X/920/1/012022

Simulation on Water Behavior in a Retention Pond at UNITEN

N. M. Zahari[1*], M. Amieroul Mirza Mohamad[1], M. H. Zawawi[1], Aizat Abas[2], Nurhanani A. Aziz[1] and F. Nurhikmah[1]

[1]Department of Civil Engineering, Universiti Tenaga Nasional (UNITEN), Kajang 43000, Selangor Darul Ehsan, Malaysia
[2]School of Mechanical Engineering, Universiti Sains Malaysia, Engineering Campus, 14300 Nibong Tebal, Pulau Pinang, Malaysia

[*]Email: mubinzahari@gmail.com

Abstract. Retention pond have been used widely to manage stormwater runoff to avoid flooding and downstream erosion and even improve water quality. It is an artificial pond that have vegetation around the perimeter and a permanent pool of water. This paper presents the study on water flow behaviours in UNITEN's retention pond by using ANSYS Fluent. Findings shows that the velocity of water in wet condition is higher than the dry condition because of an additional inlet in wet condition to imitate the heavy rainfall phenomenon. It is also found that the water pressure in wet condition is higher compare to dry condition. Therefore, the conditions that have a higher movement of water can be determined through this simulation. Based on the findings, it can be concluded that the wet condition have a high movement of water behaviour compare to dry condition.

1. Introduction

Detention pond is water runoff from streets and parking areas carry all pollutants such as dirt and oil from roadways and neighbourhood areas. Two situations would be involved in this pond which is wet and dry. Wet detention pond act to treat storm water runoff to prevent flash flood and protect stream and creeks while dry detention pond generally only hold water during a rainstorm and their main purpose is to prevent a flood. The primary function designation of detention pond is to estimate the pond's normal detention volume and it is important to know the inflow hydrograph for the detention pond that may cause worst damage under the design standard [1]. The variables or parameters that need to be considered in designing detention pond are several pipes for the service outlet, the pond storage and the diameters of pipe [2]. However, the population equivalent or population capacity also becomes a main concern on constructing a detention pond in order to estimate the size and depth will be use.

Detention pond has become widely used around the world because it provides flow control through decreasing of stormwater runoff. Any wide range of rainfall events can be stored to avoid the neighbourhood or landscaped area from a flash flood, and it is simple to construct and design and easy to maintain. Involved pollutants by settling and elimination of suspended solids are needed to improve the stormwater quality for stormwater ponds to use expansively. However, this study is to know the water behaviour and sediments or particle flow of detention pond because there is no fluid particle movement investigation on that area which is in UNITEN. The methods presently used to determine sediments flow settling in stormwater ponds are not reasonable and are based on two methods which are the computational fluid dynamics (CFD) models and ideal settling tank concept [3, 4]. Water flow of sediments or particle modelling and water flow modelling that shows the simulations of the effects of solid particles on the sparkling behaviour in a liquid-solid flow system by applying this CFD method [5-10].

2. Methodology

Figure 1 show the flow chart of methodology:

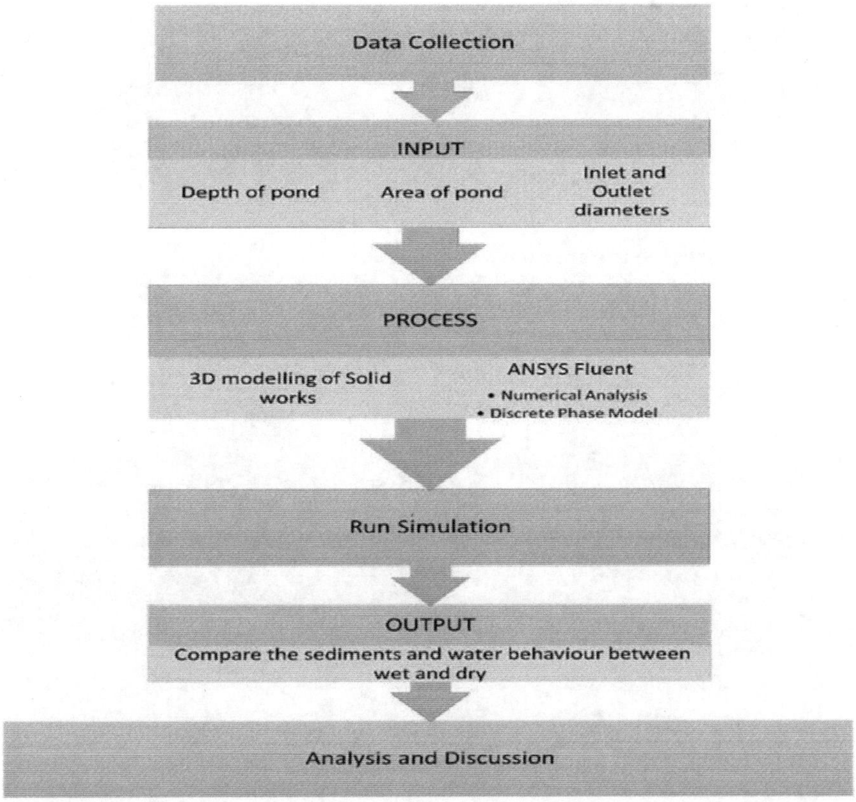

Figure 1. Flow chart of methodology.

2.1. Data Collection

The water level of the pond is measured with two different conditions which are wet and dry. The wet condition occurs during heavy rain while the dry condition is during the normal day. The depth of the pond during wet and dry is 0.7m and 1.0m respectively. The area is measured using google earth to estimate the area or manually measured at site personally. The length of the pond is 548m while the total area is 0.836 m^2 or 8360 m^2. The diameter of the inlet and outlet also measured at site or detention pond.

2.2. Modelling Setup

From the top view, the calculated and measured detention pond length was 548 m by using Google Earth. There were 5 other minor inlets adding from wet condition pond. This was due when raining heavily the water flow through the drainage system entering detention pond. The flow stops during dry condition because heavy water raining will flow to drain that causing another inlet to put out their functions.

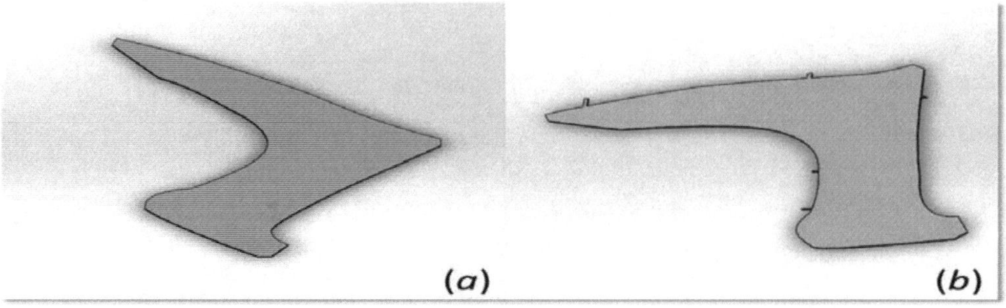

Figure 2. (*a*) Dry detention pond model and (*b*) Wet detention pond.

2.3. *Meshing*
Meshing generated for both models are shown as below:

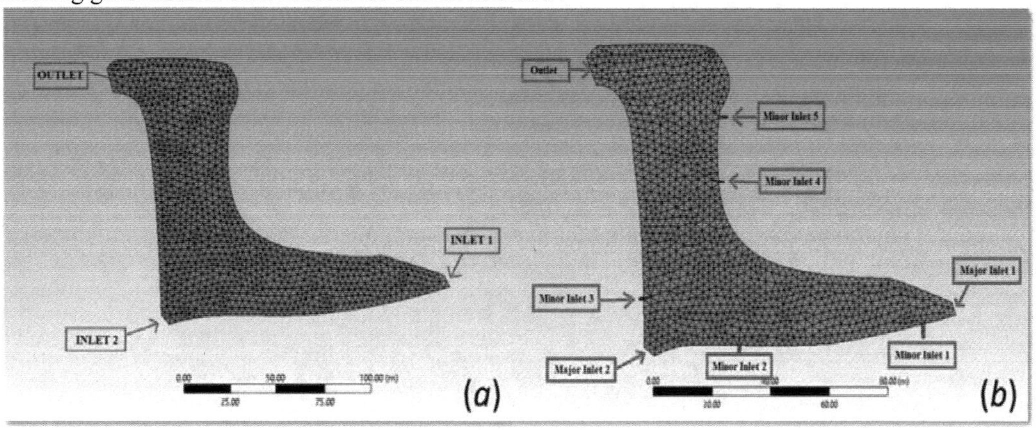

Figure 3. Meshing of (*a*) dry condition and (*b*) wet condition.

3. Results and discussion
3.1. *Velocity contour*
Theoretically, the water behaviour can be affected by velocity of water. Whenever the water flows at high speed, turbulence will occur and there would be a high movement of water behaviour can be observed. Figure 4 shown a velocity contour for dry and wet conditions. Based on the figure, during dry condition it shown that the velocity of water is lower than the wet condition. The maximum velocity for dry condition can reach until 1.631e-2 m/s while wet condition can reach until 8.21e-2 m/s but both conditions still in steady state and in laminar flow range. The major different in maximum velocity for both conditions is because of an additional minor and major inlet in wet condition due to heavy rainfall. Based on the findings, the maximum velocity of minor and major inlet is 1.5m/s and 2.5m/s respectively and the minimum velocity for each inlet is 0.2m/s and 2m/s respectively.

IOP Conf. Series: Materials Science and Engineering **920** (2020) 012022 doi:10.1088/1757-899X/920/1/012022

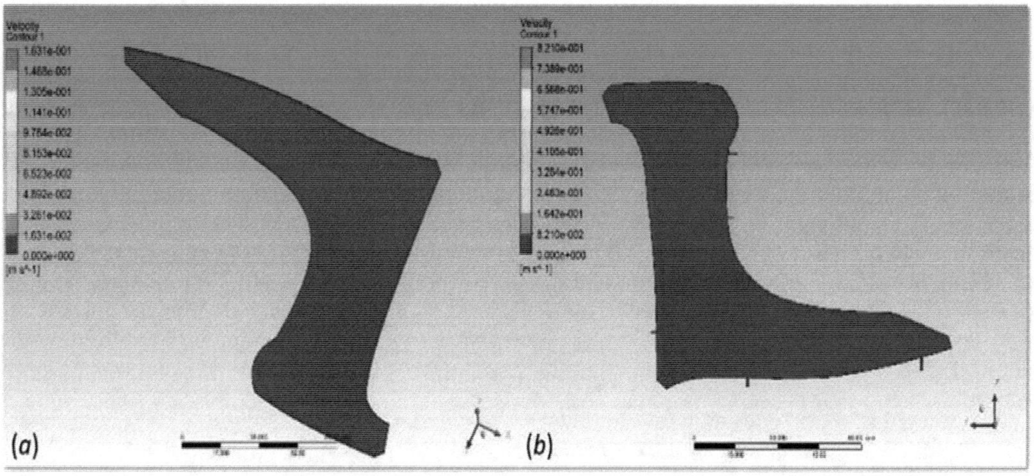

Figure 4. Velocity contour: (*a*) dry condition and (*b*) wet condition.

3.2. *Velocity vector*

Figure 5 shows the velocity vector of dry and wet conditions. Based on the figure below, the lowest water flow for dry condition was 4.077e-2 m/s which was located at the middle of the pond heading towards the outlet while the maximum reading was 1.631e-1 m/s located at the border of the pond which heading to Inlet 2. There was 2 result that was ranged between (1.223e-1 to 1.631e-1) m/s and arrow for both direction orange in colour. One was lighter and opposite for the other arrow which was darker. As per calculated the velocity for both arrow 1.427e-1 m/s. While for wet conditions it shows that the direction of water velocity vector wet condition which very different from dry condition. The maximum value for velocity vector was 8.210e-1 m/s which was in front minor inlet 2 and the direction for the flow was heading towards major inlet 2 while for the minimum value was 6.158e-1 m/s. The result for minimum value like that due arrow which yellow in colour and located between minor inlet 1 and minor inlet 2 while the direction of the arrow heading towards minor inlet 1. There were 2 arrows facing in the same direction, one with orange in colour and the other one lighter orange. The velocity vector for both arrows was 7.184e-1 m/s.

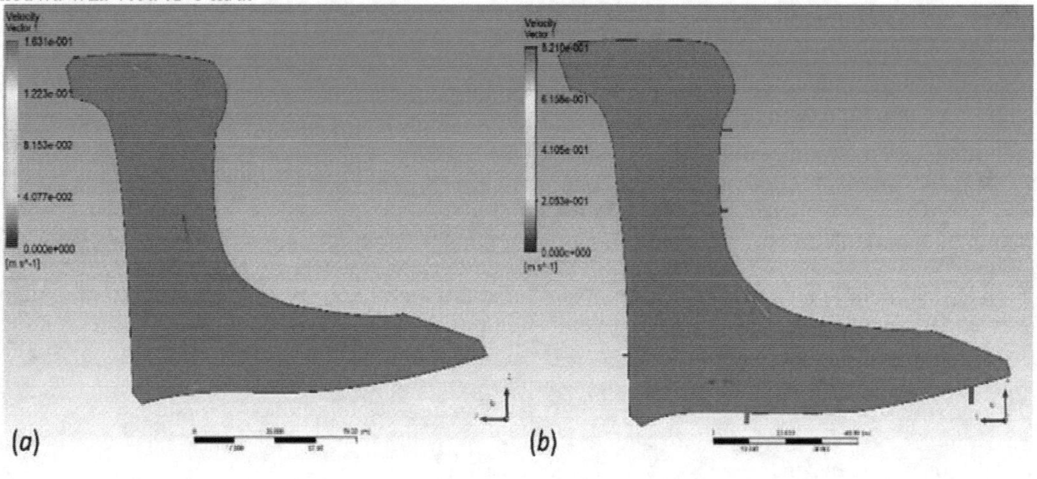

Figure 5. Velocity vector: (*a*) dry condition and (*b*) wet condition.

3.3. Pressure contour

Figure below shows the pressure contour for dry and wet conditions. The pressure in dry conditions shows that at both inlet and some spot showed a light blue zone which could be referred from the graph at the side between (-1.212e+001 – 3.568e+001) Pa. But there was a spot where dark blue zone which the pressure between (-1.077e+002 to -5.992e+001) Pa very low. The result indicates that there was very low water pressure in that area because the water keeps the flow to the outlet. But the green zone area pressure in between (8.348e+001 – 1.791e+002) Pa, showed a little bit higher. This area was approaching the outlet so the colour contour was accepted because the water would flow in that part to reach the outlet. However, there were a small spot of the yellow zone and red zone ranged between (2.269e+002 – 3.703e+002) Pa at the end of the detention pond before reaching the outlet but there was no such situation happened around inlet 2. This could be proven the water flow from inlet 1 approaching wall at inlet 2 but there was also water flow at the water inlet 2 which avoiding the water from hitting the wall hard. Then all the water flows right into the outlet and before reaching it the water may be accumulated at that spot.

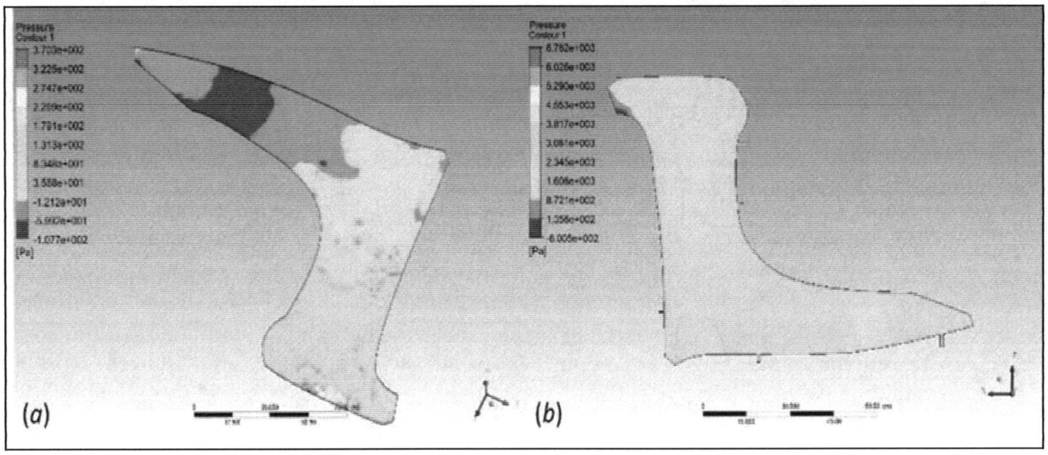

Figure 5. Pressure contour: (*a*) dry condition and (*b*) wet condition.

Figure 6(*b*) shows the pressure contour for wet condition. Wet condition has 5 minor inlets, 2 major outlets and an outlet. From the result above, the highest pressure was located at minor inlet 4 valued 6.762e3 Pa. This was due to water that flow in that area came from higher ground with high velocity down to the detention pond, which is only that spot was in the red zone. The lowest pressure for this condition was ranged between (-6.005e2 – 1.358e2) Pa which was located at the corner of the outlet. It is normal to have negative static pressure since the pressure energy gets converted into kinetic energy. After some distance, the pressure would be recovered and increased pressure but still, the result remains negative. The negative value was the difference between actual pressure and operating pressure. For light blue zone, the area was only spotted at 2 different location, one at the outlet and the other one was between minor inlet 1 and minor inlet 2. Blue zone only appears at the outlet which was (8.721e2 – 1.608e3) Pa, majorly blue area only seems to appear at the outlet. This happens because the water flow velocity decreasing from time to time as well as kinetic energy. Yellow spot only appears on minor inlet 1, 4 and 5 only. These were due to high water velocity coming from drainage area to the detention pond, while the other 2 water only flows from lower ground level from the other 3 minor inlet that was why the water flow from both inlets slow. The yellow zone pressure was ranged between (4.553e3 – 5.290e3) Pa. However, majorly water pressure contour for pond oscillated between (2.345e3 – 3.817e3) Pa which appears on the green zone as shown in the Figure 6(*b*).

4. Conclusions

Understanding the water behaviour of retention pond during dry and wet condition is necessary in order to come out with a better design for the pond in worst case. Therefore, a simulation on water behaviour in a retention pond is needed to visualize the water behaviour of the pond. The water behaviour has been simulated using ANSYS Fluent. Accordingly, the water velocity, velocity vector and water pressure can be obtained through this simulation. These three variables able to explain and elaborate more the behaviour of water in UNITEN's retention pond. Based on the findings, it shows that wet condition has high movement of water behaviour compare to the dry condition. This phenomenon occurs because of an additional inlet for the wet condition during heavy rainfall. Water movement is a bit different due to everyday forecast. The water flow is affected by weather condition such as windy, sunny or even rainy day. These data will be helpful in improving the scheme of designing detention pond that can produce the optimum result.

Acknowledgement

This research was supported by the internal grant under Universiti Tenaga Nasional (UNITEN). The authors acknowledge Civil Engineering Department, College of Engineering, UNITEN for the facilities. Special thanks to those who contributed to this project directly or indirectly.

References

[1] Hong, Y. M. (2008). Numerical simulation of laboratory experiments in detention pond routing with long rainfall duration. *International Journal of Sediment Research*, *23*(3), 233–248. https://doi.org/10.1016/S1001-6279(08)60021-9.

[2] Yang, D., Gao, P., Liu, X., Zhao, Y., & Zhang, Y. (2018). A method for determining steady velocity of shallow water fl ow on hill-slope and the distance when water fl ow reaches stability. *Journal of Hydrology*, *565*(April), 214–223. https://doi.org/10.1016/j.jhydrol.2018.08.018

[3] Krishnappan, B. G., & Marsalek, J. (2002). Modelling of flocculation and transport of cohesive sediment from an on-stream stormwater detention pond. *Water Research*, *36*(15), 3849–3859. https://doi.org/10.1016/S0043-1354(02)00087-8

[4] Burlutskii, E. (2017). CFD study of oil-in-water two-phase flow in horizontal and vertical pipes. *Journal of Petroleum Science and Engineering*, *162*(August 2017), 524–531. https://doi.org/10.1016/j.petrol.2017.10.035

[5] Shi, J., Gourma, M., & Yeung, H. (2017). CFD simulation of horizontal oil-water flow with matched density and medium viscosity ratio in different flow regimes. *Journal of Petroleum Science and Engineering*, *151*(August 2016), 373–383. https://doi.org/10.1016/j.petrol.2017.01.022

[6] Tarpagkou, R., & Pantokratoras, A. (2013). CFD methodology for sedimentation tanks: The effect of secondary phase on fluid phase using DPM coupled calculations. *Applied Mathematical Modelling*, *37*(5), 3478–3494. https://doi.org/10.1016/j.apm.2012.08.011

[7] Zahari, N. M. et al. 2018. "Introduction of Discrete Phase Model (DPM) in Fluid Flow: A Review." *AIP Conference Proceedings* 2030.

[8] Zawawi, M. H., A. Saleha, et al. 2018. "A Review: Fundamentals of Computational Fluid Dynamics (CFD)." *AIP Conference Proceedings* 2030.

[9] Zawawi, M. H., W. N. Yusairah, et al. 2018. "Computational Fluid Dynamic Analysis for Solar Powered Water Treatment Device." *AIP Conference Proceedings* 2030(November).

[10] Zawawi, M. H., N. H. Hassan, et al. 2018. "Fluid-Structure Interactions Study on Hydraulic Structures: A Review." *AIP Conference Proceedings* 2030.

IOP Conf. Series: Materials Science and Engineering **920** (2020) 012023 doi:10.1088/1757-899X/920/1/012023

Vibration Control and Smart Integrated Transportation Strategy of Crane Systems

Kazuhiko Terashima and Ho Duc Tho

Mechanical Engineering Department, Toyohashi University of Technology, 1-1 Hibarigaoka, Tempaku-cho, Toyohashi 441-8580, Japan

E-mail: terasima@tut.ac.jp, ho_duc_tho@ise.me.tut.ac.jp

Abstract. This paper presents number of vibration control schemes and smart integrated transportation strategies for crane systems. First, robust control problem of payload's skew rotation in crane systems is discussed. A robust integral sliding mode controller is proposed to realize skew transportation without residual vibration. Robust stability conditions with respect to parametric uncertainties of the closed-loop system are imposed on control gains. Second, transportation of bulk material using an overhead crane system is considered, in which the transferred material can be dropped/discharged while in the air. This paper introduces a new concept, named tossing control methodology, to enhance transportation productivity. A specific type of tossing controller is established, which relies on the phenomenon of linear resonance. It will be shown that the resonance-based tossing control is able to break the time limitation of the minimum-time vibration suppression control. Simulation and experimental results are provided to verify the effectiveness of the proposed control systems. Finally, a diffusion-based obstacle-avoidance path planning method for crane systems is outlined to realize a fully autonomous crane in future.

1. Introduction

Cranes are the most important means of transportation in various places such as factories, shipyards, construction sites, and especially harbors, to transfer heavy and hazardous material point-to-point within the workspace. The 21st century witnessed a revolution of maritime industry, where the containerization played a key role. Every year, billion tons of merchandise and containers are transported by sea and it is generally accepted that more than 90% of global trade belongs to seaborne commerce. In order to transload such an enormous and annually increasing amount of goods, crane systems are getting larger in size and in lifting capacity. They are also being designed to operate at a much higher productivity than before to shorten transloading time. As a result, transportation cost can be significantly reduced. The main focus factor in the pursuit of improving productivity of cranes is the oscillation of the payload around its equilibrium. The vibration phenomenon is a natural characteristic of every crane. The central idea here is to control that vibration as our will to suit for different engineering purposes. At present, the main stream of studies focuses on the vibration suppression control, in which the vibration is seen as an unwanted phenomenon that needs to be eliminated. However, the vibration might be undesirable for some applications but favorable for others (e.g., bulk material transportation described in Section 3). Crane vibration control should therefore also be understood in the sense of vibration excitation control, not just in the direction of vibration suppression control. Due to its significance in fostering both theory and application advancements, vibration control of cranes attracted considerable attention from both control theorists and control engineers. The last 50 years have seen a tremendous interest and effort in modeling and control of crane systems. Since the accumulative literature is too vast to list every single detail of all efforts made on the topic, only swing suppression control of the overhead crane system will be subsequently reviewed. It can be classified into two main classes: feedback and feedforward controls. These types can be combined to obtain a two degrees of freedom control system which could give a better control performance.

An extensive background can be found for the feedback control category [1]. Naturally, vibration suppression control of the overhead crane evolved along with the development of control theory field. New achievements of the linear controls and model predictive controls are promptly employed on crane systems in [2–4] and [5] respectively. Feedback linearization and differential-flatness based methodology [6], which aims at transforming the nonlinear model to a linear one to utilize well-established linear controls, are also widely applied [7, 8]. To deal with parametric uncertainties and unmodeled dynamics, sliding-mode based controls [9–12], adaptive controls [13–15], energy and passivity-based nonlinear controls [16, 17]. Intelligence—fuzzy and neural network—controls are also successfully implemented on the crane systems [18, 19]. To sum up, feedback control techniques showed their excellent properties in handling parametric uncertainties and external disturbances. However, they demand a sophisticated and costly sensor structure which requires regular maintenance. In numerous circumstances, feedforward controls are preferable in industry due to their sensorless nature and ease of implementation. In the feedforward control category, the motion trajectory of actuators is planned such that the payload arrives at the target position without residual vibration. The swing oscillation information will not be used during the transfer process. In order to do so, one can utilize the input shaping technique [20–23], S-curve commands [24–28], optimal controls [29–32], linear/nonlinear programming algorithms [33, 34]. The main disadvantage of the feedforward control is that the robustness to parametric uncertainties and disturbances are difficult to be assured.

Skew orientation control of the payload is crucial in practical cranes when the payload (e.g. a container) cannot be considered a point mass. It is also important for an obstacle avoidance application when crane systems work in a narrow workspace. Despite urgent requirements from the harbor industry and in contrast to the well-established anti-swing controls, there are only a few studies paying attention to design payload anti-skew oscillation controllers. Therefore, we will introduce a robust integral sliding mode control for the skew rotation process of crane systems, which is elaborated in Section 2. This paper will also focus on the development of a tossing controller to further enhance transfer productivity of crane systems in the context of bulk material transportation, in which the transferred material can be dropped/discharged while in the air. By imposing upon such a special property, one can break the time limitation of the minimum-time swing suppression control—the fastest controller in the swing suppression control group. Theoretical developments and results of the tossing controller are given in Section 3. In Section 4, to realize a fully autonomous crane in future, a diffusion-based path planning method is outlined to address the obstacle avoidance task.

2. Skew rotation control of crane systems

Skew orientation control of a payload plays an important role in crane systems. For instance, consider the overhead crane system shown in Fig. 1 where the payload's center of gravity G has arrived at the correct *position* in the workspace without any oscillation. However, when the target *orientation* is taken into account, the transferring process is said to be incomplete since there exists a skewing mismatch angle β between the payload's orientation and its desirable pose. Therefore, it is essential to rotate the payload at a skew angle of β along the Z axis in the counter-clockwise direction to accomplish the transferring task. A similar issue is also

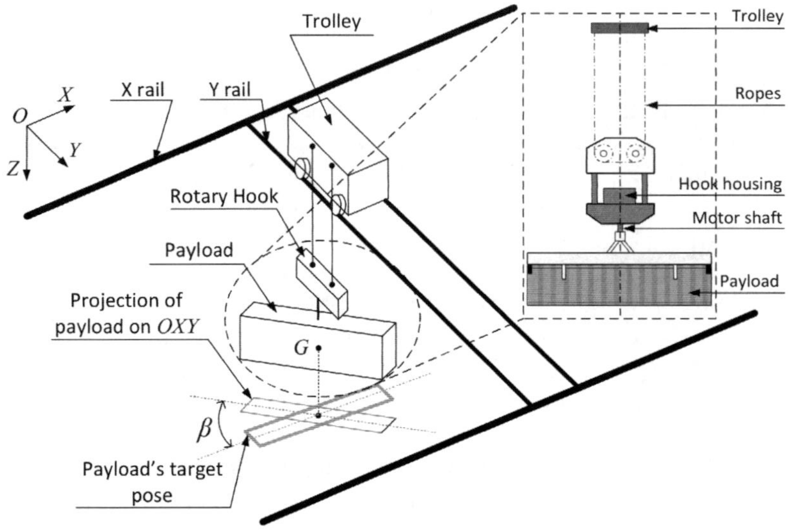

Figure 1. Necessity of the payload orientation control in an overhead crane system.

encountered in boom crane systems in use at harbors where the skew orientation of the container must be adjusted to comply with that of the truck or vessel [11].

In order to fully adjust the skew orientation of the payload, crane systems utilize a simple device called rotary hook. This device is able to directly control the skew angle of the payload by a servo motor which is mounted inside its housing. As shown in the zooming area of Fig. 1, the payload is directly attached to the motor shaft and the entire suspended load is hung below the trolley through two flexible ropes. By using this kind of direct mechanism, the skewing orientation can be controlled to any desired value (> 360 degrees). However, along with the above advantage, after reaching the desired angle, the payload will continue to oscillate rather than stop at the designated skew angle due to the law between inertia force and rope tensions. In practice, the payload inertia is large and the damping ratio of the system is low hence an artificial oscillation suppression needs to be imposed. Therefore, an automatic controller should be developed to reduce the workload for the crane operator as well as drastically enhance productivity of the system. Specifically, two control objectives need to be accomplished simultaneously: as well as drastically enhance productivity of the system. Specifically, two control objectives need to be accomplished simultaneously as well as drastically enhance productivity of the system. Specifically, two control objectives need to be accomplished simultaneously: as well as drastically enhance productivity of the system. Specifically, two control objectives need to be accomplished simultaneously: driving the payload to the desired skew angle and suppressing the residual skew oscillation caused by flexible ropes.

2.1. Mathematical modeling and validation

The skew residual oscillation resulting from the rotation of the payload is depicted in Fig. 2, where $\varphi(t)$ denotes the relative angle between the payload and the hook. Moreover, $\theta(t)$ and $\gamma(t)$ represent the absolute angles between the hook and the payload versus x axis respectively. Therefore, two aforementioned control objectives can be translated to $\varphi(t) \to \varphi_d$ and $\theta(t) \to 0$, where φ_d is a reference skew angle of the payload. The dynamics of the skew rotation process was established by using the Lagrangian mechanics under a holonomic constraint, which can be

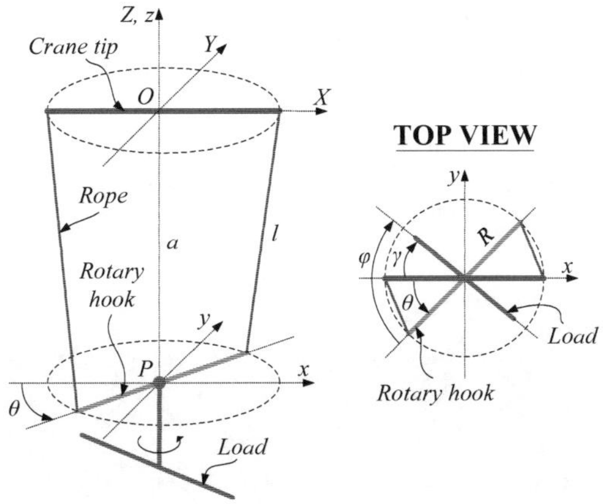

Figure 2. Modeling of the rotary hook system.

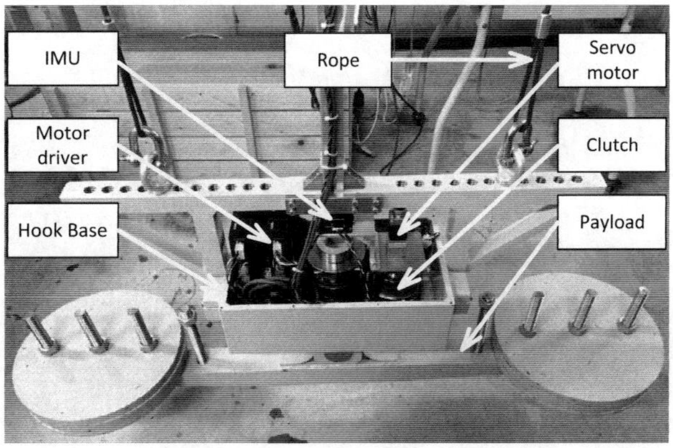

Figure 3. Experimental system.

elaborated as follows

$$\ddot{\theta} = f + bu, \quad (1)$$

where

$$f = \frac{1}{I_L + I_H + \dfrac{mR^4 \sin^2\theta}{a^2}} \times \left[\frac{-mR^2 \sin\theta}{a} \left(\frac{R^4\dot{\theta}^2 \sin^2\theta}{a^3} + \frac{R^2\dot{\theta}^2 \cos\theta}{a} + g \right) - c\dot{\theta} \right]$$

$$b = \frac{I_L}{I_L + I_H + \dfrac{mR^4 \sin^2\theta}{a^2}}$$

$$u = \ddot{\varphi},$$

and $a = \left(l^2 - 2R^2\left(1 - \cos\theta\right)\right)^{1/2}$ is the holonomic constraint due to inelastic rope. In (1), I_L is the inertia of the payload: $I_L \in [2.64,\ 41.89]$ kgm^2; I_H is the inertia of the hook: $I_H \in [1,\ 1.5]$ kgm^2; m is the total mass of the payload and the hook: $m \in [77.39,\ 207]$ kg; R is the skewing radius: $R = 0.25$ m; l is the rope length: $l \in [2.5,\ 5.5]$ m; c is the viscous friction coefficient: $c \in [0.1,\ 0.5]$ Ns/rad; and g is the gravity acceleration: $g = 9.81$ m/s^2.

The formulated mathematical model is now validated on an experimental apparatus, which is described in Fig. 3. A representative example of the model validation process is illustrated in Fig. 4. It can be observed that the experimental data are in a complete agreement with the mathematical model, which verifies the correctness of the established dynamical model.

The actual specification of the rotary hook system in use at the harbor is characterized by wide parametric uncertainties; hence, without a robustness consideration, the controller will likely cause instability to the closed-loop system. Therefore, a robust nonlinear integral sliding mode control (ISMC) is subsequently proposed as a vibration controller for the payload's skew rotation process to cope with parametric uncertainties in the system parameters.

2.2. Integral sliding mode control design

Denoting the reference trajectory of the payload angle as $\varphi_d(t)$. Thus, the tracking error term can be expressed by $e(t) = \varphi(t) - \varphi_d(t)$. The coupled integral sliding function is proposed as

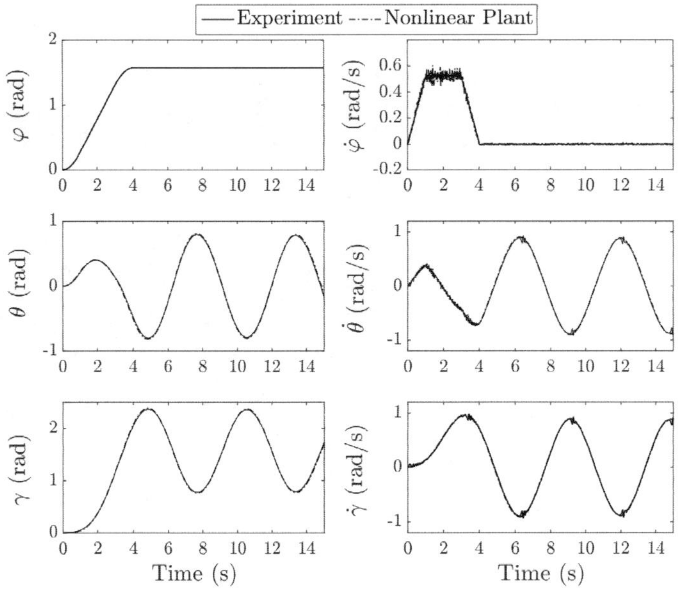

Figure 4. Validation result of the mathematical model.

follows

$$\sigma = \beta_1 \dot{e} + \beta_2 e + \int_0^t (\beta_3 e(\tau) + \beta_4 \theta(\tau))\, d\tau + \beta_5 \theta + \beta_6, \tag{2}$$

where $\beta_1, \ldots, \beta_6 \in \mathbb{R}$. By adding the constant term β_6 into the sliding function, it is expected to completely cancel the reaching phase (i.e. force the system to slide on the sliding surface at the very beginning of time). Therefore, the constraint $\sigma(t = 0) = 0$ should be imposed, which implies that

$$\beta_6 = -\beta_1 \dot{e}(0) - \beta_2 e(0) - \beta_5 \theta(0). \tag{3}$$

By letting $\dot{\sigma} = -K\text{sgn}(\sigma)$ where K is a strictly positive real number satisfying $K > |\beta_1||\ddot{\varphi}_d|_{\max} + \eta$ and $\eta > 0$ (η can be arbitrarily small), the sliding mode control law can be obtained as

$$u_I = -\beta_1^{-1} \left(\beta_2 \dot{e} + \beta_3 e + \beta_4 \theta + \beta_5 \dot{\theta} + K\text{sgn}\,(\sigma) \right). \tag{4}$$

By choosing a Lyapunov function Lyapunov function $V = \frac{1}{2}\sigma^2$, it is easy to verify that the control law (4) can render the sliding condition $\dot{V} = \dot{\sigma}\sigma \leq -\eta|\sigma|$. Therefore, the sliding surface $\sigma = 0$ becomes an invariant set [35, p. 280]. Note that, with the choice of β_6 as in (3) and the usage of the sliding mode control law (4), the system will be attracted to the invariant set $\sigma = 0$ at $t = 0$, and thus the reaching phase is actually eliminated.

On the sliding surface $\sigma = 0$, the equation $\dot{\sigma} = 0$ also holds, hence by defining the state variables as $x_1 = \theta$, $x_2 = \dot{\theta}$, $x_3 = \varphi - \varphi_d$, $x_4 = \dot{\varphi} - \dot{\varphi}_d$, the closed-loop system under sliding mode is governed as follows

$$\begin{cases} \dot{x}_1 = x_2 \\ \dot{x}_2 = f - b\beta_1^{-1}(\beta_4 x_1 + \beta_5 x_2 + \beta_3 x_3 + \beta_2 x_4) \\ \dot{x}_3 = x_4 \\ \dot{x}_4 = -\beta_1^{-1}(\beta_4 x_1 + \beta_5 x_2 + \beta_3 x_3 + \beta_2 x_4). \end{cases} \tag{5}$$

It is observed that when the integral sliding surface (2) is used, under the sliding mode, the closed-loop system (5) remains the same order with the original system (1). Next, define the state vector as $\boldsymbol{x} = [x_1 \ x_2 \ x_3 \ x_4]^\top$, linearized model of the nonlinear autonomous system (5) around the equilibrium point $\boldsymbol{x} = 0$ can be described as

$$\dot{\boldsymbol{x}} = \boldsymbol{A}\boldsymbol{x}, \tag{6}$$

where

$$\boldsymbol{A} = \begin{bmatrix} 0 & 1 & 0 & 0 \\ -\mu_1 & -\mu_2 & -\mu_3 & -\mu_4 \\ 0 & 0 & 0 & 1 \\ -\beta_1^{-1}\beta_4 & -\beta_1^{-1}\beta_5 & -\beta_1^{-1}\beta_3 & -\beta_1^{-1}\beta_2 \end{bmatrix}, \tag{7}$$

therein $\mu_1 = \omega_n^2 + \bar{b}\beta_1^{-1}\beta_4$, $\mu_2 = c_N + \bar{b}\beta_1^{-1}\beta_5$, $\mu_3 = \bar{b}\beta_1^{-1}\beta_3$, $\mu_4 = \bar{b}\beta_1^{-1}\beta_2$, where $c_N = c/(I_L + I_H)$, $\bar{b} = I_L/(I_L + I_H)$, and $\omega_n^2 = (mgR^2)/(l(I_L + I_H))$. It is easy to check that the linearized closed-loop system is stable regardless of parametric uncertainties in all system parameters if the following conditions are made

$$\beta_1 > 0, \beta_2 > 0, \beta_3 > 0, \beta_4 > 0, \beta_5 > 0, \beta_2\beta_4 - \beta_3\beta_5 > 0. \tag{8}$$

Therefore, based on the indirect (or linearization) Lyapunov method [35, p. 55], the nonlinear autonomous system (5) is robustly asymptotically stable at the equilibrium point $\boldsymbol{x} = 0$ if the control gains adhere the constraints in (8).

In order to obtain a good control performance while ensuring the robust stability, an optimization routine based on the metaheuristic particle swarm optimization mechanism can be employed, which adopts the robust stability conditions (8) as inequality constraints [11]. The optimized gains of ISMC are

$$\beta_1 = 1.988,\ \beta_2 = 10.258,\ \beta_3 = 0.0058, \beta_4 = 14.136,\ \beta_5 = 1.058,\ K = 3.726. \tag{9}$$

To reduce the chattering in the control signal, the signum function $\mathrm{sgn}(\sigma)$ in (4) should be replaced by a saturation function $\mathrm{sat}(\sigma)$ which is defined in (10), therein ε is a constant related to the thickness of the boundary layer [35]

$$\mathrm{sat}(\sigma) = \begin{cases} 1 & \text{if} \quad \sigma/\varepsilon > 1 \\ -1 & \text{if} \quad \sigma/\varepsilon < -1 \\ \sigma/\varepsilon & \text{if} \quad |\sigma/\varepsilon| < 1. \end{cases} \tag{10}$$

2.3. Simulation and experimental results

The designed robust controllers will now be applied on the experimental apparatus in three cases: minimum load case, nominal (or average) load case, and maximum load case. The control structure of ISMC is implemented in the C programming language. The Motion Designer software platform provided by TAMAGAWA SEIKI CO., LTD. is utilized to compose the executing C code. Note that the controller gains of ISMC used in the simulation will be directly employed on the experimental testbed. In the other words, there is not any gains tuning process in our experiments. It is observed that the experimental data of ISMC (see Figs. 5–7) strongly agree with the simulation results in all cases. Moreover, despite the fluctuation in the control input, the sliding functions are kept inside the neighborhood of zero bounded by the designated boundary layer with a thickness of ε. Due to the effect of inertia, peak of the skew vibration $\theta(t)$ is smallest in the minimum case and largest in the maximum case, which adheres to the common physical sense.

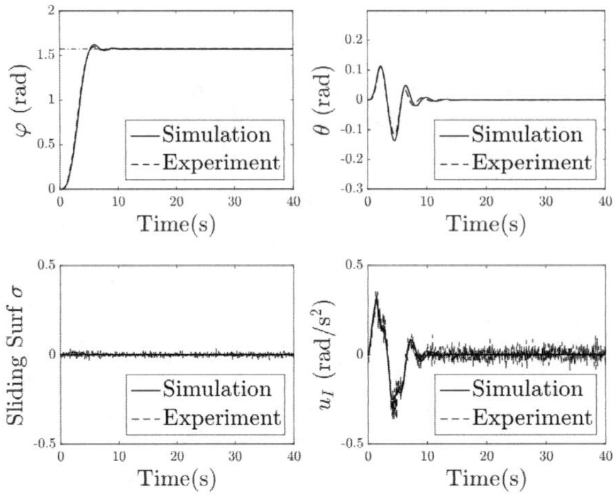

Figure 5. Experimental result of ISMC in the minimum load case.

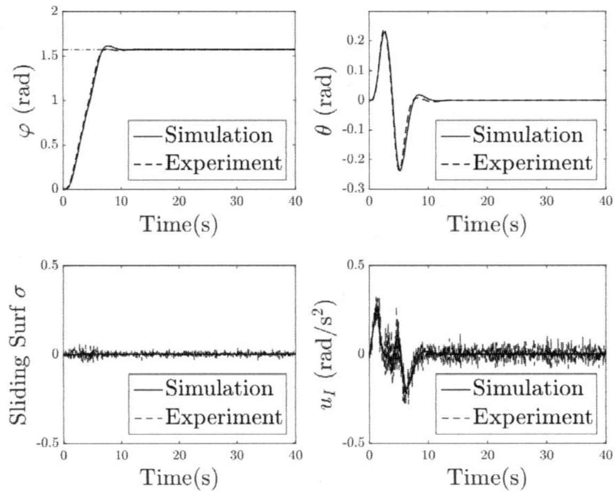

Figure 6. Experimental result of ISMC in the nominal load case.

3. Resonance-based tossing control of the overhead crane system

Transportation of bulk material using an overhead crane system possesses a special property, that is, the transferred material can be dropped/discharged right in the air. This paper will show that, under similar transportation requirements, conditions, and constraints, it is possible for the tossing control methodology, which aims at exciting a large swing angle, provides a considerably faster transferring time than the conventional minimum-time swing suppression control, whose objective focuses on suppressing/minimizing the vibration. In order to induce the oscillation with periodically increasing amplitude, the linear resonance phenomenon is employed. Consequently, a resonance-based tossing controller—comprising of an optimal linearization law and a resonance evoking component—is introduced to transfer the payload to the target discharging destination, at which velocities of the payload in both horizontal and vertical directions are zero, the trolley completely stops, and a large swing angle is obtained. For a specific configuration, in one cycle of transportation, it will be shown that the proposed resonance-based tossing control is 6.3 seconds faster than the minimum-time swing suppression control. Both simulation and experimental results are presented to verify the effectiveness of the proposed method.

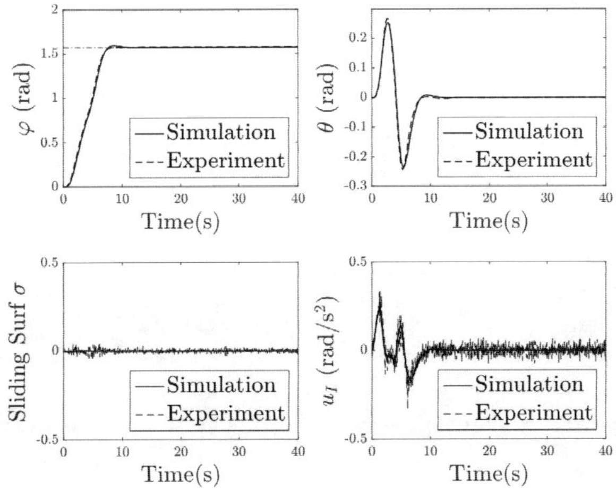

Figure 7. Experimental result of ISMC in the maximum load case.

3.1. Introduction to the tossing control and resonance controller design

The fundamental property of the bulk material transport is that the material can be dropped/discharged from the air and they naturally fall down toward their intended destination by gravity. This is completely different to a large block transfer, in which the payload must be eventually laid down on the ground by the crane. Due to the fact that the rope is flexible, the payload will exhibit vibration during and after the transportation [36]. The residual oscillation is usually undesirable when transferring a large block payload, since, for safety, the swinging motion needs to cease before the payload can be safely landed. Therefore, in such circumstances, vibration suppression control must be emphasized. In the context of bulk material transfer, nevertheless, the vibration suppression control is not necessary owning to its distinct transportation property, namely the material can be discharged right in the air. This indeed opens a possibility in further enhancing the transfer productivity by exciting the payload vibration rather than suppressing/minimizing it. Such a type of control is termed tossing control. Tossing control and vibration suppression control methodologies have a common objective in driving the payload (grab with material inside) above the desired destination, at which the payload's horizontal and vertical velocities are zero, meanwhile the trolley must stop. The only difference between them is the tossing control aims at achieving a large terminal swing angle, whereas the swing angle must be zero in the vibration suppression control. This dissimilarity is depicted in Fig. 8.

A mathematical description of the overhead crane system in the case of bulk material transportation is illustrated in Fig. 9, in which $x(t)$ (m) and $\theta(t)$ (rad) are the trolley position and the swing angle. The net force driving the trolley is denoted as $F(t)$. In addition, m (kg), M (kg), l (m), and g (= 9.81 m/s^2) represents the payload mass, trolley mass, rope length, and gravitational acceleration respectively. From Fig. 9, horizontal and vertical positions of the payload can be computed by

$$x_p(t) = x(t) - l\sin\theta(t) \tag{11}$$

$$y_p(t) = l\cos\theta(t). \tag{12}$$

By differentiating (11) and (12), horizontal and vertical velocities of the payload are

$$\dot{x}_p(t) = \dot{x}(t) - l\dot{\theta}(t)\cos\theta(t) \tag{13}$$

$$\dot{y}_p(t) = -l\dot{\theta}(t)\sin\theta(t). \tag{14}$$

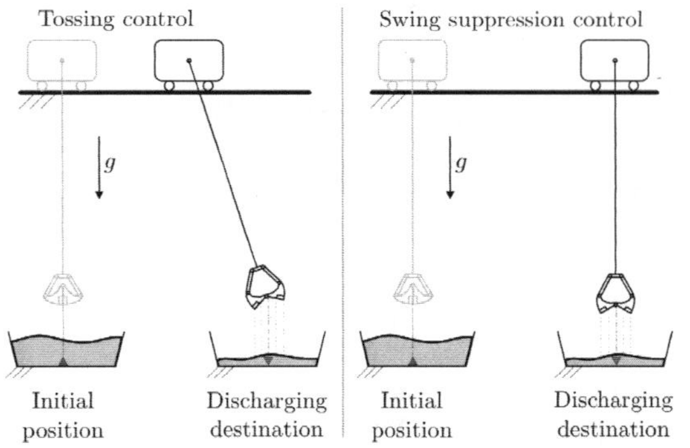

Figure 8. Difference between tossing control and swing suppression control.

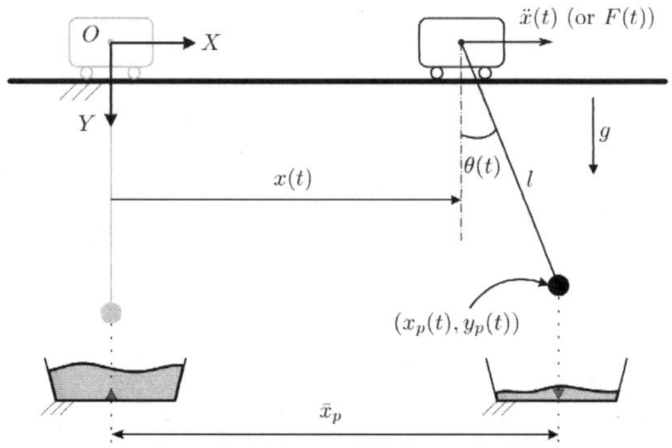

Figure 9. Mathematical model of the overhead crane system.

By utilizing the Euler–Lagrange method, dynamical equations of the overhead crane system are described by

$$(M + m)\,\ddot{x} - ml\ddot{\theta}\cos\theta + ml\dot{\theta}^2\sin\theta = F \tag{15}$$
$$l\ddot{\theta} + g\sin\theta = \ddot{x}\cos\theta. \tag{16}$$

The control input of the system is chosen to be $\ddot{x}(t)$ and only (16) is employed in the subsequent tossing control design. Note that the nonlinear differential equation (16) will be directly used to design the tossing controller without linearizing around the equilibrium point. The reason is that the small angle approximation is not longer valid in the tossing control methodology, as a large swing angle is supposed to be generated.

In order to realize the tossing control requirements, the following terminal conditions must be fulfilled

$$\dot{x}(t_f) = \dot{x}_p(t_f) = \dot{y}_p(t_f) = 0,\ x_p(t_f) = \bar{x}_p,\ \theta(t_f) < 0. \tag{17}$$

In (17), t_f is the terminal time and $\bar{x}_p$ represents the horizontal desired destination of the payload (see Fig. 9). The terminal conditions (17) of the tossing control can easily be transformed to

$$\dot{x}(t_f) = \dot{\theta}(t_f) = 0,\ x(t_f) - l\sin\theta(t_f) = \bar{x}_p,\ \theta(t_f) < 0. \tag{18}$$

Due to the physical limitation of the trolley actuator, the following maximum velocity and maximum acceleration constraints must be strictly complied with

$$|\dot{x}(t)| \le v_{\max} \text{ and } |\ddot{x}(t)| \le a_{\max}, \tag{19}$$

where $v_{\max} > 0$ and $a_{\max} > 0$. By using the information of the servo motor's maximum velocity and maximum torque, as well as the structure of the transmission system, the values of $v_{\max}$ and $a_{\max}$ can be easily estimated

The following exact linearization resonance control law is proposed

$$\ddot{x}(t) = \underbrace{g\left(\tan\theta - \frac{k\theta}{\cos\theta}\right)}_{:=L(t)} \overbrace{-\frac{a}{\cos\theta}\sin(\omega_0 t) - \frac{b}{\cos\theta}\cos(\omega_0 t)}^{:=R(t)}, \tag{20}$$

where $L(t)$ is the linearization law to exactly linearize the nonlinear system (16) and $R(t)$ is the control input for generating the resonance. It is therefore chosen that $\omega_0 = \sqrt{kg/l}$ to invoke resonance. Furthermore, the constant k in (20) is designed to minimize the magnitude of the linearization law $L(t)$. It can be computed by [36]

$$k = \frac{1}{\cos\theta^\star + \theta^\star \sin\theta^\star}, \tag{21}$$

where $\theta^\star$ is determined by

$$\sin\left(\bar{\theta} + \theta^\star\right) - \theta^\star \cos\left(\bar{\theta} + \theta^\star\right) = \bar{\theta}. \tag{22}$$

Note that, the parameters $\bar{\theta}$, $\theta^\star$ in (21)–(22), a, b in (20), and the terminal time t_f are unknown, which can be determined by solving (24).

By using the method of undetermined coefficients, a solution of the system (16) under the action of the control input (20) can be expressed by

$$\begin{aligned} \theta(t) &= -\frac{a}{2l\omega_0^2}\sin(\omega_0 t) + \frac{a}{2l\omega_0}t\cos(\omega_0 t) - \frac{b}{2l\omega_0}t\sin(\omega_0 t) \\ \dot{\theta}(t) &= -\frac{b}{2l\omega_0}\sin(\omega_0 t) - \frac{a}{2l}t\sin(\omega_0 t) - \frac{b}{2l}t\cos(\omega_0 t)\,. \end{aligned} \tag{23}$$

By exploiting (18), to solve for five unknowns $(a, b, \bar{\theta}, \theta^\star, t_f)$ in the resonance control law (20), we have the following set of five nonlinear equations

$$\begin{cases} \dfrac{-a}{2l\omega_0^2}\sin(\omega_0 t_f) + \dfrac{t_f}{2l\omega_0}\left[a\cos(\omega_0 t_f) - b\sin(\omega_0 t_f)\right] = -\bar{\theta} \\ \dfrac{b}{\omega_0}\sin(\omega_0 t_f) + t_f\left[a\sin(\omega_0 t_f) + b\cos(\omega_0 t_f)\right] = 0 \\ \displaystyle\int_0^{t_f} \ddot{x}(t)dt = 0 \\ l\sin\bar{\theta} - \displaystyle\int_0^{t_f} \ddot{x}(t)t\,dt = \bar{x}_p \\ \sin\left(\bar{\theta} + \theta^\star\right) - \theta^\star\cos\left(\bar{\theta} + \theta^\star\right) = \bar{\theta}. \end{cases} \tag{24}$$

The MATLAB routine *fsolve* can be utilized to solve (24). Note that, in order to strictly comply with the maximum velocity and maximum acceleration constraints, namely $|\dot{x}(t)| \le v_{\max}$ and $|\ddot{x}(t)| \le a_{\max}$, the initial guess of the terminal time t_f can be gradually increased until those inequalities are met.

3.2. Simulation result: comparison with minimum-time swing suppression controller

A real gantry crane in use at the harbor is utilized in the simulation study. The maximum velocity and maximum acceleration are $v_{\max} = 1.5$ m/s and $a_{\max} = 0.6$ m/s^2. The rope length is $l = 20$ m. The desired transfer distance of the payload is $\bar{x}_p = 10$ m. By using those quantities, control parameters of the resonance-based tossing control and the minimum-time suppression control [37] can be computed. Accordingly, it takes 17.72 seconds for the proposed resonance tossing control to complete one cycle of transportation, whereas it requires 24.02 seconds in the case of minimum-time swing suppression control. Therefore, under similar transportation requirements, conditions, and constraints, the resonance-based tossing controller is 6.3 seconds faster than the minimum-time swing suppression controller. Simulation results of

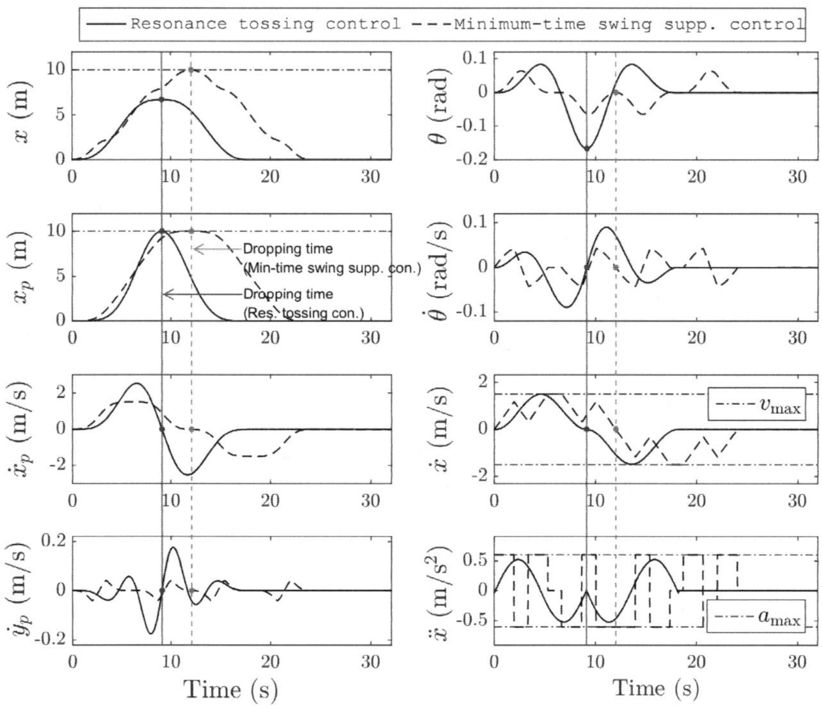

Figure 10. Comparative simulation result between the proposed resonance-based tossing control and the minimum-time swing suppression control.

two controllers are shown in Fig. 10. It can be seen that, at the material dropping/discharging time (represented by the solid blue line and the dash red line for the resonance-based tossing control and the minimum-time vibration control respectively), the payload reaches the target destination with zero horizontal and vertical velocities, whereas the trolley completely stops. Therefore, requirements of the transferring phase are fulfilled by both controllers. Moreover, at the end of the returning phase, in both control schemes, the vibration is completely suppressed. To sum up, two controllers satisfy the necessary conditions of the bulk material transport. The main difference between the proposed resonance-based tossing control and the minimum-time vibration control is explained as follows. At the dropping time, the swing angle in the case of tossing controller is $\bar{\theta} = -0.167$ rad (or -9.57 degrees), whereas it is zero degree when the minimum-time swing suppression control is employed (see θ graph in Fig. 10). This characteristic leads to the fact that, at the discharging point, the payload must not be exactly under the trolley in the tossing control case, which is completely in contrast to the minimum-time swing suppression control (see x and x_p graphs of Fig. 10). For this reason, the tossing controller can take full advantage of the long rope length to save the transferring time since the trolley is not required to travel a full distance as it does in the case of vibration control (see subfigure of x). The above discussions explain the result that the proposed resonance-based tossing control is 6.3 seconds faster than the minimum-time vibration control. To sum up, it is proven that, in the case of bulk material transportation, one could further enhance the transfer productivity by using the tossing control methodology instead of the conventional minimum-time swing suppression control.

In order to clearly illustrate the feasibility of the above-mentioned tossing controller, the images captured from an actual footage of a real-size boom crane system, which utilized a similar tossing control methodology proposed in this paper, are shown in Fig. 11.

Figure 11. Illustration of the actual discharging process: (a) Starting point of the transfer phase: the payload (grab bucket with materials inside) is at rest, (b) One second before the payload arrives at the discharging destination: the grab opens to drop the materials, (c) At the discharging destination: both horizontal and vertical velocities of the grab are zero, and a large terminal swing angle is obtained (end of the transfer phase, the return phase immediately starts), (d) One second after the return phase started: all materials are discharged from the grab, (e) At the end of the return phase: the payload reaches its initial position with no residual vibration.

4. Towards fully automated crane systems

A future crane system must be fully autonomous. In order to realize such a requirement, one need to integrate three main components, namely anti-swing control, anti-skew control, and path planning for obstacle avoidance. The first part (i.e. anti-swing control) was intensively done in the literature. The second part (i.e. anti-skew control) was discussed in Section 2. The final one (i.e. path planning for obstacle avoidance) will be introduced in the subsequent presentation.

One method can be utilized in the obstacle avoidance for crane systems is based on the diffusion process. Assuming that the motion of a real crane can be reduced to a problem of finding a path in known environments, the following physical analogy can be employed for path-planning purposes. A goal point G of the crane's collision-free motion path is considered to be the location of a virtual source like a scent or a perfume. While the concentration at point G is kept constant, the substance diffuses steadily into the surrounding space.

The diffusion process is modeled by an unsteady or dynamic diffusion equation as described by Fick's second law

$$\frac{\partial C}{\partial t} = D\left(\frac{\partial^2 C}{\partial x^2} + \frac{\partial^2 C}{\partial y^2} + \frac{\partial^2 C}{\partial z^2}\right), \tag{25}$$

where D is the diffusion coefficient. By applying the standard finite difference methods, the following time and space discretized model of (25) can be obtained for a grid point or node

$$\begin{aligned} C_{t+1,i,j,k} =& r\left(C_{t,i+1,j,k} + C_{t,i,j+1,k} + C_{t,i,j,k+1} + C_{t,i-1,j,k} + C_{t,i,j-1,k} + C_{t,i,j,k-1}\right) \\ &+ (1-6r)C_{t,i,j,k}. \end{aligned} \tag{26}$$

The boundary conditions of (26) are given by

$$C_{t,i,j,k} = \begin{cases} 1 & \text{(Goal point)} \\ 0 & \text{(Wall, Obstacles)} \end{cases} \tag{27}$$

The initial conditions of (26) are

$$C_{0,i,j,k} = 0 \quad \text{(except for Goal)}, \tag{28}$$

where $r := D\Delta t/(\Delta h)^2$, $i \in x$, $j \in y$, $k \in z$. The concentration at time t and for the space (i, j, k) is $C_{t,i,j,k}$. The sampling time and the grid size are Δt and Δh respectively. It is chosen that $\Delta h = 0.05$ m. Furthermore, the stable region of the coefficient r is $0 < r \leq 1/6$.

The outline of the diffusion-based path planning strategy can be explained as follows. First, a crane's (transfer object's) collision-free path between a start and a mission-dependent goal point is generated by off-line simulation of a diffusion process, and by evaluating the gradient of the computed concentration distribution functions. Second, for both position and swinging-suppression control, command inputs (reference trajectory given at every time-step) for the path obtained in potential fields must be given in order to implement a feedback control system. For the command inputs obtained, a time-varying controller using a fixed-pole approach is applied to obtain the position control and swinging-suppression control.

5. Conclusions

This paper introduced an integral sliding mode controller for the payload's skew orientation process in crane systems to cope with parametric uncertainties. By using the indirect Lyapunov method, algebraic inequality constraints for the ISMC gains are formulated to ensure the robust stability of the closed-loop system under the sliding mode and in a context where the reaching phase is completely eliminated. Another crucial conclusion of the paper emphasizes on the result that there are possibilities to break the time limitation of the minimum-time swing suppression controller—the fastest member of the swing suppression control family—by using the tossing control methodology. In addition, a diffusion-based obstacle avoidance method for crane systems is also introduced to realize a fully autonomous crane in future.

References

[1] Rahman E M A, Nayfeh A H and Masoud Z N 2003 *Modal Analysis* **9** 863–908
[2] Sawodny O, Aschemann H and Lahres S 2002 *Control Engineering Practice* **10** 1323 – 1338 ISSN 0967-0661
[3] Mori Y and Tagawa Y 2018 *Control Engineering Practice* **81** 256 – 263 ISSN 0967-0661
[4] Schlott P, Geise A, Grabmair G, Fritzel T and Sawodny O 2019 *Control Engineering Practice* **85** 149 – 162 ISSN 0967-0661
[5] Galuppini G, Magni L and Raimondo D M 2018 *Control Engineering Practice* **78** 56 – 64 ISSN 0967-0661
[6] Rigatos G G 2015 *Nonlinear Control and Filtering Using Differential Flatness Approaches* (Cham, Switzerland: Springer)
[7] Tuan L A, Lee S G, Hung D V, Moon S C and Kim B S 2013 *Int. J. of Control, Automation and Systems* **11** 718–727 ISSN 2005-4092
[8] Wu X and He X 2016 *ISA Trans.* **65** 361 – 370 ISSN 0019-0578
[9] Almutairi N B and Zribi M 2009 *J. of Vibration and Control* **15** 1679–1730
[10] Hieu N Q and Hong K S 2012 *IEEE/ASME Trans. on Mechatronics* **17** 201–209 ISSN 1083-4435
[11] Tho H D and Terashima K 2019 *IEEE Trans. on Control Systems Technology* **27** 1608–1621 ISSN 1063-6536
[12] Tho H D, Tasaki R, Kaneshige A, Miyoshi T and Terashima K 2017 *2017 IEEE Int. Conf. on Advanced Intelligent Mechatronics (AIM)* pp 998–1003
[13] Fang Y, Ma B, Wang P and Zhang X 2012 *IEEE Trans. on Control Systems Technology* **20** 241–248 ISSN 1063-6536
[14] Yang J H and Shen S H 2011 *J. of Intelligent & Robotic Systems* **62** 59–80 ISSN 1573-0409
[15] Zhang M, Ma X, Rong X, Song R, Tian X and Li Y 2017 *Nonlinear Dynamics* **89** 1779–1791 ISSN 1573-269X

[16] Sun N, Yang T, Fang Y, Wu Y and Chen H 2019 *IEEE Trans. on Systems, Man, and Cybernetics: Systems* **49** 1408–1418
[17] Lu B, Fang Y and Sun N 2018 *Control Engineering Practice* **76** 123 – 132 ISSN 0967-0661
[18] Lee L H, Huang P H, Shih Y C, Chiang T C and Chang C Y 2014 *J. of Vibration and Control* **20** 749–760
[19] Zhao Y and Gao H 2012 *IEEE Trans. on Fuzzy Systems* **20** 181–186 ISSN 1063-6706
[20] Singer N C and Seering W P 1990 *ASME J. of Dynamic Systems, Measurement, and Control* **112** 76–82 ISSN 0022-0434
[21] Singer N, Singhose W and Kriikku E 1997 An input shaping controller enabling cranes to move without sway Tech. rep. United States wSRC-TR–97-0049
[22] Singhose W, Porter L, Kenison M and Kriikku E 2000 *Control Engineering Practice* **8** 1159 – 1165 ISSN 0967-0661
[23] Tho H D, Takemori K, Tasaki R, Miyoshi T and Terashima K 2016 *2016 55th Annu. Conf. of the Soc. of Instrument and Control Engineers of Japan (SICE)* pp 596–602
[24] Sun N, Fang Y, Zhang X and Yuan Y 2012 *IET Control Theory Applications* **6** 1410–1423 ISSN 1751-8644
[25] Hoang N Q, Lee S G, Kim H and Moon S C 2014 *J. of Mechanical Science and Technology* **28** 2879–2888 ISSN 1976-3824
[26] Béarée R 2014 *Control Engineering Practice* **28** 112 – 120 ISSN 0967-0661
[27] Tho H D, Uchiyama N and Terashima K 2019 *2019 IEEE 58th Conf. on Decision and Control (CDC)* pp 7963–7969
[28] Tho H D, Kaneshige A and Terashima K 2020 *Control Engineering Practice* **98** 104390 ISSN 0967-0661
[29] Terashima K, Shen Y and Yano K 2007 *Control Engineering Practice* **15** 1179 – 1192 ISSN 0967-0661
[30] Auernig J and Troger H 1987 *Automatica* **23** 437 – 447 ISSN 0005-1098
[31] Tho H D, Tasaki R, Miyoshi T and Terashima K 2018 *2018 IEEE Conf. on Control Technology and Applications (CCTA)* pp 729–735
[32] Tho H D, Suzuki K, Tsume M, Tasaki R, Miyoshi T and Terashima K 2019 *Control Engineering Practice* **84** 247 – 260 ISSN 0967-0661
[33] Cruz J J D and Leonardi F 2013 *Optimal Control Applications and Methods* **34** 191–201
[34] Zhang X, Fang Y and Sun N 2014 *IEEE Trans. on Industrial Electronics* **61** 6915–6925 ISSN 0278-0046
[35] Slotine J J E and Li W 1991 *Applied Nonlinear Control* (Englewood Cliffs, N.J., USA: Prentice Hall)
[36] Tho H D, Uchiyama N and Terashima K 2020 *IEEE Trans. on Industrial Electronics* 1–1
[37] Yoshida Y and Tabata H 2008 *2008 IEEE/ASME Int. Conf. on Advanced Intelligent Mechatronics* pp 1114–1119 ISSN 2159-6247

Modal identification of cylindrical shell with high damping using reduction method focusing on symmetry of mode shape

Daiki Tajiri[1*], Masaki Ojiro[1], Masami Matsubara[2] and Shozo Kawamura[2]

[1] Graduate School, Toyohashi University of Technology, 1-1 Hibarigaoka, Tempaku Toyohashi Aichi 441-8580, Japan

[2] Department of Mechanical Engineering, Toyohashi University of Technology, 1-1 Hibarigaoka, Tempaku Toyohashi Aichi 441-8580, Japan

*E-mail: d189103@edu.tut.ac.jp

Abstract. The viscoelastic structures such as tires have high damping characteristics, so that resonance peaks do not appear clearly and they are difficult to identify the existence of modes in the high frequency region. Previous studies have proposed methods that make it easier to understand modes. In that method, the number of modes are reduced by using the fact that the basis function of the mode shape of the cylindrical structure can be expressed by a periodic function. However, a large number of measurement points are placed on the structure and data is collected, so the workload of the experiment is large. In this study, we conducted a numerical experiment of the mode reduction method using some measurement point data for the purpose of reducing the workload, and evaluate the influence on the identification accuracy. As a result, it was confirmed that the identification results of the natural frequency and the structural damping coefficient had sufficient accuracy even if some measurement points were used.

1. Introduction

The experimental modal analysis is widely used to understand the vibration characteristics of mechanical structures. This is a method for identifying modal characteristics by curve fitting with the theoretical formula based on the frequency response function (FRF) obtained from vibration tests [1][2]. Recently, highly accurate mode identification is required. This is because the problem of vibration and noise has shifted to the high frequency side including many modes [3], making mode identification difficult. Furthermore, viscoelastic structures such as tires have high damping characteristics, so that resonance peaks do not appear clearly and they are difficult to identify the existence of modes in the high frequency region [4]. Therefore, tire mode identification is generally targeted at the low frequency region [5][6][7].

From such a background, a method of identifying after reducing the number of modes from FRF has been proposed. In the complex mode indicator function (CMIF) method [8], only the one-degree-of-freedom mode component is extracted by performing singular value decomposition on the response obtained from the measurement points installed in the entire machine structure. In this method, the extraction accuracy of the mode component depends on the position and number of measurement points. Therefore, there is a problem that the number of experiment man-hours increases to obtain the necessary identification accuracy. There is also a method (mode expansion reduction method) that reduces the number of modes by focusing on the mode shape of the mechanical structure [9]. This method reduces the FRF by utilizing the fact that the mode shape can be expressed by a periodic function in the circumferential direction when a cylindrical shell such as a tire is used. In previous studies, it has been reported that mode identification in high damping systems and high frequency regions is possible by

applying the reduction method and the conventional curve fitting method. On the other hand, the experimental load is not reduced because a large number of measurement points are placed on the entire cylindrical shell. In addition, although the mode identification results have been reported, the accuracy of the reduction method has not been fully discussed.
Therefore, in this study, we conduct a numerical experiment of the mode reduction method by using some measurement points data for the purpose of reducing the workload, and evaluate the influence on the identification accuracy. About the target structure, we set the cylindrical shell with high damping and used a periodic function as the basis function. Here, we define a high damping system as a system with a structural damping coefficient of 5.0×10^{-2} or higher. Beside, in order to improve the identification accuracy, the linear fit method [10][11] is used for modal identification.

2. Circumferential mode reduction method for cylindrical structures

In the previous study [9], the response points were set on multiple virtual rings that arranged in the whole cylindrical shell as shown in figure 1. The response of each ring could be expressed using a periodic function. In this study, only the response of the measurement points set in the range of 0 to 180 ° of the virtual ring is used for the purpose of reducing the workload of the experiment. The reason is that the case of a cylindrical shell, the positions of the antinodes and nodes of the mode shape excited by the position of the input point can be easily grasped. Here, the basic theory of circumferential mode reduction when using the response of the whole measurement point is explained.

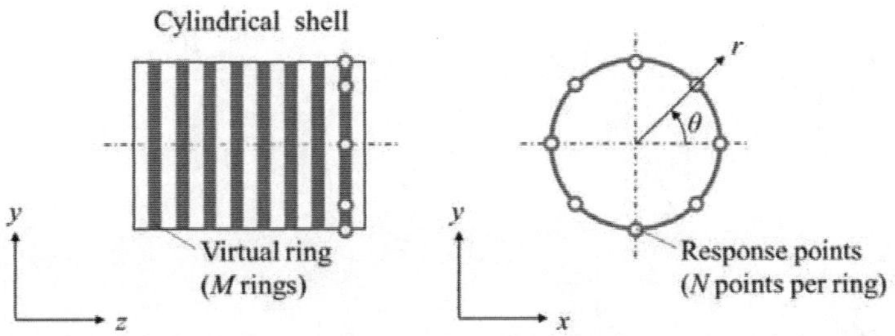

Figure 1. Input points and response points set on the virtual ring

The excitation points P are arbitrarily arranged on the cylindrical shell. The response points are arranged with M virtual rings at equal intervals in the axial direction and N points at equal intervals ($2\pi/N$) in the circumferential direction of each ring. Therefore, the total of the response points are $M \times N$ points. The FRF of all response points are represented by the following matrix $\mathbf{H}(\omega)$. ω is an angular frequency.

$$\mathbf{H}(\omega)=\begin{bmatrix}\mathbf{H}_1(\omega)\\ \mathbf{H}_2(\omega)\\ \\ \mathbf{H}_M(\omega)\end{bmatrix} \quad (1)$$

$$\mathbf{H}_{m,p}(\omega)=\begin{bmatrix}H_1(\omega)\\ \vdots\\ H_N(\omega)\end{bmatrix} \quad \begin{cases} m=1,\cdots,M \\ p=1,\cdots,P\end{cases} \quad (2)$$

$\mathbf{H}(\omega)$ is $M\times N\times P$ matrix in equation (1). $\mathbf{H}_{m,p}(\omega)$ is N matrix between the p-th input point and the response point on the m-th virtual ring.
Here, the bending vibration of the virtual ring is expressed as a basis function using a periodic function. The bending vibration of the cylindrical shell is excited in the radial direction. Specifically, it is excited

every wave numbers in the circumferential direction and every half wave numbers in the axial direction. Therefore, the wave number i in the circumferential direction is defined as the circumferential i-th mode, and the half wave number j in the axial direction is defined as the j-th mode in the axial direction. Figure 2 shows an example of the circumferential 0th to 4th modes.

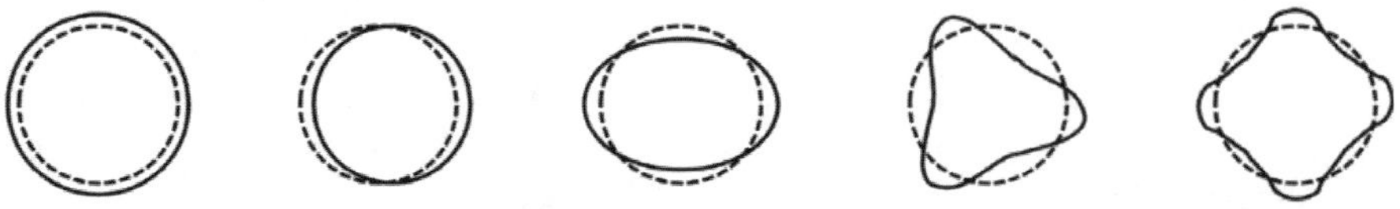

Figure 2. Relationship between circumferential mode shape and circumferential mode order

Now, $H_n(\omega)$, which is the n-th FRF in the circumferential direction in equation (2), is expanded in the Fourier series in the circumferential direction. The FRF $H_n(\omega)$ between the n-th response point and the p-th input point on the m-th virtual ring, is rewritten as follows by the Fourier series expansion of $H_n(\omega)$ in equation (2). $A_i(\omega)$ and $B_i(\omega)$ mean the amplitude representing the circumferential mode shape of the m-th virtual ring.

$$H_n(\omega)=\frac{A_0(\omega)}{2}+\sum_{i=1}^{N/2-1}\left[A_i(\omega)\cos\left(i\frac{2\pi n}{N}\right)+B_i(\omega)\sin\left(i\frac{2\pi n}{N}\right)\right]+\frac{A_{N/2}(\omega)}{2}\cos(\pi n) \quad (n=1,\cdots,N) \tag{3}$$

$$A_i(\omega)=\frac{2}{N}\sum_{n=1}^{N}H_n(\omega)\cos\left(i\frac{2\pi n}{N}\right), \quad B_i(\omega)=\frac{2}{N}\sum_{n=1}^{N}H_n(\omega)\sin\left(i\frac{2\pi n}{N}\right) \quad (i=0,\cdots,N/2) \tag{4}$$

equations (3) and (4) are the sums of the circumferential direction 0 to $N/2$-th mode components. So the circumferential i-th mode component can be extracted as following equation.

$$H_n^{(i)}(\omega)=A_i(\omega)\cos\left(i\frac{2\pi n}{N}\right)+B_i(\omega)\sin\left(i\frac{2\pi n}{N}\right) \tag{5}$$

From the above, $\mathbf{H}_{m,p}(\omega)$, which have N degrees of freedom, is reduced to a matrix with two degrees of freedom for each circumferential mode order as shown in the following equation.

$$\mathbf{H}_{m,p}^{(i)}(\omega)=\begin{bmatrix}A_i(\omega)\\B_i(\omega)\end{bmatrix} \tag{6}$$

Therefore, the FRF matrix $\mathbf{H}(\omega)$ that had $M\times N\times P$ degrees of freedom is reduced to an FRF matrix with $2\times P\times M$ degrees of freedom for each circumferential mode order. At this time, only the circumferential i-th mode component is extracted, so that modes with different axial orders exist as eigenmodes, and the mode density is greatly reduced.

IOP Conf. Series: Materials Science and Engineering **920** (2020) 012024 doi:10.1088/1757-899X/920/1/012024

3. Modal identification using finite element model of cylindrical shell

3.1. Cylindrical shell model and analysis conditions

Figure 3 shows the cylindrical shell model created by commercial software MSC.Patran. Table 1 shows the design parameters. FRF is calculated by a hysteretic damping system with a structural damping coefficient set to 5.0×10^{-2}. The cylindrical shell is simply supported at both ends. The input point is the point of $\theta=0$ in the second ring (marked ○ in figure 3), and the input direction is perpendicular to the cylindrical shell surface. The response points are 80 points (8 virtual rings × 10 points on the ring) distributed at equal intervals in the range of $0\leqq\theta\leqq\pi$ on the cylindrical shell. This method is effective for measuring the response of cylindrical structures. The reasons are that the positions of the antinodes and nodes of the excited mode shape are determined by the positions of the input points, and the positions can be easily grasped.

Table 1. The design parameters

Young's modulus	200 [GPa]
Poisson's ratio	0.3 [-]
Radius	0.5 [m]
Length	1.26 [m]
Thickness	0.01 [m]

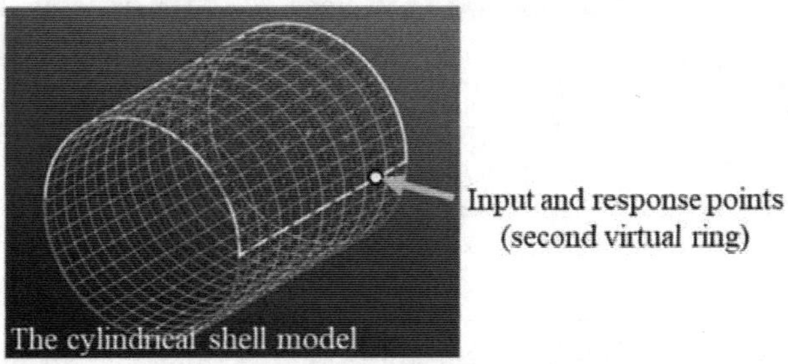

Figure 3. Input and response points on virtual ring in the cylindrical shell model

3.2. Effects of circumferential mode reduction and results of modal identification

As an example of numerical analysis, using the driving point FRF in the cylindrical shell model, the circumferential 3rd mode and 5th mode components are extracted based on the circumferential mode reduction method. Figure 4 shows the effect of circumferential mode reduction. Figure 4 (a) shows the original FRF (Original; solid black line) and all mode components calculated in the cylindrical shell model. Figure 4 (b) shows the original FRF and the reduced circumferential mode components of all orders. It can be seen that modes that could not be clearly identified from original FRF were separated by circumferential mode reduction. This effect makes it relatively easy to identify the mode characteristics.

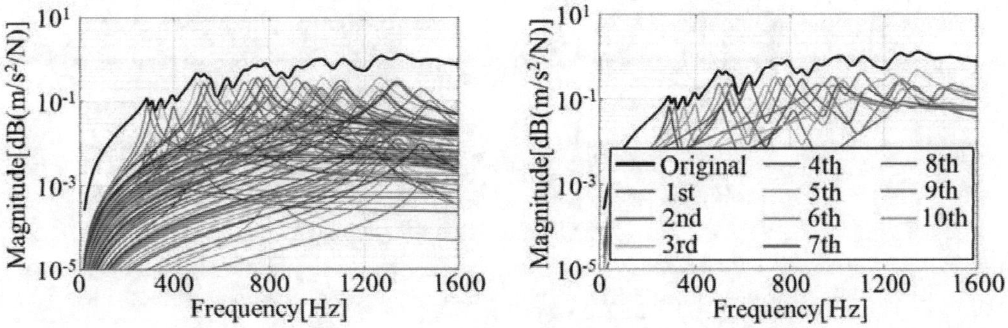

(a) The original FRF and all mode components (b) The original FRF and mode components obtained by circumferential mode reduction

Figure 4. The effect of circumferential mode reduction

Subsequently, the modal characteristics are identified by using the FRF of the extracted circumferential 3rd. mode component and 5th. mode component shown figure 5 (a)(b) (Extracted FRF; blue solid line). The FRFs in figure 5 (a)(b) has five resonant peaks. These are the axial 1st to 5th mode from the low frequency side. An example of a mode shape representing the coupling in the circumferential direction and the axial direction is also shown in figure5.

In this study, the modal characteristics are identified using the linear fit method [10][11], which is one of the curve fitting methods. This method uses simultaneous equations of the real and imaginary parts of the FRF. In these simultaneous equations, the natural frequency and damping characteristic are unknown quantities, and they are obtained using the least squares method (LSM).

Table 2 shows the identified natural frequency f_{np}, structural damping coefficient η_p, and their relative errors with respect to their exact values. The identification results of natural frequency and structural damping coefficient are close to exact values, and it can be seen that both can be identified within a very small error range. In addition, it was confirmed that there was no difference in identification accuracy when using the response of the entire cylindrical shell ($0 \leqq \theta \leqq 2\pi$). Similarly, the mode characteristics can be identified with high accuracy for other circumferential modes. Figures 5(a)(b) shows the reconstructed FRF (Reconstructed FRF; red broken line) using the identified mode characteristics. Since all the modal characteristics were identified with high accuracy, the blue solid line and the red broken line coincide.

Table 2. The exact values and identification values of modal characteristics

Axial order	Exact value		Identification value		Relative error [%]	
	f_{np} [Hz]	η_p [-]	f_{np} [Hz]	η_p [-]	f_{np} [Hz]	η_p [-]
(a) Circumferential 3rd mode						
1	350.45	5.00×10^{-2}	350.46	5.00×10^{-2}	2.59×10^{-6}	4.30×10^{-6}
2	696.44	5.00×10^{-2}	696.44	5.00×10^{-2}	4.24×10^{-7}	2.07×10^{-5}
3	1023.92	5.00×10^{-2}	1023.92	5.00×10^{-2}	4.35×10^{-6}	3.77×10^{-5}
4	1268.98	5.00×10^{-2}	1268.98	5.00×10^{-2}	2.88×10^{-6}	2.91×10^{-5}
5	1453.26	5.00×10^{-2}	1453.26	5.00×10^{-2}	2.98×10^{-6}	3.01×10^{-5}
(b) Circumferential 5th mode						
1	311.60	5.00×10^{-2}	311.60	5.00×10^{-2}	9.00×10^{-7}	1.98×10^{-5}
2	500.33	5.00×10^{-2}	500.33	5.00×10^{-2}	2.54×10^{-6}	3.69×10^{-5}
3	752.76	5.00×10^{-2}	752.76	5.00×10^{-2}	5.65×10^{-6}	2.78×10^{-6}
4	1017.51	5.00×10^{-2}	1017.52	5.00×10^{-2}	4.09×10^{-6}	3.24×10^{-6}
5	1271.93	5.00×10^{-2}	1271.94	5.00×10^{-2}	2.49×10^{-6}	3.63×10^{-5}

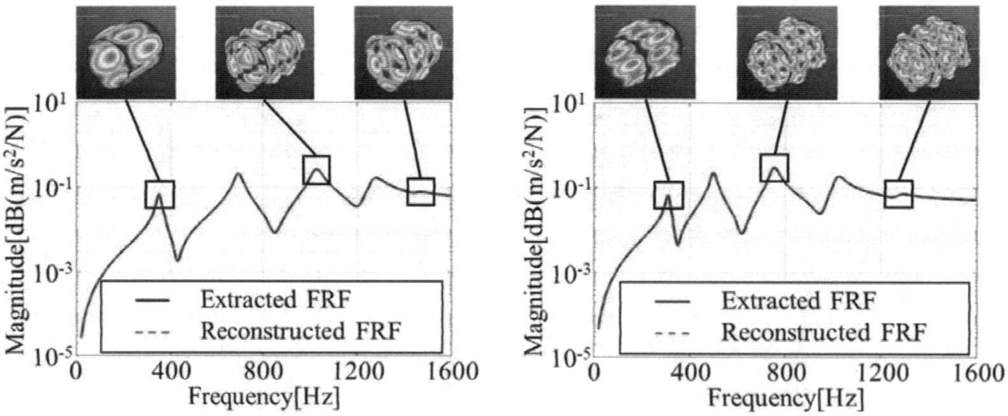

(a) The circumferential 3rd mode (b) The circumferential 5th mode

Figure 5. Comparison of the original and the reconstructed FRF

4. Conclusion

In this study, we conducted a numerical experiment of the mode reduction method using some measurement point data for the purpose of reducing the workload. Then, the influence on the identification accuracy was evaluated. The target structure was a cylindrical shell with high damping, and a periodic function was used as the basis function.

Although the circumferential 3rd mode and 5th mode were described as examples, the accuracy of the identification results of the natural frequency and the structural damping coefficient was very high in the circumferential modes of all orders.

From the above, it was understood that the circumferential mode reduction method had sufficient accuracy and could reduce the experimental work load even when only the data of some measurement points were used.

5. Future works

In the future, the influence on the identification accuracy will be grasped by numerical analysis considering measurement noise. Then, we plan to conduct experimental verification on cylindrical structures such as automobile tires.

6. References

[1] Ewins D J 2000 *Modal testing: theory, practice, and application* (Hertfordshire: Research Studies Press Ltd)

[2] Jimin H and Zhi-Fang F 2001 *Modal Analysis* (Milton Keynes: Butterworth Heinemann)

[3] Waki Y, Kozuki K, and Heguri H 2011 Estimation of noise radiating parts of a tire using the wave finite element method. *Proc. of the Internoise 2011* (1) 577-584

[4] Vercammen S, Diaz C G, Kindt P, Thiry C, Middelberg J and Leyssens J 2012 Dynamic behavior of rolling tires under different operating conditions *Proc. of the Internoise 2012 / ASME NCAD meeting* 301-307

[5] Y T Wei L Nasdala and H Rothert 2009 Analysis of forced transient response for rotating tires using REF models *J. of Sound and Vibration* **320** 145-162

[6] Matsubara M, Tsujiuchi N, Koizumi T, Ito A. and Bito K 2015 Natural frequency analysis of tire vibration using a thin cylindrical shell model *SAE Technical Paper* 2015-01-2198 DOI: https://doi.org/10.4271/2015-01-2198.

[7] Matsubara M and Kawamura S 2019 Parameter identification of a three-dimensional flexible ring-based model of a tire using experimental modal analysis *Int. J. of Automotive Engineering* **10**(2) 133-138 DOI: https://doi.org/10.20485/jsaeijae.10.2_133

[8] Shih C Y, Tsuei Y G, Allemang R J and Brown D L 1989 Complex mode indication function and its applications to spatial domain parameter estimation *Proc. of the 7th Int. Modal Analysis Conf. (IMAC VII)* 533-540

[9] Kitahara A and Yoshimura T 2015 Modal identification of cylindrical shell using circumference reduction method. *Transactions of the JSME* **81**(822) DOI: 10.1299/transjsme.14-00461 (in Japanese)

[10] Kawamura S, Kita M, Matsubara M and Ise T 2016 Study of the effect of specimen size and frequency on the structural damping property of beam *Mechanical Engineering J.* **3**(6) DOI: 10.1299/mej.16-00446

[11] Matsubara M, Ise T and Kawamura S 2018 Application of modal properties identification to multi-degree-of-freedom system using simultaneous equations of the real and imaginary parts of frequency response function *Transactions of the JSME* **84**(860) DOI: 10.1299/transjsme.17-00540 (in Japanese)

IOP Conf. Series: Materials Science and Engineering **920** (2020) 012025 doi:10.1088/1757-899X/920/1/012025

The Impact of Vortex Formation Due to The Operational Dam Condition: A Review

Aqil Azman[1], N.M. Zahari[2], M.H. Zawawi[2], M.H. Mansor[3], F.C. Ng[1], Aizat Abas[1*], F. Nurhikmah[2], and Nurhanani A. Aziz[2]

[1]School of Mechanical Engineering, Universiti Sains Malaysia, Engineering Campus, 14300 Nibong Tebal, Pulau Pinang, Malaysia.
[2]Department of Civil Engineering, Universiti Tenaga Nasional (UNITEN), Kajang 43000, Selangor Darul Ehsan, Malaysia.
[3]Department of Electrical Power Engineering, Universiti Tenaga Nasional (UNITEN), Kajang 43000, Selangor Darul Ehsan, Malaysia.

[*]Email: aizatabas@usm.my

Abstract. There are various sources of energy throughout the world, renewable as well as non-renewable through which renewable energy sources are considered more environmentally friendly. Encompassed by all the sources of renewable energy, hydropower is considered the most beneficial source of energy. Proper operation of hydropower plant is very important for generating maximum energy by utilizing the available. However, formation of vortices at power intake can cause number of problems. The stronger vortices have more negative effects on the performance of hydropower plant which can also, draw debris and air into an intake causing vibration and damage to turbines. The present study addresses the vortices formation at the intake dam, several types of vortices and anti-vortex applications in order to overcome the vortex formations.

1. Introduction

Energy is very important for a developing country and the common energy used is electrical energy. There are various ways to harvest electricity energy by constructing power plant such as nuclear power plant, charcoal power plant, solar harvester, and hydropower. The role of hydropower has been changed from constant energy supplier to the highly flexible one to maintain the grid stability which leads to operate the hydropower under transient operations [1, 2]. Dam intakes divided into two types which are horizontal intakes and vertical intakes, that can be observed in Figure 1 while Table 1 summarize the study of vortex formation for two types of dam intake which horizontal and vertical.

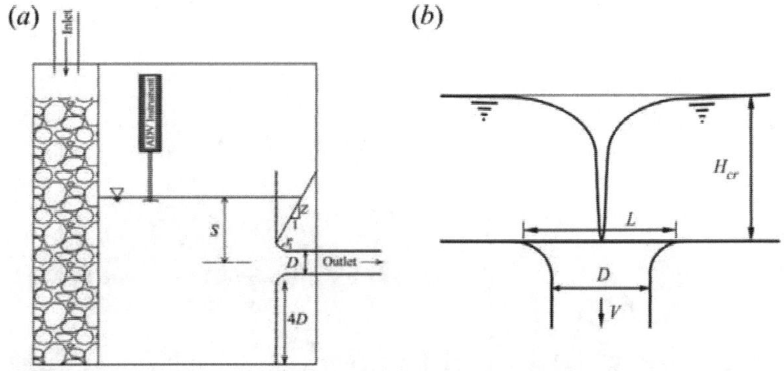

Figure 1. Types of intake: (*a*) horizontal intake and (*b*) vertical intake.

Table 1. types of dam intake for vortex formation study

Author	Types of intake study	Year
Padmanabhan and Hecker. [3]	Horizontal intake	1984
Cheng and Ying. [4]	Horizontal intake	2007
Roshan et al. [5]	Horizontal intake	2009
Sarkardeh et al. [6]	Horizontal intake	2010
Kerem and Yildirim. [7]	Vertical intake	2010
Trivellato. [8]	Vertical intake	2010
Amiri et al. [9]	Horizontal intake	2011
Wang et al. [10]	Horizontal intake	2011
Yang et al. [11]	Vertical intake	2014
Keller et al. [12]	Horizontal intake	2014
Taştan and Yildirim. [13]	Vertical intake	2014
Azarpira et al. [14]	Horizontal intake	2014
Sarkardeh et al. [15,16]	Horizontal intake	2014
Suerich-Gulick et al. [17–19]	Horizontal intake	2014
Sarkardeh. [20, 21]	Horizontal intake	2017

Usually, horizontal intake was more preferred as compared to vertical intakes due to construction cost and fabrication materials [9]. In this review, the impact of vortex in the literature are reviewed with a focus on their formation in dam intakes and the anti-vortex applications in order to reduce the vortex formation.

2. The impact of vortex formation

In general, intake vortices can be recognized as a prevailing phenomenon which must be considered in hydraulic engineering applications and the formation of vortices leads to flow unsteadiness and non-uniformity at hydro machines [8, 9, 22]. Vortex formation at power intakes may occur at lower reservoir levels which can draw debris and air into an intake causing vibration and damage to turbines [6, 23–25]. Eventually, Yang et al., (2018) mentioned that the vortex strength increased accordingly, and the air-core diameter was around 3 – 4 m [26]. Instead, Zhang et al., (2018) stated that vortex is a typical flow structure in the hydro turbines [27], leading to various kinds of instabilities such as large pressure fluctuation [28, 29], significant noise, prominent vibrations, rotating stall [30], cavitation erosion [31], and material fatigue.

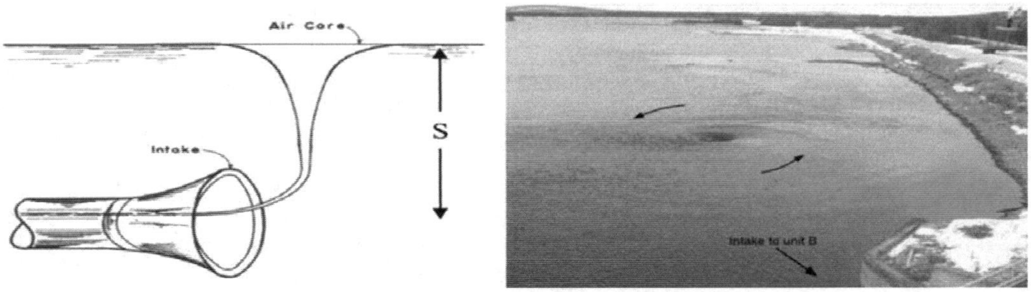

Figure 2. Occurrence of intake vortex [5, 26].

On the whole, Walker, (2016) discussed that there are no information was found on the loss of power generation with vortex suppressors, however many documents acknowledged that the design of a trash rack can create head loss which could reduce power generation of a turbine [32]. In other book section by Ratnayaka et al., (2009) mentions that, radial vanes are generally incorporated in the top of a bell mouth shaft to minimize the occurrence of vortex action, which can otherwise reduce the discharge capacity and cause problematic flow conditions in the shaft and tunnel [33].

In the past, study by Padmanabhan and Hecker, (1984) using horizontal intake approached to study the vortex formation with three types of condition: one pipe operation, two pipe operation and screen blockage test [3]. Based on findings, considerable swirl in the pipe flow was induced, and the swirl meter indicated higher swirl angles at lower submergences when the subsurface vortices were stronger [3]. In the book by Paul, (1998) stated that, for intakes feeding directly to a power tunnel, the exclusion of air is an important design consideration and the intake may require submergence to prevent vortex formation and air entrainment [34]. In contrast, findings by Azarpira et al., (2014) found that, velocity measurements in the reservoir showed that increasing the submerged depth from 1.6 to 3.2 has effect on reducing plane velocities, whereby the maximum plane velocity decreased about 30% [14].

Meanwhile, the study by Khanarmuei et al., (2018) encountered the strength of vortices that form in single and dual pipe intakes decreases by changing the intake direction from vertical to 45° and horizontal [35]. Based on their findings, the circulation number of vortices that formed at horizontal single and dual pipe intakes as compared to vertical single and dual pipe intakes was decreased on average 31% and 35%, respectively. In addition, the critical submergence was considerably affected by changing the intake direction at single and dual pipe intakes, and by changing the angle of pipe intakes from vertical to horizontal, the critical submergence at single and dual pipe intakes decreases 16% to 30% and 19% to 40%, respectively [35].

3. Vortex classifications

Alden Research Laboratory [36] classified vortices into 6 types of classification as shown in Figure 2(*a*) as compared to Sarkadesh et al., (2010) which classified the vortex formation into three class vortex which vortex class A, B and C in Figure 2(*b*). The vortices of class C are considered to be safe while Class B surface vortices are stronger and the rotation in the water surface extends down to the intake [6]. Lastly, the strongest vortices which should be avoided are categorised in class A which is in this class, air bubbles are trapped and conveyed down from the water surface to the intake.

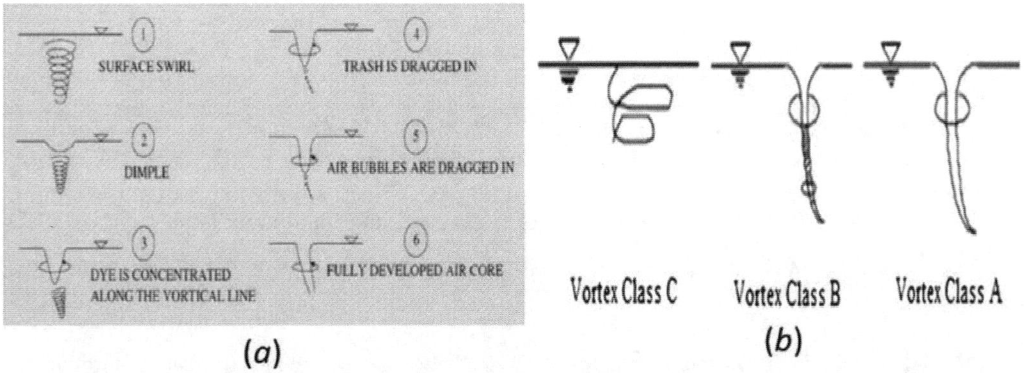

Figure 3. Vortex classification: (*a*) Alden Research Laboratory [36] and (*b*) Sarkadesh et al. (2010) [6].

4. Anti-vortex applications

Roshan et al., (2009) found that the vortices can be reduced by introduced anti-vortex walls (Figure 4) whereby anti-vortex wall was then recommended as a cheap method to prevent vortex formation in the prototype [5]. Based on their results of experiments, the type of vortices reduced from type 4 to type 2 and 1. Similarly, Sarkardeh et al., (2010) also proposed an anti-vortex which trash rack to reduce the vortices formation [6]. However, since water level in a reservoir should not be reduced to levels below the critical submergence, the volume of water in this zone cannot be used for power generation [9]. Thus, increasing the submergence of the intake to prevent the formation of air-entraining vortices may not always be an economical solution and be cost prohibitive.

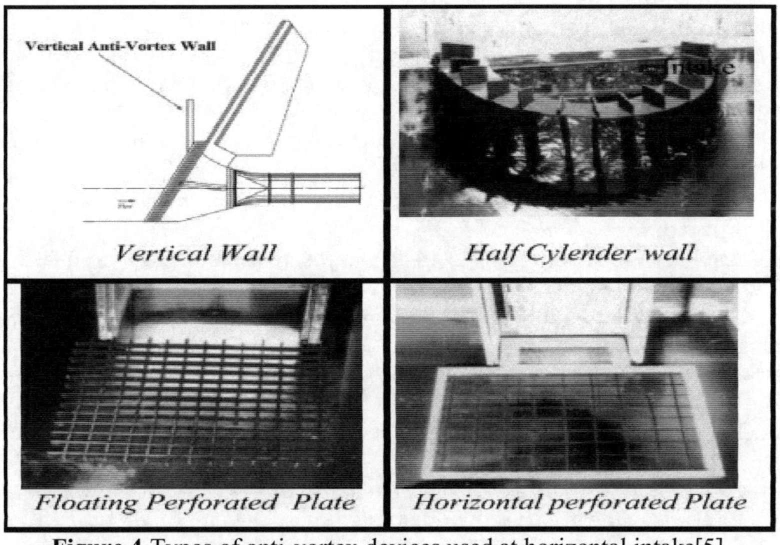

Figure 4. Types of anti-vortex devices used at horizontal intake[5].

5. Conclusions

Hydropower plant is the most energy harvester used throughout world due to higher energy output as well as renewable energy compared to other types of power plant. Consequently, formation of vortex at dam intake reduce the productivity of the collected electrical energy. The present study has been undertaken to review the earlier research work carried out in the area of dam intakes which the occurrence of vortex formation. It has been found that the water depth, *S* affect the vortex formation at dam intakes and increasing the submergence of the intake will be cost prohibitive which not always be an economical solution. In the current status, the impact of vortex reduces the turbine capabilities in generating electricity due to the air-entrain which lower the mass flow and increase void inside the penstock. Based on the review, anti-vortex can reduce the vortex formation but there are limited studies of the design for the anti-vortex. Thus, more studies for the design of anti-vortex are needed to optimally reduce the vortex formation.

Acknowledgments
This research is partly funded by Tenaga Nasional Berhad seeding fund under grant No. U-TG-RD-19-14.

References

[1] Goyal R, Gandhi B K and Cervantes M J 2018 PIV measurements in Francis turbine – A review and application to transient operations *Renew. Sustain. Energy Rev.* **81** 2976–91
[2] Singh V K and Singal S K 2017 Operation of hydro power plants-a review *Renew. Sustain. Energy Rev.* **69** 610–9
[3] Padmanabhan M and Hecker G E 1984 Scale effects in pump sump models *J. Hydraul. Eng.* **110** 1540–56
[4] Cheng Y and Ying B 2007 Numerical Simulation and Comparison of Water Intake-Outlet Methods in Power Plants *J. Hydrodyn.* **19** 623–9
[5] Roshan R, Sarkardeh H and Zarrati A R 2009 Vortex study on a hydraulic model of Godar-e-Landar Dam and hydropower plant *WIT Trans. Eng. Sci.* **63** 217–25
[6] Sarkardeh H, Zarrati A R and Roshan R 2010 Effect of intake head wall and trash rack on vortices *J. Hydraul. Res.* **48** 108–12
[7] Kerem T and Yildirim N 2010 Effects of dimensionless parameters on air-entraining vortices *J. Hydraul. Res.* **48** 57–64
[8] Trivellato F 2010 Anti-vortex devices: Laser measurements of the flow and functioning *Opt. Lasers Eng.* **48** 589–99
[9] Amiri S M, Zarrati A R, Roshan R and Sarkardeh H 2011 Surface vortex prevention at power intakes by horizontal plates *Proc. Inst. Civ. Eng. Water Manag.* **164** 193–200
[10] Wang Y, Jiang C and Liang D 2011 Comparison between empirical formulae of intake vortices *J. Hydraul. Res.* **49** 113–6
[11] Yang J, Liu T, Bottacin-Busolin A and Lin C 2014 Effects of intake-entrance profiles on free-surface vortices *J. Hydraul. Res.* **52** 523–31
[12] Keller J, Möller G and Boes R M 2014 PIV measurements of air-core intake vortices *Flow Meas. Instrum.* **40** 74–81
[13] Taştan K and Yildirim N 2014 Effects of Froude, Reynolds, and Weber numbers on an air-entraining vortex *J. Hydraul. Res.* **52** 421–5
[14] Azarpira M, Sarkardeh H, Tavakkol S, Roshan R and Bakhshi H 2014 Vortices in dam reservoir: A case study of Karun III dam *Sadhana - Acad. Proc. Eng. Sci.* **39** 1201–9
[15] Sarkardeh H, Jabbari E, Zarrati A R and Tavakkol S 2014 Velocity field in a reservoir in the presence of an air-core vortex *Proc. Inst. Civ. Eng. Water Manag.* **167** 356–64
[16] Sarkardeh H, Zarrati A R, Jabbari E and Marosi M 2014 Numerical simulation and analysis of flow in a reservoir in the presence of vortex *Eng. Appl. Comput. Fluid Mech.* **8** 598–608
[17] Suerich-Gulick F, Gaskin S J, Villeneuve M and Parkinson É 2014 Free surface intake vortices: Theoretical model and measurements *J. Hydraul. Res.* **52** 502–12
[18] Suerich-Gulick F, Gaskin S J, Villeneuve M and Parkinson É 2014 Free surface intake vortices: Scale effects due to surface tension and viscosity *J. Hydraul. Res.* **52** 513–22
[19] Suerich-Gulick F, Gaskin S J, Villeneuve M and Parkinson É 2014 Characteristics of free surface vortices at low-head hydropower intakes *J. Hydraul. Eng.* **140** 291–9
[20] Sarkardeh H 2017 Minimum Reservoir Water Level in Hydropower Dams *Chinese J. Mech. Eng. (English Ed.* **30** 1017–24
[21] Sarkardeh H 2017 Numerical calculation of air entrainment rates due to intake vortices

Meccanica **52** 3629–43
[22] Monshizadeh M, Tahershamsi A, Rahimzadeh H and Sarkardeh H 2018 Vortex Dissipation Using a Hydraulic-Based Anti-Vortex Device at Intakes *Int. J. Civ. Eng.* **16** 1137–44
[23] Rabe B K, Najafabadi S H G and Sarkardeh H 2018 Numerical simulation of anti-vortex devices at water intakes *Proc. Inst. Civ. Eng. Water Manag.* **171** 18–29
[24] Ofosu E A, Amo-Boateng M, Domfeh M K and Andoh R 2018 *Re-engineering Hydropower Plant for Improved Performance* (Elsevier Inc.)
[25] Naghian M, Lashkarbolok M and Jabbari E 2017 Numerical Simulation of Turbulent Flows Using a Least Squares Based Meshless Method *Int. J. Civ. Eng.* **15** 77–87
[26] Yang J, Andreasson P, Högström C M and Teng P 2018 The tale of an intake vortex and its mitigation countermeasure: A case study from akkats hydropower station *Water (Switzerland)* **10** 1–14
[27] Zhang Y, Liu K, Xian H and Du X 2018 A review of methods for vortex identification in hydroturbines *Renew. Sustain. Energy Rev.* **81** 1269–85
[28] Chen T, Zhang Y and Li S 2016 Instability of large-scale prototype Francis turbines of Three Gorges power station at part load *Proc. Inst. Mech. Eng. Part A J. Power Energy* **230** 619–32
[29] Zhang T, Meng Q, Chen C, Zheng W, Liu W, Yu Q and Tong L 2017 A model-free CAF fringe search algorithm with wavelet boosting for VLBI observation *Publ. Astron. Soc. Pacific* **129** 74501
[30] Zhang Y and Wu Y 2017 A review of rotating stall in reversible pump turbine *Proc. Inst. Mech. Eng. Part C J. Mech. Eng. Sci.* **231** 1181–204
[31] Zhang Y, Zhang Y, Qian Z, Ji B and Wu Y 2016 A review of microscopic interactions between cavitation bubbles and particles in silt-laden flow *Renew. Sustain. Energy Rev.* **56** 303–18
[32] Walker K 2016 *Intake vortex formation and suppression at hydropower facilities*
[33] Ratnayaka D D, Brandt M J and Johnson K M 2009 Dams, Reservoirs and River Intakes *Water Supply* pp 149–93
[34] Paul J H 1998 *Extending the life of concrete repairs* vol 20
[35] Khanarmuei M, Rahimzadeh H and Sarkardeh H 2018 Effect of dual intake direction on critical submergence and vortex strength *J. Hydraul. Res.* **57** 272–9
[36] Knauss J 1987 Swirling flow problems at intakes *IAHR Hydraul. Struct. Des. Man.* **1** 57 – 76

Cubic Spline-Based Genetic Algorithm Optimization for Millimeter and Microwave Integrated Circuits

N H Mansor and R Abdul-Rahman
School of Mechanical Engineering, Universiti Sains Malaysia, 14300 Nibong Tebal, Pulau Pinang, Malaysia.

*Corresponding author: menurhidayah@usm.my

Abstract. A clamped cubic spline profile taper of an air-filled substrate integrated waveguide (SIW) transition was optimized to search for the optimal design of taper in this paper. The optimization technique involved a combination of the multi-objective genetic algorithm (MOGA) and full-wave analysis of the waveguide. The data from the return and transmission losses provides a basis for the optimization of the transition tapers design. The simulation results show the efficacy of our proposed strategy, where the optimal taper geometry yields an improvement of the return loss while maintaining its original transmission loss at Ka-band (26–40 GHz) frequencies.

Keywords: Millimeter-wave, air-filled substrate integrated waveguide (SIW), substrate integrated waveguide (SIW), clamped cubic spline method, genetic algorithm optimization, multi-objective genetic algorithm.

1. Introduction

The genetic algorithm (GA) optimization method is widespread in engineering electromagnetics design [1–4]. The electromagnetic optimization problems generally involve several parameters, thus making the design process inefficient and complicated [5]. Meanwhile, spline would be useful for a design process because of the simplicity of their construction, the accuracy of evaluation, and their capacity to approximate complex shapes [6]. Previous studies using the spline optimization approach include optimization of a spline-profile horn with the quasi-Newton method and optimization of microstrip antennas using cubic splines with particle swarm optimization [7, 8]. This paper will develop an optimization procedure based on the combination of GA with the clamped cubic spline. The developed procedure used an air-filled SIW transition design as a case study.

Recently, there has been increasing interest in the manufacturing of millimetre and microwave devices components based on substrate integrated waveguide (SIW) due to the demand for compact, light, flexible, low-cost, and high-performance integrated circuits. Until today, various components based on the SIW technology have been proposed and applied [10, 11]. One of the significant issues with the SIW circuit application in the high-quality microwave millimetre wave component is its losses. Dielectric loss is the primary source of loss in SIW and is significantly larger than the ohmic and radiation losses [9]. Therefore, many efforts had taken to improve the performance of SIW circuits [12–17].

The modified SIW with the dielectric-to air-filled transition taper has been discovered by [16]. The effects of the transition taper geometry on losses were reported in [18]. In parallel, an optimization of the transition taper design of modified SIW also was reported in [15–17]. Therefore, the specific geometry of the transition taper is required to go through to find an optimal design of an air-filled SIW transition taper. Hence, a cubic clamped spline-based genetic algorithm optimization method is applied to re-design the transition taper geometry to minimize the SIW losses in this study.

2. Dielectric-to-air filled SIW transition

The design of the taper for the transition using the characteristic equation is useful in [16]. Based on the full-wave analysis results in [18], the raised cosine taper, as depicted in Figure 1, has an influence on the transition losses compared to others.

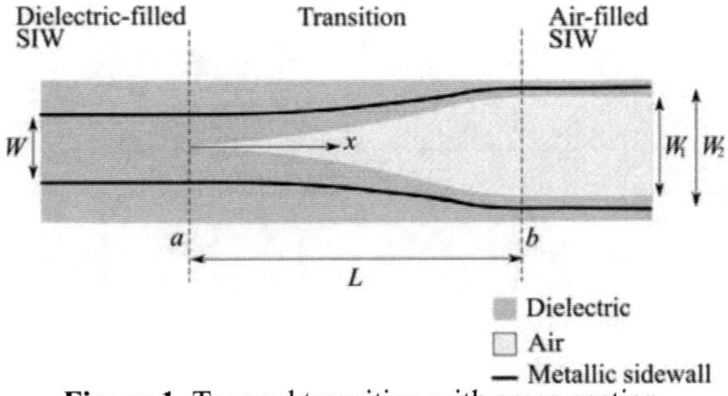

Figure 1. Tapered transition with cross-section.

By referring to the taper designs in Figure 1, the clamped cubic spline function is applied to define the shape of the transition taper in order to minimize the transition losses further. The W_1 is the total width in the air-filled region, W is the width of the dielectric-filled SIW and W_2 is the width of the air-filled SIW. The width of the taper W_2 increases along the transition length L with the initial width a and the final width b of the transition taper are constants to ensure the transition continues as presented in Figure 1. All the structural dimensions of the transitions are described in Table 1, with L being 20 mm, and height h are 0.508 mm. For our computational experiment, the five fixed number of nodal points are used to form a set of splines that define the taper geometry, as presented in Figure 2. As stated in [16], a transition length of 20 mm is commonly accepted to provide acceptable losses. The dielectric-to air-filled SIW transition is designed for the Roger RT/Duroid 6002 substrate ε_r in the Ka-band frequencies.

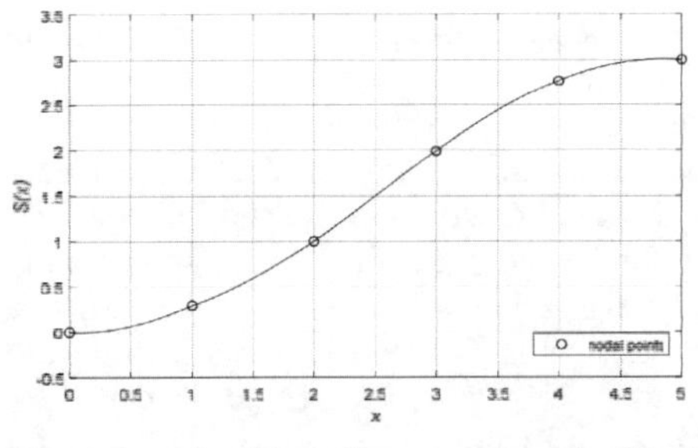

Figure 2. Taper profile with clamped cubic splines: x= length, $S(x)$= the clamped cubic spline function

Table 1. Structural parameters.

Band Freq.	Dimension Properties		
	Parameters	*Air-filled SIW*	*Dielectric-filled SIW*
Ka-band	W_1	7.05 mm	–
	W_2	6.034mm	–
	W	–	4.11mm
	ε_r	2.94	2.94
	h	0.508mm	0.508mm
	W	0.508mm	0.508mm

3. Optimization Procedures

3.1 Taper Design

The first step is to form a spline profile of the taper. The spline profile of transition taper is constructed to have zero slopes at the endpoints. By constructing the equation of the taper with the clamped cubic splines as shown in Figure 2, the internal nodes provide four degrees of freedom $\{\boldsymbol{x_1}, \boldsymbol{x_2}, \boldsymbol{x_3}, \boldsymbol{x_4}\}$ as the design variables, later, these nodes can serve as a finite number of unknowns to be optimized to find an optimal design of taper. The upper and lower boundaries of each node, i.e., $\boldsymbol{x} \pm \mathbf{0.16}$ mm, are also specified so that the optimized variables do not exceed the range for the shape construction. From the geometric construction, a Visual Basic Script (VBS) code is generated for the optimization.

3.2 Optimization Procedure

The second step is setting up the interface code for real-time data transfer between VBS and MATLAB for GA optimization. In order to re-design the transition geometry, the losses along the transition taper need to be optimized. So, GA is run to optimize a spline profile of taper on return $\boldsymbol{S_{11}}$ and transmission $\boldsymbol{S_{21}}$ losses. Those losses are indicated by scattering parameters S-parameters. The GA procedure is used for the computation, where it generates the design variables associated with each taper structure. The calculation of fitness values is then evaluated based on the full-wave analysis results. The calculated fitness values are then used in the GA computation for the subsequent iterations to find the optimal geometry. The transition is validated by comparison with an evaluation of a fitness function. If the fitness meets the requirement, the procedure ends. Otherwise, new structures are generated, which are used in the next iteration for full-wave analysis to validate their performance concerning the expectations. A strategy of the analysis is illustrated in Figure 3.

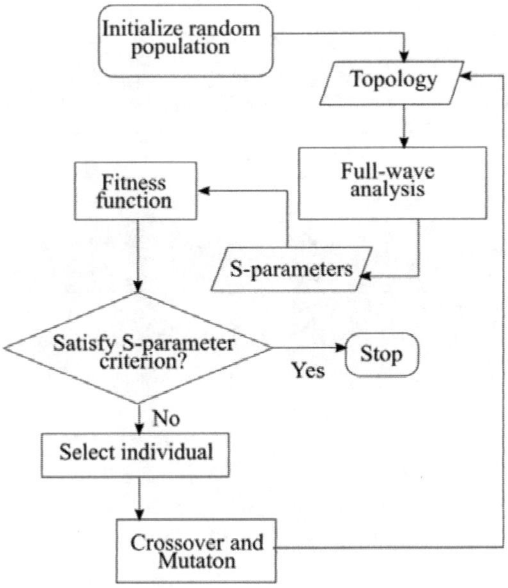

Figure 3. Flowchart of GA optimization with full-wave analysis.

The transition design goal is to minimize the return and transmission loss by optimizing the tapered design. Since there are two objectives, the multi-objective genetic algorithm (MOGA) was performed. The fitness functions for the multi-objective GA are defined as the average of those S_{11} and S_{21} values over the frequencies band of interest as

$$Fitness = \frac{1}{N}\sum_{i=1}^{N}|S_{11}(f_i)| \leq -20\ dB \quad (1)$$

$$Fitness = \frac{1}{N}\sum_{i=1}^{N}|S_{21}(f_i)| \geq -0.2\ dB \quad (2)$$

where f_i is the sampling frequency, and N is the number of the sample.

4. Results and discussion

The ranges of the optimization parameters to be optimized and the optimal results are displayed in Table 2. The formation of an optimal taper is shown in Figure 4.

Table 2. Optimized parameters obtained using MO optimization at Ka-band.

	Parameters(mm)			
$S(x)$	$S_1(1)$	$S_2(2)$	$S_3(3)$	$S_4(4)$
Original value	0.2865	1.0365	1.9635	2.7135
Optimal value	0.2900	1.0187	1.8497	2.6667

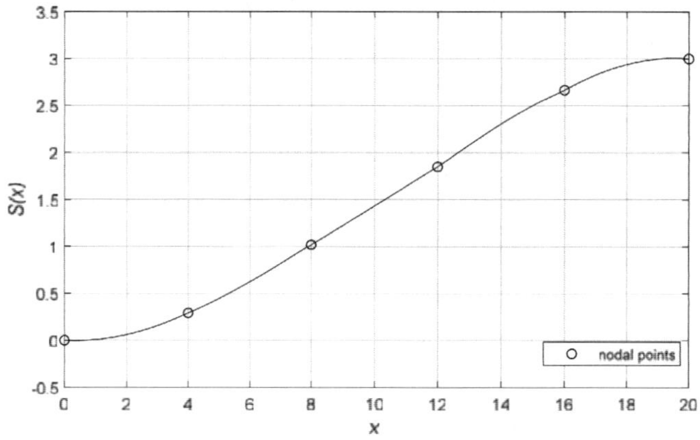

Figure 4. Taper geometry with a clamped cubic spline for L = 20 mm: x= length, S(x)= the clamped cubic spline function

In the design specifications, the centre operating frequency is expected to be 33 GHz. In contrast, after the design was simulated, the results displayed that the centre operating frequency is 33 GHz, as shown in Figures 5 and 6. The return loss comparison is depicted in Figure 5. It can be observed that for the Ka-band frequency, the reflection coefficient, S_{11} indicates that the optimal transition taper yields the best performance with a return loss of 20.88 to 52.60 dB at the frequency range from 26 to 40 GHz. However, for the original transition taper the S_{11} of 20.78 to 47.60 dB from 26 to 40 GHz is obtained. The percentage of return loss improvement, ΔS_{11} of the optimized taper design compared to the original-taper design can be calculated by

$$\Delta S_{11} = \frac{\max S_{11}^{o} - \max S_{11}^{rc}}{\max S_{11}^{rc}} \times 100\% \tag{3}$$

where S_{11}^{o} is the return loss of the optimized taper and S_{11}^{rc} is the return loss of the original taper. From equation (3), the optimal transition taper showed a 17% improvement concerning the return loss.

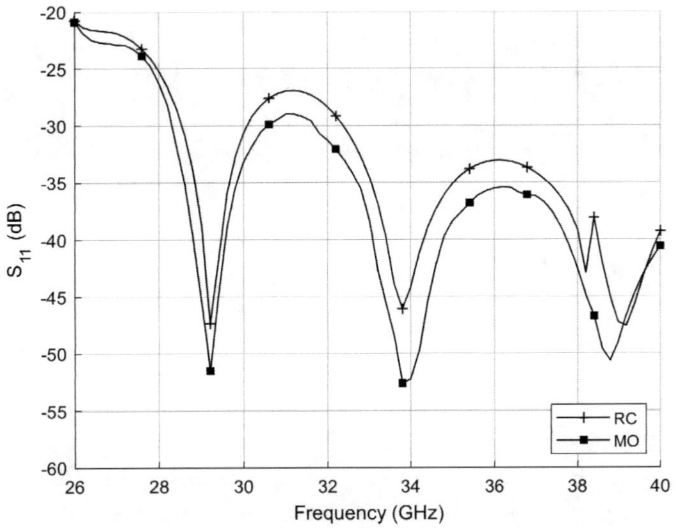

Figure 5. Comparison of the optimized $|S_{11}|$ between MO and RC taper in Ka-band frequency at 33 GHz: RC = original taper, MO = optimal taper.

From figure 6, the transmission loss S_{21} of 0.0745 to 0.14 dB from 26 to 40 GHz are obtained for the optimal transition taper. In comparison, the original transition taper yields the transition loss 0.149 to 0.28 dB at the frequency range from 26 to 40 GHz. The percentage of insertion loss improvement, ΔS_{21} of the optimized design compared to the original-taper design is calculated as

$$\Delta S_{21} = \frac{\max S_{21}^{o} - \max S_{21}^{rc}}{\max S_{21}^{rc}} \times 100\% \tag{4}$$

where S_{21}^{o} is the transmission loss of the optimized taper and S_{21}^{rc} is the transmission loss of the original taper. Referring to equation (4), the improvement in transmission loss of the optimal taper is 50%.

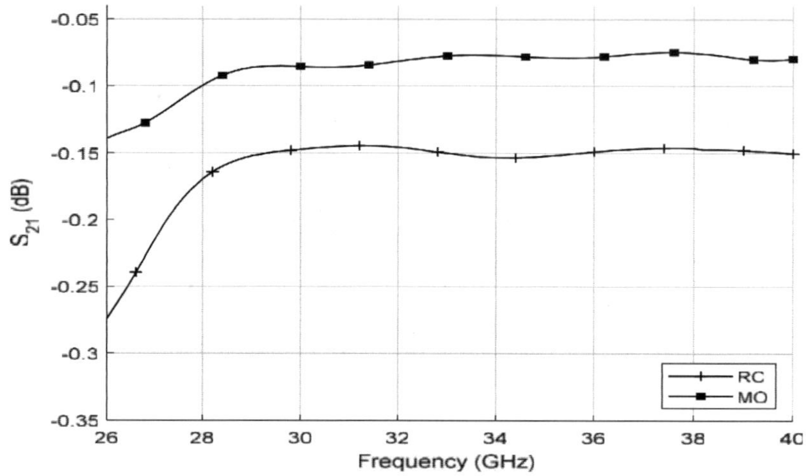

Figure 6. Comparison of the optimized $|S_{21}|$ between MO and RC taper in Ka-band frequency at 33 GHz: RC = original taper, MO= optimal taper.

5. Conclusion

The numerical studies in specific cases of the transition optimization show an improvement of 17% return loss and 50% transmission loss at Ka-band frequencies. Based on the results, the above procedure is viable to find an optimal transition taper, where the new taper transition geometry is shown to be the best choice to minimize the return and transmission losses further. The successful optimization strategy of the present study may be used to the development of the more optimal design of a waveguide component for any frequency band with better performance.

6. References

1. Choo H and Ling H 2003 Design of Broadband and Dual-Band Microstrip Antennas on a High-Dielectric Substrate Using a Genetic Algorithm *IEE Proc.-Microwave Antennas Propagation* ***150(3),*** 137–142
2. Santhosi B V S R N, Ramji K and Mohan Rao N B R 2019 Optimization of double-layered graphene-based microwave absorber in X—band using Pareto genetic algorithm *Materials Research Express* **10(6)**
3. Jungyoul L and Kok-Meng L 2015 Design of Electromagnetic Actuators Using Layout Optimization With Distributed Current Source Models *IEEE/ASME Transactions on Mechatronics* **20(6)** 2726–2735
4. Kalyan S K, Debasis M, Jyotirmay T, Ved P R and Anup K. B 2017 Optimum Design of Thin Wideband Multilayer Electromagnetic Shield Using Evolutionary Algorithms *Advanced Electromagnetics* **6(2)** 59–63
5. Rahmat-Samii Y and Michielssen E 1999 *Electromagnetic optimization by genetic algorithms* (New York: John Wiley & Sons)
6. John M and Ammann M J 2007 Spline-based geometry for printed monopole antennas. *Electronics Letters* **43(6)** 317–319

7. Granet C and James G L 2004 A smooth-walled spline-profile horn as an alternative to the corrugated horn for wideband millimetre-wave applications *IEEE Trans. Antennas Propagation* **52(3)** 848–854
8. Moniruzzaman M and Islam M R et al. 2011 Cubic spline-based particle swarm optimization for microstrip antennas *Australian J. Basic Appl. Sci.* **5(12)** 804–807
9. Bozzi M, Perregrini L, and Wu K 2008 Modeling of conductor, dielectric, and radiation losses in substrate integrated waveguide by the boundary integral-resonant mode expansion method IEEE Trans. Microw. Theory Tech. **56(12)** 3153–3161
10. Bozzi M, Perregrini L and Arcioni P 2009 Current and future research trends in substrate integrated waveguide technology *Radioengineering* **18(2)** 201–209
11. Ghassemi N and Boudreau I et al. 2013 Millimetre-wave broadband transition of substrate integrated waveguide on high-to-low dielectric constant substrates *IEEE Trans. Compon., Packag. Manuf. Technol.* **3(10)** 1764–1770
12. Lai Q, Fumeaux C, Hong W, and Vahldieck R 2009 Characterization of the propagation properties of the half-mode substrate integrated waveguide *IEEE Trans. Microw. Theory Tech.* **57(8)** 1996–2004
13. Ranjkesh N and Shahabadi M 2006 Reduction of dielectric losses in substrate integrated waveguide *Electronics Letters* **42(21)** 1230–1231
14. Jin L, Lee R M A, and Robertson I 2014 Analysis and design of a novel low-loss hollow substrate integrated waveguide *IEEE Trans. Microw. Theory Tech.* **62(8)** 1616–1624
15. Belenguer A and Borja A L et al. 2015 High-performance coplanar waveguide to empty substrate integrated coaxial line transition *Trans. Microw. Theory Tech.* **63(12)** 4027– 4034
16. Parment F and Ghiotto A et al. 2015 Air-filled substrate integrated waveguide for low-loss and high power-handling millimetre-wave substrate integrated circuits *IEEE Trans. Microw. Theory Tech.* **63(4)** 1228–1238
17. Isapour A and Kouki AB 2017 Empty LTCC Integrated Waveguide with Compact Transitions for Ultra-Low Loss Millimeter-wave Applications *IEEE Microwave and Wireless Components Letters* **27(2)** 144–146
18. Mansor H and Abdul-Rahman R 2016 Optimal transition in air-filled substrate integrated waveguide Proc. *IEEE Stud. Conf. Res. Dev* (Kuala Lumpur, Malaysia: IEEE) 172–176

IOP Conf. Series: Materials Science and Engineering **920** (2020) 012027 doi:10.1088/1757-899X/920/1/012027

Application of electromagnetic induction for impact load measurement

T Adachi* and Y Ishii

Department of Mechanical Engineering, Toyohashi University of Technology, 1-1 Tempaku, Hibarigaoka, Toyohashi, 441-8580, Japan

*Corresponding author: adachi@me.tut.ac.jp

Abstract. In this paper, a small impact testing machine was developed as an application of the method for measuring impact load based on electromagnetic induction phenomena. The experimental results were shown to determine the effectiveness of the method. The electromotive forces generated in two coils near a specimen were measured when a small impactor containing a neodymium magnet, having a mass of 0.36 g, was dropped freely onto a 1.2-mm-thick glass plate. The histories of the impact load and the relation between the impact load and specimen deflection were evaluated from the measured electromotive forces because the electromotive force per impactor velocity is dependent on only the position of the impactor. The method and developed machine with the measurement system were determined to be effective by considering the experimental impact test results for the glass plates.

1. Introduction

Impact load [1] is very important to evaluate the dynamic strength and stiffness of a material or structure. Generally, the impact load cannot be easily measured, especially when a small impactor collides with a structure or specimen. There are mainly two reasons for this difficulty: attaching an impact load sensor to the impactor, and balancing impact load and reaction force. First, the sensor, such as a load cell, a strain gage, or an accelerometer, cannot be attached to the impactor due to lack of space. Even if the sensor could be attached, the motion of the impactor is disturbed by the cable from the sensor to a measuring instrument. Second, because the inertia of the specimen or structure is generated under dynamic loading, the impact load often does not coincide with the reaction force at supporting positions. Namely, dynamic equilibrium is not always satisfied. Therefore, in previous studies, impact test results were estimated using the impact velocity or kinetic energy of the impactor, which was determined from the impactor behavior before the tests. [2–8].

In our previous works [9-11], impact load and specimen deflection were evaluated accurately from the measured electromotive force induced in coils near a specimen when a small impactor containing a magnet collided with the specimen. In the present work, the principle and method for measuring impact load were explained and a small impact testing machine with the measurement system was developed.

2. Measurement principle

We have suggested a novel method based on electromagnetic induction phenomena [12] to measure impact load when a small impactor collides with a structure or specimen [9-11]. The measurement principle is explained as follows.

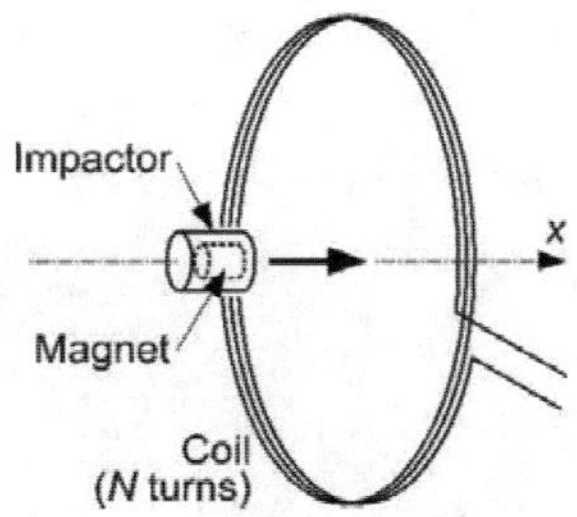

(a) Magnet-coil system.

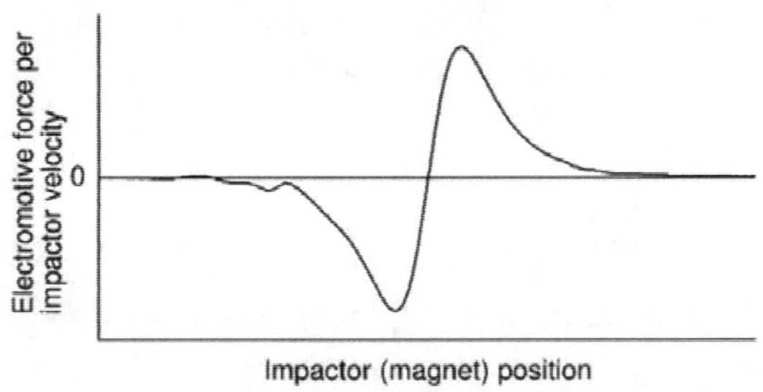

(b) Relation of electromotive force per impactor velocity and impactor position.

Figure 1. Electromagnetic induction phenomena.

Electromotive force is generated in a coil when an impactor containing a magnet passes through the coil with N turns (Fig. 1(a)). The electromotive force V is evaluated by Faraday's law of induction [12].

$$V(t) = -N\frac{d\Phi(t)}{dt}, \tag{1}$$

where Φ and t are the magnetic flux in the area surrounded by the coil and time, respectively. The induced electromotive force is proportional to the turns of the coil and the time derivative of Φ, namely the velocity of the impactor. The electromotive force becomes zero when the magnet arrives at the coil. The electromotive force per impactor velocity relation (Fig. 1(b)) is dependent on only the magnet position regardless of impactor velocity. Impact load and displacement of the impactor can be evaluated on the basis of the impactor's initial position and velocity.

Two groups of coils are used to measure the impact load and displacement of the impactor as shown in Fig. 2(a). The first group is composed of Coils No. 1 and 2 that obtain the initial position and velocity of the impactor as the initial condition in the impact load analysis. Typical electromotive forces generated in Coils No. 1 and 2 are shown in Fig. 2(b). The impactor position coincides with the positions of Coils No. 1 and 2, x_1 and x_2, at time, t_1 and t_2, when the electromotive forces of the coils become zero. The initial velocity of the impactor v_0 can be calculated as follows:

$$v_0 = \frac{x_2 - x_1}{t_2 - t_1}. \tag{2}$$

The second coil group is composed of a single coil or multiple coils connected in series. Multiple coils are used to extend the measurable region [9]. Because the electromotive force is proportional to the impactor velocity, the velocity can be evaluated as (Fig. 2(c))

IOP Conf. Series: Materials Science and Engineering **920** (2020) 012027 doi:10.1088/1757-899X/920/1/012027

$$v(t) = \frac{V(t)}{e(x)},\ x = x(t) \tag{3}$$

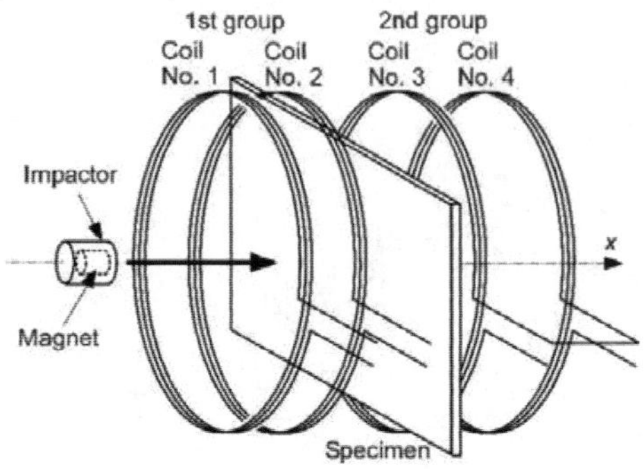

(a) System for measuring impact load.

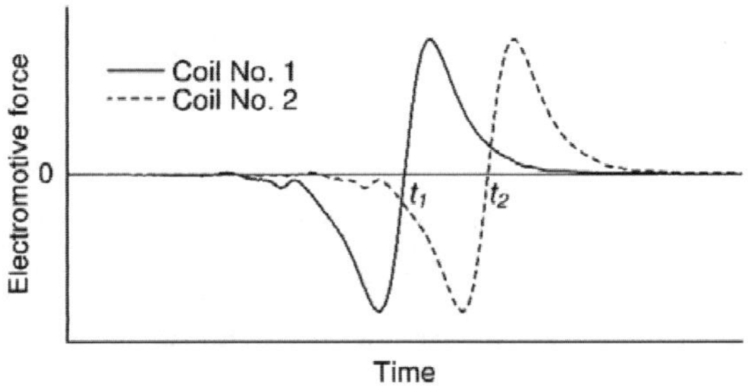

(b) Histories of electromotive forces in 1st coils.

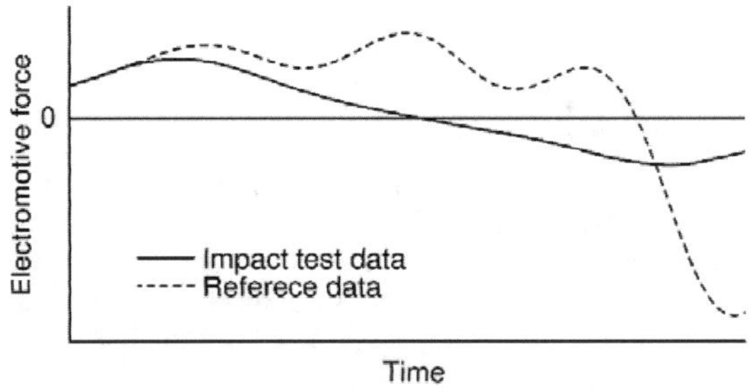

(c) Histories of electromotive forces in 2nd group coils.

Figure 2. Principle of measuring impact load.

where e, V, and x denote the electromotive force per impactor velocity (reference data), the electromotive force in the impact test, and the impactor position, respectively.

The impactor position can be determined by integrating Eq. (3) with the initial conditions.

$$x(t) = x_i + \int_{t_i}^{t} v(\tau)d\tau \quad i = 1 \text{ or } 2 \tag{4}$$

The impactor acceleration α can be determined by differentiating Eq. (1).

$$\alpha(t) = \frac{dv(t)}{dt} \quad (5)$$

The impact force F can be evaluated by multiplying Eq. (5) by the mass of the impactor m on the assumption that the impactor is a mass point. This assumption is valid when the natural frequency of the impactor is much higher than that of the specimen.

$$F(t) = m\alpha(t) \quad (6)$$

The impactor displacement x can be regarded as the deformation of the specimen because the impactor is in contact with the specimen during the impact loading.

3. Application to small impact testing machine

3.1. Impact testing machine

We developed the impact testing machines for penetration testing of rubber sheets [9-11]. In this paper, a novel impact testing machine developed for a small impactor is denoted. The machine and impactor are illustrated in Fig. 3. The machine was designed and manufactured on the basis of a drop weight testing machine (Fig. 3(a)).

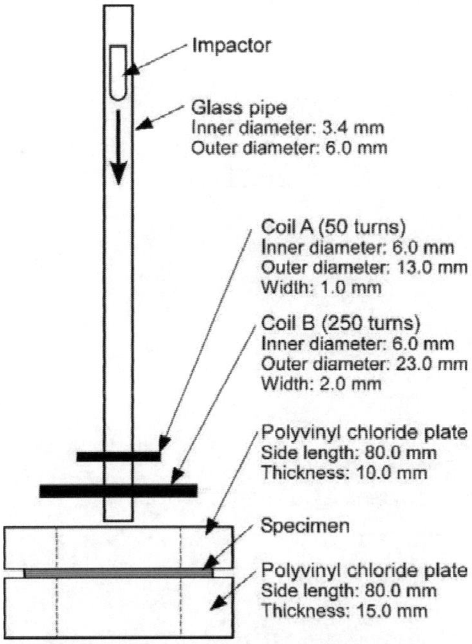

(a) Small drop weight testing machine.

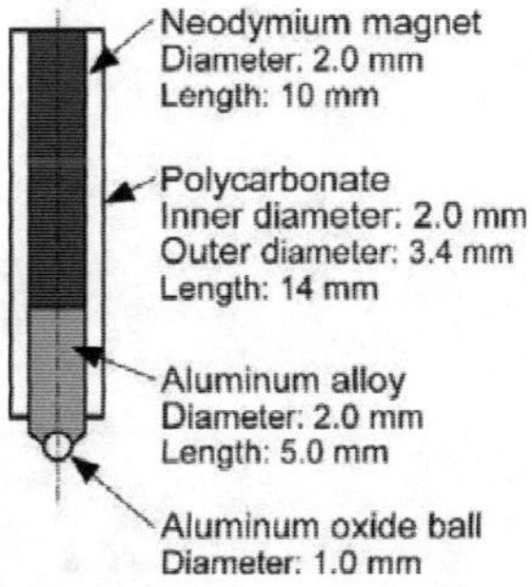

(b) Impactor (Cross section).

Figure 3. System of small impact testing machine.

The impactor (Fig. 3(b)) contained a cylindrical neodymium magnet with 2 mm in diameter and 10 mm in length. The tip of the impactor was an aluminum oxide ball on the aluminum-alloy cylinder having 2.0 mm in diameter and 5.0 mm in length, and the sides of the cylinder and magnet were covered by a polycarbonate pipe. The impactor mass was 0.36 g.

The impactor was dropped inside a glass pipe with inner and outer diameters of 3.4 mm and 6.0 mm respectively and collided with a specimen clamped between two thick polyvinyl chloride plates having holes with a diameter of 30 mm. Two coils, A and B, were placed in front of the specimen. A

0.5-mm polyurethane-covered copper wire (2UEW-0.16, JIS C3202) was wrapped around the external area of the circular pipe to prepare both coils. Coil A had 50 turns with an outer diameter of 13 mm and a width of 1 mm, and Coil B had 250 turns with an outer diameter of 23 mm and a width of 2 mm. Coils A and B (first group) were used to determine the initial conditions and Coil B (second group) was also used to measure the impact load and specimen deformation. n contact with the specimen during the impact loading.

3.2. Experimental results

The impactor was dropped freely from a height of 250 mm and collided with the glass plate specimen. The glass plate was 1.2 mm thick, 26 mm wide, and 76 mm long. No damage was observed in the plate after the impact test.

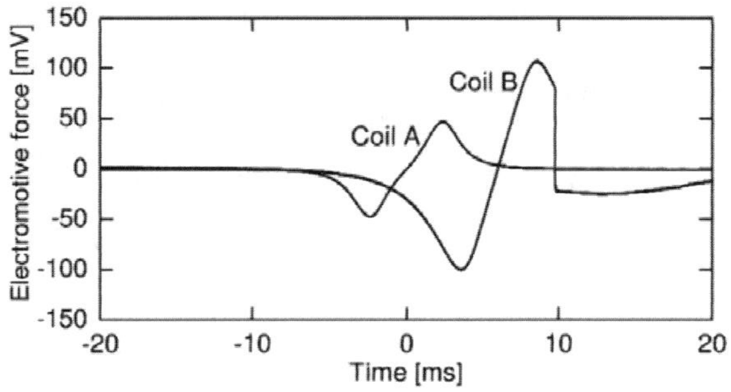

(a) Histories of electromotive forces in the impact test.

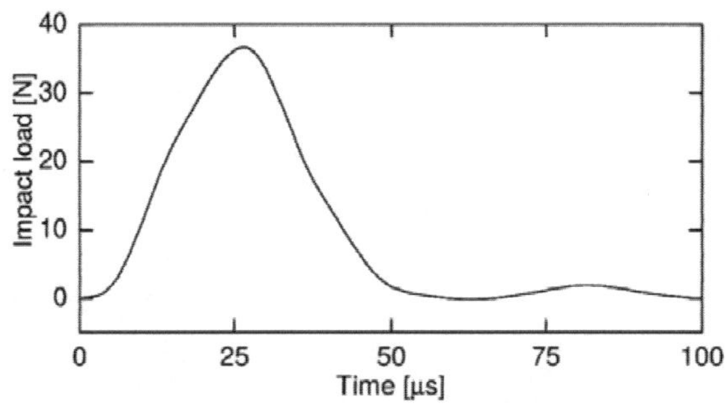

(b) History of impact load.

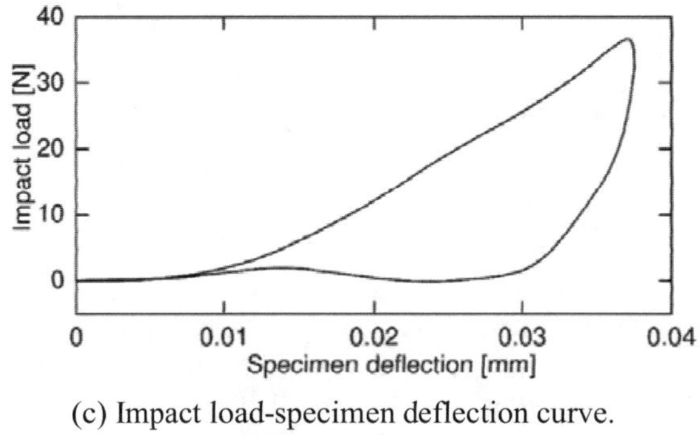

(c) Impact load-specimen deflection curve.

Figure 4. Experimental results.

Figure 4 shows the history of the measured electromotive forces in the impact test. The time of the horizontal axis in Fig. 4 is relative. The impactor collided with the specimen at about 10 ms when the electromotive force of Coil No. B decreased drastically.

The impact load and specimen deflection were evaluated from the histories of the electromotive forces in Fig. 4(a) with Eqs. (2)-(6) and the reference data measured before the impact test. The history of the impact load is shown in Fig. 4(b). Because the specimen was not penetrated, the impact load increased and decreased linearly after the maximum load of 37 N. The duration of the impact load was approximate 50 μs. The minor fluctuation after 50 μs was caused by the natural vibration of the impactor. The impact load-specimen deflection curve was determined by evaluating the deflection. The curve is shown in Fig. 4(c). The curve after the collision was linear to the deflection until the maximum value, and the curve decreased drastically after the maximum value.

The momentum variation of the impactor due to the collision was evaluated as 0.93 gm/s from the velocities before and after the impactor collision of 1.99 and –0.58 m/s. The impulse of the impact load in Fig. 4(b) was also calculated as 0.93 mNs. The validity of the analysis for the impact load and specimen deflection was confirmed by the agreement of the momentum and impulse.

The impact load and specimen deflection were determined on the basis of the electromagnetic induction even though an impactor weighting less than 1 g collided with the specimen.

4. Conclusion

In this paper, the principle and system for measuring impact load and specimen deformation based on the electromagnetic induction phenomena was explained, and a small impact testing machine was developed as an application of the method. The impact load and specimen deflection, which were difficult to measure with conventional methods, were determined when a small impactor with a mass of less than 1 g collided with a 1.2-mm-thick glass plate. The method and developed machine with the measurement system were determined to be effective by considering the experimental impact test results for the glass plates.

References

[1] Adachi T and Ishii Y 2017 *J. Jpn. Soc. Exper. Mech.* **17** 167 (in Japanese).
[2] Abrate S 1998 *Impact on Composite Structures* (Cambridge: Cambridge University Press)
[3] Zukas J A, Nicholas T, Swift H F, Greszczuk L B and Curran D R 1982 *Impact Dynamics* (New York: John Wiley & Sons).
[4] Walker J D 2001 *Int. J. Impact Eng.* **26** 809
[5] Bourne N K 2012 *Exp. Mech.* **52** 153
[6] Warren T L 2016 *Int. J. Impact Eng.* **91** 6
[7] Antonucci V, Caputo F, Ferraro P, Langella A, Lopresto V, Pagliarulo V, Ricciardi M R, Riccio A and Toscano C 2016 *Prog. Aerosp. Sci.* **81** 26
[8] Omidvar M, Iskander M and Bless S 2014 *Int. J. Impact Eng,* **66** 60
[9] Adachi T, Matsukawa N, Takamizo C and Ishii Y 2017 *J. Jpn. Soc. Exper. Mech.* **17** 231 (in Japanese).
[10] Adachi T, Matsukawa N, Takamizo C and Ishii Y2018 *Int. J. Impact Eng.* **121** 172
[11] Adachi T, Mochizuki Y and Ishii Y 2019 Measurement system of impact force and specimen deflection based on electromagnetic induction phenomena. *Dynamics and Control of Advanced Structures and Machines* ed V. P. Matveenko V P, Krommer M, Belyaev A and Irschik H (Wien: Springer) Chapter 1 pp 1-9, Springer.
[12] Ida K 2015 *Engineering electromagnetics 3rd ed* (New York: Springer) p 515

IOP Conf. Series: Materials Science and Engineering **920** (2020) 012028 doi:10.1088/1757-899X/920/1/012028

The Electronic States of Copper Oxides Photoactive Layers Prepared by Electrodeposition followed by Annealing

P L Khoo*, K Satou and M Izaki

Department of Mechanical Eng., Toyohashi University of Technology, 1-1 Hibari Gaoka, Tenpaku, Toyohashi, Aichi 441-8580, Japan

*E-mail: khoo@tf.me.tut.ac.jp

Abstract. The copper oxides are best performing metal oxide materials regarding solar energy conversion, and due to their abundance, scalability, ease of fabrication and subsequently being low-cost, Cu_2O and CuO have attracted extensive attraction. In this research, Cu_2O layers were electrodeposited on Ga:ZnO (GZO) and annealed to form directly-stacked GZO/Cu_2O/CuO. The external quantum efficiency (EQE) of the devices was evaluated which demonstrated peculiar negative regions around 500 nm besides the typical 410 nm originating from the Cu_2O. Reversed bias voltages were applied to investigate the changes, and along with the disappearance of negative regions, a peak around 550 nm and an absorption edge at 850 nm appeared, which increased with the increment of bias voltage. To elucidate the mechanism, the electronic states were investigated by x-ray photoelectron spectra (XPS). Species identification was also successful which showed the presence of CuO on the topmost layer and the underlying Cu_2O, which was difficult to analyze by XRD due to its thinness. The schematics of the band-alignments were drawn based on the calculations from the XPS measurements. The conduction band of the Cu_2O appeared elevated, which contributed to the speculated two-way flow of the charge when illuminated, and explains the appearance of the negative EQE regions. The charge transportation was successfully controlled and aligned when bias voltages were applied, which caused the disappearance of negative regions and the appearance of the absorption edge of the CuO.

1. Introduction

One ultimate goal of researches in the solar energy conversion field is to reach grid parity for world-wide deployment to reduce greenhouse gas emissions and cut down dependence on fossil fuels. As such, the cost of the material and fabrication of photovoltaics or photocathodes should be minimized, and a shift of focus to earth-abundant photovoltaic materials is imperative in realizing sustainability. Being environmental-friendly (nontoxic) is also a huge advantage. Photovoltaic materials satisfying all mentioned demands are the copper oxides, which are unique among oxides being intrinsically p-type, and also are among the best performing metal oxide materials. The copper (I) oxide (Cu_2O) and copper (II) oxide (CuO) possess bandgap energies of 2 eV and 1.5 eV respectively, with Cu_2O possessing a theoretical hydrogen conversion efficiency of 18% in water splitting and a theoretical power conversion efficiency of 20% as a solar cell. CuO possesses lower bandgap energy, absorbing corresponding light wavelengths below ~850 nm. As one single light-absorbing material cannot satisfy the wide span of the sunlight spectrum, a strategy to combine Cu_2O and CuO to utilize multiple bandgaps was reported,[1] which showed that both oxides could simultaneously act as the light-absorbing layer.

Our previous research has demonstrated the fabrication of Cu_2O/CuO layer by annealing in air, and also the external quantum efficiency (EQE) behavior of the directly stacked GZO/Cu_2O/CuO under reversed bias voltage which resembles the working condition during water splitting.[2] However, the species identification was insufficient due to the thinness of the CuO film which made the characterization of X-ray diffraction difficult. The need for detailed X-ray photoelectron spectroscopy (XPS) for the directly stacked Cu_2O/CuO arose not only to more accurately determine the chemical state and identification of fabricated copper oxide but also to investigate their energy states. Negative regions of EQE were observed in some of the devices when reversed bias voltage was not applied, which indicated the presence of two distinct charge transportations which might originate from two separate electric fields, presumably at the heterointerfaces of GZO/Cu_2O and Cu_2O/CuO, and XPS characterization could give insight regarding the carrier transport mechanism.

2. Experimental

2.1. Device Preparation: Electrodeposition and Annealing

Conductive gallium-doped zinc oxide (GZO)-coated glass was partly used as substrates due to its ability to demonstrate the EQE behavior of Cu_2O without any intermediary layer. Electrodepositions of Cu_2O layers were carried out from alkaline aqueous solution (pH 12.5) containing 0.4 mol L^{-1} copper (II) acetate monohydrate ($Cu(CH_3COO)_2 \cdot H_2O$, Nacalai Tesque, Inc.) and 3 mol L^{-1} lactic acid (Kanto Chemical, Co., Inc.). A standard 3-electrode cell was used for the potentiostatic electrodeposition with a Pt plate as the counter electrode and Ag/AgCl in saturated KCl as the reference electrode. The electrodepositions were carried out with a potentiostat (Hokuto Denko, HAL-3001) and a Coulomb meter (Hokuto Denko, HF-301) with the potential and electrical charge set at -0.4 V and 1.5 C cm^{-2} respectively, and the water bath temperature set at 313 K. ZnO electrodepositions on FTO were deposited with the similar method from a solution of 80 mmol L^{-1} zinc nitrate hexahydrate ($Zn(NO_3)_2 \cdot 6H_2O$, Nacalai Tesque, Inc.) in a water bath of temperature 336 K.[3]

Annealing of the devices was carried out on an alumina plate in a quartz holder, annealed at various temperatures of 523, 573 and 673 K in a rapid thermal annealing furnace (MILA 5000) for GZO-based devices. The annealing rate was set at 10 K s^{-1} and the heating time was set at 60 min after the target temperature had been reached.

2.2. Characterizations

X-ray photoelectron spectroscopy (XPS) measurements were performed with a scanning X-ray microprobe (Quantera SXM-CI) made by ULVAC-PHI Inc, with a 24.8 W monochrome Al $K\alpha$ (1486.6 eV) with a 100-μm-beam. To lessen the drift caused by charge-up, the neutralizer was operated at 1.0 V of 20 μA, and peak corrections of C 1s were made to 284.8 eV. The software used for the analyses was CasaXPS (Casa Software Ltd.)

The external quantum efficiency (EQE) measurements were performed using a monochromator (Bunko Keiki, SM-250 KB) with approximate monochromatic light intensity above 5 mW cm^{-2}. Reversed bias voltages were applied to investigate the carrier transportation and also simulate approximate conditions when the photoactive layers were to be used as photocathodes for water splitting.

3. Results and Discussions

3.1. The Behavior of External Quantum Efficiency (EQE) of Cu_2O/CuO

Figure 1 (a) shows the EQE of the 523K-annealed device without and under bias voltage of −0.01 to −0.1 V, with −0.01 V increments. Observed is the appearance of a new absorption edge around 850 nm with the increase of voltage. Without any biased voltage applied (0 V), the 523-K-prepared directly-stacked photoactive layer showed a peak at approximately 410 nm and a dip in the spectra towards the negative region around 500 nm, indicating two distinct charge transportations from different light-absorbing layers. The peak at 410 nm can be attributed to the characteristic EQE peak of electrodeposited Cu_2O as already reported.[4] The overall low performance may be attributed to the formation of scattering defects such as pores during the oxidation of the Cu_2O, which involves the transportation of Cu ions.[1,4,5]

Before annealing, all charge collected can be attributed to the flow from photo-induced (starting from 650 nm) electrons from the Cu_2O layer towards the direction of the source of irradiation and extracted via the GZO layer. [4] For the 523-K-device, the dip towards the negative region in the EQE spectra within 470 and 580 nm then reveals the presence of photo-induced electrons moving in the opposite direction. Increment of reverse biased voltages by −0.001 V towards −0.01 V caused a dramatic increase in the 550-nm-peak, while the increase in the peak at around 410 nm was gradual in comparison. This also indicates that the charge collection as represented by these peaks were constituted by charges originating from more than one light-absorbing layer, which were the Cu_2O and CuO layer. Also, a distinct absorption edge for the peak of 550 nm at around 850 nm emerged. This absorption edge can be attributed to the characteristic absorption edge of CuO, as can clearly be seen from the absorption spectra of the 673 K device as already reported, which among the devices, possesses the thickest layer of CuO at the surface.[2] This behavior of dipping of the EQE curve towards the negative region was not observed when annealed in vacuum at the same temperature, even though the Cu_2O was prepared similarly, as previously reported.[4] This means that the EQE behavior is specific to devices annealed in a normal atmosphere which can be attributed by the appearance of Cu_2O/CuO internally-stacked layer. Also shown are the EQE spectra of 523, 573 and 673 K annealed devices as already reported.[2]

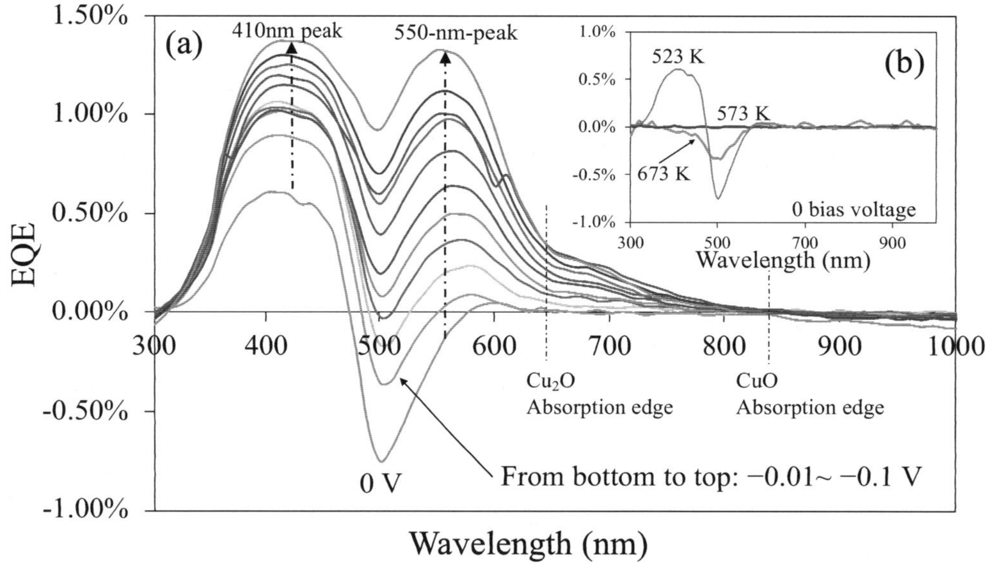

Figure 1. EQE spectra of (a) 523-K annealed device without and under reversed bias voltage, (b) 523, 573 and 673-K annealed devices.

3.2. Surface X-ray Photoelectron Spectroscopy (XPS)

Figure 2 shows the Cu $2p_{3/2}$ and $CuL_3M_{4,5}M_{4,5}$ spectra of the as-deposited and annealed devices. One significant change observed for the Cu $2p_{3/2}$ spectra is the overall shifts of the main $2p_{3/2}$ peak from 932.3 eV (as-deposited) to 933.1 eV (523 K), 933.4 eV (573 K), and 993.5 eV (673 K), which also showed broaden peaks. These values seem to increase from reported literature average values of the binding energies of 932.4 eV belonging to Cu_2O, to 933.6 eV of CuO, based on the NIST Database,[6] as compiled by Biesinger, who reported an extensive analysis of copper XPS.[7] Another significant change observable from the Cu $2p_{3/2}$ spectra is the emergence of satellite peaks at approximately 941.5 eV and 944 eV for the annealed devices. These evident peaks are characteristic shake-up satellites which indicate the presence of Cu(II) species, originating from the surface CuO layer.[8–13] The $CuL_3M_{4,5}M_{4,5}$ spectra as shown in figure 2 (b) indicates a shift in peak position from 916.7 eV to 917.6 eV. The peak value of the as-deposited device at 916.7 eV was close to the reported value of 916.5 eV belonging to Cu_2O.[14] While the Auger main peaks of Cu(I) and Cu(II) species are close to each other, the $CuL_3M_{4,5}M_{4,5}$ spectral line-shape can provide information about species differentiation.[15] The shoulder peaks at around 912-914 eV are usually observed for the spectra of CuO. In short, the spectra of the annealed devices showed XPS characteristics leaning to Cu_2O for the as-deposited device and CuO for the annealed layers.

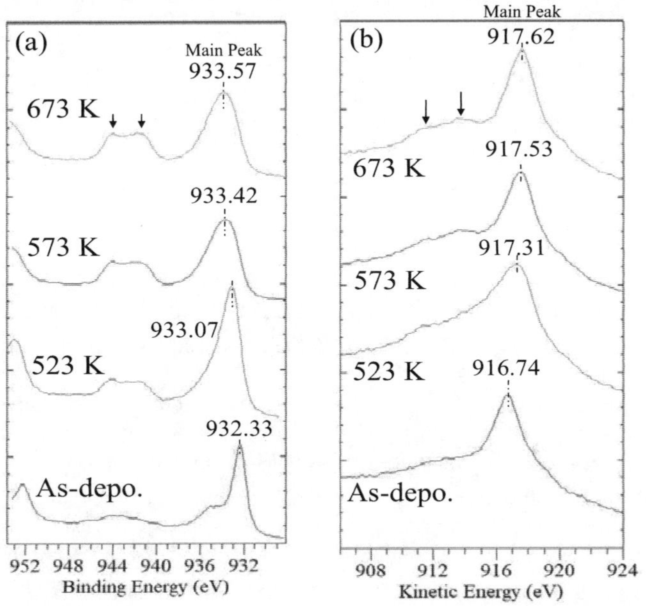

Figure 2. The surface (a) Cu $2p_{3/2}$ and (b) $CuL_3M_{4,5}M_{4,5}$ spectra with horizontal axes as binding and kinetic energy respectively, for the as-deposited GZO/Cu_2O device, and devices annealed at 523, 573, and 673 K. Along with peak shifts, the appearance of characteristic shake-up satellites in (a) and changes in spectral line-shapes in (b) can be observed here, as denoted with arrows.

Wagner has shown the chemical shifts for Auger transitions involving electrons in the atomic core and changing of the line shape for Auger transition involving valence electrons and introduced the concept of Auger parameter based on the idea that two line energies (Auger and photoelectron) of the same element in the same measurement is fixed, and the charge corrections to individual peak measurements can be canceled off and the work function corrections are not necessary.[16–18] The Auger parameter α' is defined as,[19]

$$\alpha' = \alpha + h\upsilon = E_k(C'C''C''') + E_b(C) \quad (1)$$

where $E_k(C'C''C''')$ is the kinetic energy of the Auger transition involving electrons from the C', C'', and C'''core levels, and $E_b(C)$ is the binding energy of the photoelectron from core level C. In other words, the sum of Cu $2p_{3/2}$ peak value in binding energy and the $CuL_3M_{4,5}M_{4,5}$ peak value in kinetic energy gives the value of a characteristic modified Auger parameter which could be used to identify species. This method was effectively used by Goh *et al.* in assigning Cu_2S and CuS,[14] besides Poulston *et al.* who have used CuLMM and the Auger parameter to distinguish Cu(0), Cu(I) and Cu(II)[8]. Calculated values are summarized in Table 1 below.

Table 1. Auger Parameter calculation of as-deposited, 523 K, 573 K, and 673 K devices.

	Binding Energy Cu 2p $_{3/2}$	Kinetic Energy $CuL_3M_{4,5}M_{4,5}$	Auger Parameter
As-deposited	932.33	916.74	1849.07
523 K	933.07	917.31	1850.38
573 K	933.42	917.53	1850.95
673 K	933.57	917.62	1851.19

The literature average compiled values from the NIST database[6] are 1849.2 for Cu_2O and 1851.5 for CuO[7,20], which are near to the calculated Auger Parameter values of 1849.1 for as-deposited and 1851.2 for 673 K. Both 523 K and 573 K devices possess Auger Parameter approaching the value for CuO. Thus, it is safe to assume that the topmost layers of all annealed devices consist of CuO layers, consistent with the FESEM observations as previously reported.

3.3. Band-alignment Calculation of Cu_2O/CuO

The difference in valence band maximums (VBM) of the Cu_2O and CuO which is the valence band offset (VBO, denoted as ΔE_V), is needed to understand the band-alignment of the heterojunction which was formed after annealing. By using the core-level alignment, a more accurate band offset calculation that does not rely on shift correction could be done as shown in the following equation,[21,22]

$$\Delta E_V = \left(E_{Cu\,2p}^{CuO} - E_{Cu\,2p}^{Cu2O}\right) + \left(E_{Cu\,2p}^{Cu2O} - E_{VBM}^{Cu2O}\right) - \left(E_{Cu\,2p}^{CuO} - E_{VBM}^{CuO}\right) \tag{2}$$

where $E_{Cu\,2p}^{CuO}$ is the 2p main peak value of the surface CuO, $E_{Cu\,2p}^{Cu2O}$ is the 2p main peak value of the underlying Cu_2O, $E_{Cu\,2p}^{Cu2O}$ is the Cu 2p main peak value of the underlying Cu_2O, E_{VBM}^{Cu2O} is the measured valence band maximum (VBM) of the underlying Cu_2O, and E_{VBM}^{CuO} is the measured VBM of the surface CuO. After the initial surface CuO measurement, to expose the underlying Cu_2O, sputtering was carried out for 18 seconds from the top surface until 180 seconds and the XPS spectra recorded for each annealed device.

The Cu 2p binding energies for the main peaks of the surface CuO ($E_{Cu\,2p}^{CuO}$) were 933.1, 933.5, and 933.7 eV, which were close to the reported value of 933.6 eV.[7] It was found that sputtering during the first 18 seconds had sufficiently removed the topmost CuO layer for all annealed devices, and the average Cu 2p binding energies for the main peaks of the underlying Cu_2O ($E_{Cu\,2p}^{Cu2O}$) were 932.6, 932.4, and 932.6 eV for 523, 573, and 673 K respectively, which were close to the reported value of 932.4 eV. Also, the E_{VBM}^{Cu2O} and E_{VBM}^{CuO} values were extracted from the XPS spectra in the vicinity of 0 eV binding energy, along with all summarized values needed for the VBO calculations, are shown in figure 3. The above VBO calculation is the typical method calculating p-n junction band alignment, where all the peaks may appear on a single spectrum. However, for our measurement, the first sputtering had completely exposed the underlying layer, and by assuming that the spectra do not shift during continuous measurement after sputtering, the VBM could be easily calculated by calculating the difference between measured E_{VBM}^{Cu2O} and E_{VBM}^{CuO}.

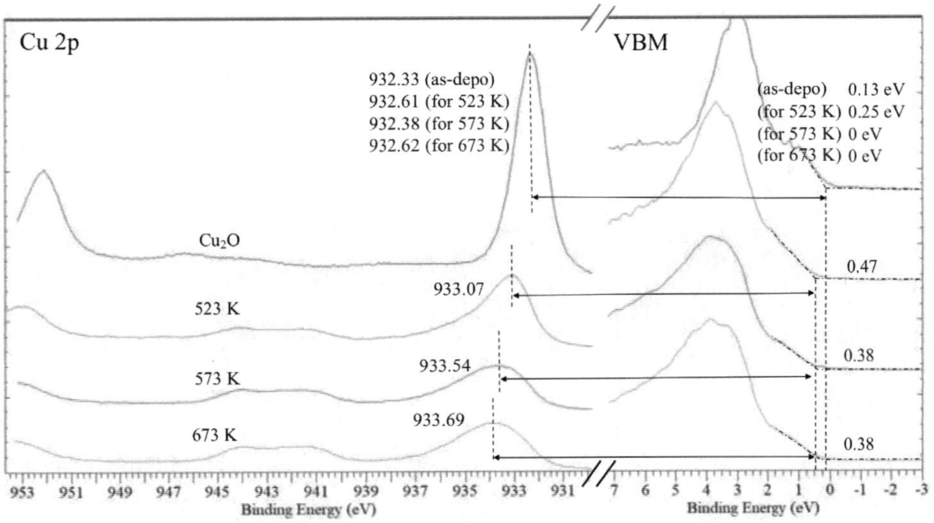

Figure 3. The XPS spectra showing all measurements needed to determine the VBO for the 523, 573 and 673 K annealed devices. The as-deposited values are for reference.

Figure 4 shows the Tauc plots for the as-deposited device and for the 673 K-annealed devices which showed a clear absorption edge for the CuO layer at 850 nm. These bandgap energies for the band offset calculations were estimated from previously measured absorbance spectra.[2] The relationship between optical bandgap energy, E_g and the absorption coefficient α calculated from the absorbance and thickness can be represented by equation 3 as follows[23]:

$$\alpha h\upsilon = A(h\upsilon - E_g)^{n/2} \quad (3)$$

A is a constant, $h\upsilon$ is photon energy, for thin films n=1 for direct transition, and n=4 for an indirect allowed transition. For the as-deposited device, the direct bandgap estimated was 2.05 eV, corresponding to reported values of 2.1 eV belonging to Cu_2O.[24] For the 673 K-annealed devices, the direct bandgap estimated from this figure was 1.56 eV, while the indirect bandgap was 1.47 eV which were close to the reported CuO direct bandgap values by Nakaoka *et al.* of 1.56 eV and 1.45 eV which were heat-treated at 573 and 773 K respectively.[23] However, the actual type of bandgap energy of CuO remains controversial.[25–30]

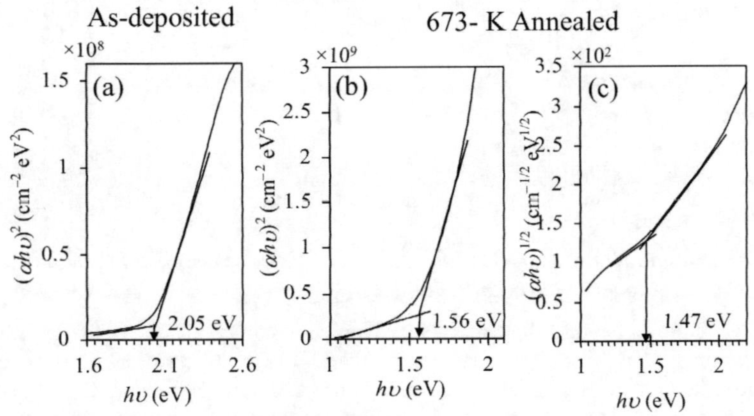

Figure 4. Tauc plots of (a)$(\alpha h\nu)^2$ versus $h\nu$ for as-deposited device and (b,c)$(\alpha h\nu)^2$, $(\alpha h\nu)^{1/2}$ versus $h\nu$ for the 673-K annealed device.

Using the direct bandgap for Cu_2O as 2.05 eV and indirect bandgap for CuO as 1.47 eV from the Tauc plots, the calculated VBO values were then 0.22, 0.38, and 0.38 eV for the 523, 573 and 673 K devices, respectively. The conduction band offsets (CBO) calculated were 0.8, 0.96 and 0.96 eV. The schematic band alignments are shown in figure 5.

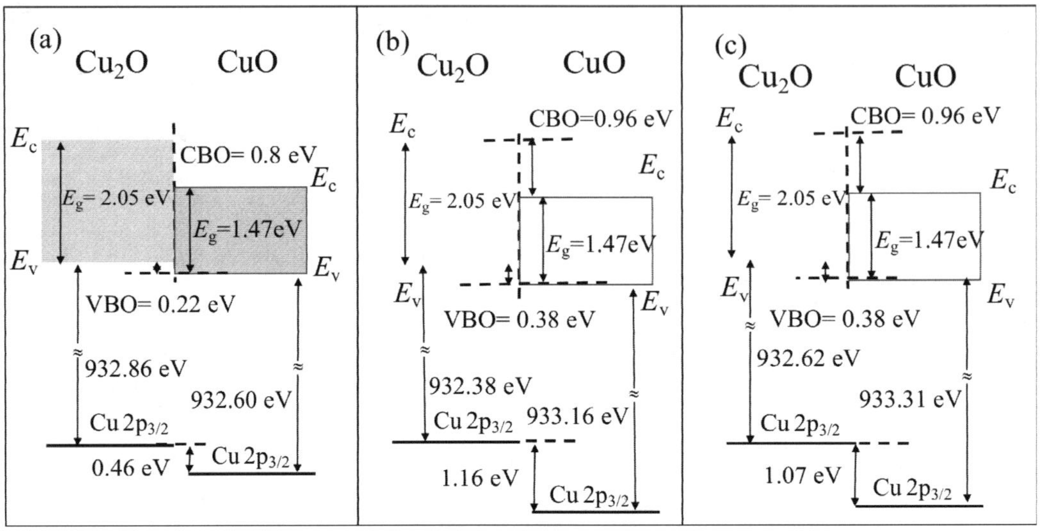

Figure 5. The schematic band alignment of Cu_2O/CuO based on calculated VBO and bandgap energy estimated from Tauc Plots. (a) is 523 K, (b) is 573 K, and (c) 673 K.

The CBO of the 573 and 673 K-annealed devices were higher than the 523 K-annealed device, contributing to dimmed EQE performances due to higher recombination rates. In an attempt to further comprehend the charge collection mechanism of GZO/Cu_2O/CuO which is closely related to the appearance of EQE spectra, reference is made to the following figure 6, which shows the band alignments at equilibrium and under reversed bias voltage.

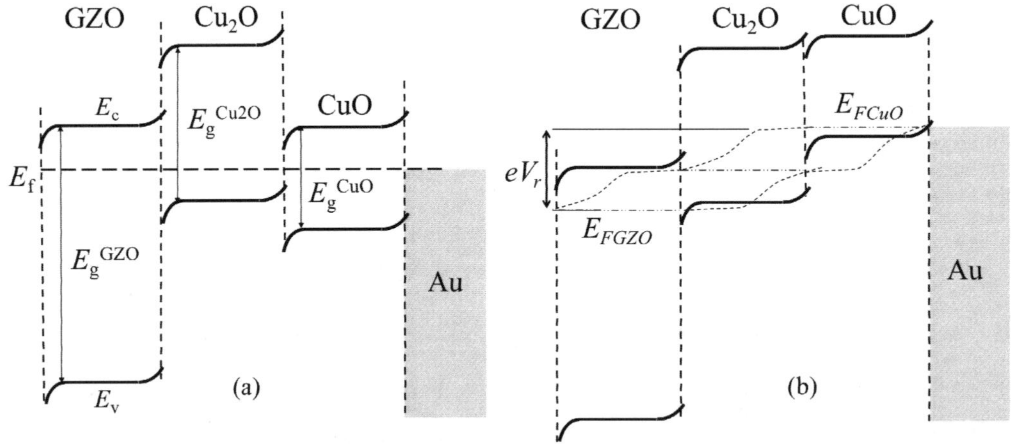

Figure 6. The schematic band alignment of GZO/Cu_2O/CuO speculated at (a) equilibrium and (b) under reversed bias voltage.

At equilibrium (0 V), the conduction band of the Cu_2O should appear elevated against both the GZO and CuO layers. Here, charges originating from the Cu_2O nearby the GZO/Cu_2O heterojunction flowed towards the GZO layer and appeared as the EQE peak at around 410 nm. However, the charge originating from the CuO, or deep in the Cu_2O layer near the vicinity of Cu_2O/CuO can favorably flow in the opposite direction towards the Au electrodes, which explains the negative regions in the EQE around 500 nm. When reversed bias voltage was applied, the potentials across the heterojunctions were increased, as denoted with eV_r, which exaggerated depiction is shown in figure 6(b). Consequently, along with the formation of quasi-Fermi levels, the band energy at the CuO side was relatively raised, facilitating movements of photo-induced electrons in the CuO layer towards the Cu_2O layer when irradiated. This should enable more charge collection via the GZO layer, which mechanism corresponds to the EQE measurement. The depletion width and the peak electric field at the junctions were also increased in this condition,[31] which explains the heightened peak at 410 nm. On the other hand, the higher rate of increase in peak 550 nm along with the reversed bias voltage suggests not only that charges originating from the Cu_2O which once flowed in two directions were corrected and aligned, but also the contribution of charge originating from the CuO layer, which is also evidenced by the appearance of absorption edge at 850 nm.

4. Conclusions

The EQE of the directly-stacked Cu_2O/CuO photovoltaic layers and their changes under reversed bias voltage was shown, giving insights into the change in charge transportation. The EQE demonstrated the ability of both Cu_2O and CuO layers to act as light-absorbing layers, which showed an augmented light-absorption span. The band-alignments were calculated and the mechanism of the EQE behavior was elucidated by investigation of the energy states. The results show that Cu_2O/CuO may work well as photocathodes in water splitting under applied biased voltage, and better optimization could be carried out to further improve the Cu_2O/CuO as a photovoltaic device to utilize its ability for a superior light-absorption span. This research also highlights the importance of minimal defects for a better semiconductor quality and may provide insights for an improved band-alignment for future fabrication of the Cu_2O/CuO photoactive layer.

Acknowledgments

This work was supported by a Grant-in-Aid for Scientific Research (19H02810) from the Japan Society for the Promotion of Science.

References

[1] Izaki M, Fukazawa K, Sato K, Khoo P L, Kobayashi M, Takeuchi A and Uesugi K 2019 Defect Structure and Photovoltaic Characteristics of Internally Stacked CuO/Cu_2O Photoactive Layer Prepared by Electrodeposition and Heating *ACS Appl. Energy Mater.* **2** 4833–40

[2] Loon Khoo P, Satou K, Nozaki K and Izaki M 2018 External Quantum Efficiency of Electrochemically Prepared and Annealed Directly Stacked SLG/GZO/Cu_2O/CuO Photoactive Layer *J. Phys. Conf. Ser.* **1083** 012030

[3] Izaki M and Omi T 1995 Transparent zinc oxide films prepared by electrochemical reaction *Appl. Phys. Lett.* **68** 2439

[4] Khoo P L, Kikkawa Y, Shinagawa T and Izaki M 2017 Effects on external quantum efficiency of electrochemically constructed n-ZnO/p-Cu_2O photovoltaic device by annealing *AIP Conf. Proc.* **1865** 050005

[5] Yuan L, Wang Y, Mema R and Zhou G 2011 Driving force and growth mechanism for spontaneous oxide nanowire formation during the thermal oxidation of metals *Acta Mater.* **59** 2491–500

[6] Wagner C D, Naumkin A V, Kraut-Vass A, Allison J W, Powell C J and Rumble J R J 2004 *NIST Standard Reference Database 20 version 3.4*

[7] Biesinger M C 2017 Advanced analysis of copper X-ray photoelectron spectra *Surf. Interface Anal.* **49** 1325–34

[8] Poulston S, Parlett P M, Stone P and Bowker M 1996 Surface Oxidation and Reduction of CuO and Cu2O Studied Using XPS and XAES *Surf. Interface Anal.* **24** 811–20

[9] Velásquez P, Leinen D, Pascual J, Ramos-Barrado J R, Grez P, Gómez H, Schrebler R, Del Río R and Córdova R 2005 A Chemical, Morphological, and Electrochemical (XPS, SEM/EDX, CV, and EIS) Analysis of Electrochemically Modified Electrode Surfaces of Natural Chalcopyrite ($CuFeS_2$) and Pyrite (FeS_2) in Alkaline Solutions *J. Phys. Chem. B* **109** 4977–88

[10] Salvador P, Fierro J L G, Amador J, Cascales C and Rasines I 1989 XPS study of the dependence on stoichiometry and interaction with water of copper and oxygen valence states in the $YBa_2Cu_3O_{7-x}$compound *J. Solid State Chem.* **81** 240–9
[11] Meda L and Cerofolini G F 2004 A decomposition procedure for the determination of copper oxidation states in Cu-zeolites by XPS *Surf. Interface Anal.* **36** 756–9
[12] Li F, Zhang L, Evans D G and Duan X 2004 Structure and surface chemistry of manganese-doped copper-based mixed metal oxides derived from layered double hydroxides *Colloids Surfaces A Physicochem. Eng. Asp.* **244** 169–77
[13] Otamiri J C, Andersson S L T and Andersson A 1990 Ammoxidation of toluene by $YBa_2Cu_3O_{6+x}$ and copper oxides. Activity and XPS studies *Appl. Catal.* **65** 159–74
[14] Goh S W, Buckley A N, Lamb R N, Rosenberg R A and Moran D 2006 The oxidation states of copper and iron in mineral sulfides, and the oxides formed on initial exposure of chalcopyrite and bornite to air *Geochim. Cosmochim. Acta* **70** 2210–28
[15] Deroubaix G and Marcus P 1992 X-ray photoelectron spectroscopy analysis of copper and zinc oxides and sulphides *Surf. Interface Anal.* **18** 39–46
[16] Moretti G 1998 Auger parameter and Wagner plot in the characterization of chemical states by X-ray photoelectron spectroscopy: A review *J. Electron Spectros. Relat. Phenomena* **95** 95–144
[17] Satta M and Moretti G 2010 Auger parameters and Wagner plots *J. Electron Spectros. Relat. Phenomena* **178–179** 123–7
[18] Wagner C D 1978 X-ray photoelectron spectroscopy with x-ray photons of higher energy *J. Vac. Sci. Technol.* **15** 518-523
[19] Wagner C D, Gale L H and Raymond R H 1979 Two-dimensional chemical state plots: a standardized data set for use in identifying chemical states by X-ray photoelectron Spectroscopy. *Anal Chem* **51** 466–82
[20] Biesinger M C, Lau L W M, Gerson A R and Smart R S C 2010 Resolving surface chemical states in XPS analysis of first row transition metals, oxides and hydroxides: Sc, Ti, V, Cu and Zn *Appl. Surf. Sci.* **257** 2717-2730
[21] Wong L M, Chiam S Y, Huang J Q, Wang S J, Pan J S and Chim W K 2010 Growth of Cu_2O on Ga-doped ZnO and their interface energy alignment for thin film solar cells *J. Appl. Phys.* **108** 033702
[22] Kraut E A, Grant R W, Waldrop J R and Kowalczyk S P 1980 Precise determination of the valence-band edge in X-Ray photoemission spectra: Application to measurement of semiconductor interface potentials *Phys. Rev. Lett.* **44** 1620–3
[23] Nakaoka K, Ueyama J and Ogura K 2004 Photoelectrochemical Behavior of Electrodeposited CuO and Cu_2O Thin Films on Conducting Substrates *J. Electrochem. Soc.* **151** C661
[24] Izaki M, Shinagawa T, Mizuno K-T, Ida Y, Inaba M and Tasaka A 2007 Electrochemically constructed p-Cu_2O/n-ZnO heterojunction diode for photovoltaic device *J. Phys. D. Appl. Phys.* **40** 3326–9
[25] Koffyberg F P and Benko F A 1982 A photoelectrochemical determination of the position of the conduction and valence band edges of p-type CuO *J. Appl. Phys.* **53** 1173–7
[26] Heinemann M, Eifert B and Heiliger C 2013 Band structure and phase stability of the copper oxides Cu2O, CuO, and Cu4O3 *Phys. Rev. B - Condens. Matter Mater. Phys.* **87** 3–7
[27] Rödl C, Sottile F and Reining L 2015 Quasiparticle excitations in the photoemission spectrum of CuO from first principles: A GW study *Phys. Rev. B - Condens. Matter Mater. Phys.* **91**, 045102
[28] Zhang Q, Zhang K, Xu D, Yang G, Huang H, Nie F, Liu C and Yang S 2014 CuO nanostructures: Synthesis, characterization, growth mechanisms, fundamental properties, and applications *Prog. Mater. Sci.* **60** 208–37
[29] Ekuma C E, Anisimov V I, Moreno J and Jarrell M 2014 Electronic structure and spectra of CuO *Eur. Phys. J. B* **87** 1–6
[30] Wu D, Zhang Q and Tao M 2006 LSDA+U study of cupric oxide: Electronic structure and native point defects *Phys. Rev. B* **73** 235206

[31] Mishra U K and Singh J 2007 *Semiconductor Device Physics and Design* (Dordrecht: Springer Netherlands) Chap. 4.3 156

IOP Conf. Series: Materials Science and Engineering **920** (2020) 012029 doi:10.1088/1757-899X/920/1/012029

Test Condition Optimization for Testing of Hip Protectors

S. A. Yahaya[1,2,*], J. D'raj[1], Z. M. Ripin[1], M. I. Z. Ridzwan[1]
[1]School of Mechanical Engineering, Engineering Campus, Universiti Sains Malaysia, 14300, Penang, Malaysia
[2]Department of Biomedical Engineering, University of Ilorin, Ilorin, Nigeria.

*Email: yahaya.say@student.usm.my

Abstract. Determining the efficacy of hip protectors designed to reduce the menace of the alarming rise in hip fracture due to fall among the elderly population is particularly challenging. Hip protector evaluation has been done using different soft tissue surrogate and at different energy level to determine the effectiveness of various hip protectors. However, some hip protectors that had been ranked effective in mechanical testing systems proof otherwise when deployed for clinical trials, this may be due to lack of optimization of the testing condition, most notably in testing when the soft tissue surrogate attenuates impacts that may be unduly attributed to the effectiveness of the hip protector. In this study, the performance of a surrogate soft tissue, fabricated from polyethylene, was evaluated at different energy levels (3.68 J – 37.9 J) to ascertain the optimal position for evaluating the efficacy of a hip protector to simulate a representative condition of an actual fall to the sideways by a person. A drop-weight impact testing machine was used to evaluate the impact force response of an employed femoral geometry with and without the soft tissue at various residual impact energy. The result showed that the soft tissue might be responsible for up to about 95.17 % of peak impact force attenuation if not adequately modelled. Therefore various conditions were examined to get representative sideways impact condition where the soft tissue employed do not attenuate beyond 11 - 28 % of the peak impact force. This finding demonstrates that the impact condition is as critical as all other parameters in determining the performance of a hip protector.

Keywords: Test system, soft tissue, hip protectors, fall and impact force.

1. Introduction

Epidemiological studies have estimated that there would be an exponential increase in the incidence of osteoporotic fractures in Asia so that by 2050, 50% of all hip fractures would occur in this region [1]. When falls cannot be prevented, and the coordinative response of the body is too slow to break the fall, more force will be applied to the hip, making it susceptible to fracture most especially in osteoporotic bone [2]. Hip protectors prove to be an effective strategy in reducing the impact force to the hip in a sideways fall [3]. Biomechanical test and clinical trials have proven that hip fracture can be drastically reduced by wearing a hip protector [4 - 8]. Researchers have, therefore, designed various types of hip protectors to prevent hip fracture in a sideways fall [8]. Lauritzen et al. [9] showed that hip fractures in the elderly could be reduced by 50 per cent if they wore a particular design of protector.

By using mechanical testing systems, researchers have been able to provide significant insights into the efficacy of a hip protector in the prevention of hip fracture [10]. However, test results are often affected by the test conditions, mainly when tests are not carried out at representative condition of fracture causing impact [11]. Current direction is to ensure that only clinically effective hip protectors are deployed to the market. Therefore, the protective value of a hip protector must be adequately measured with an optimized test system. The influence of soft tissue is often time overlooked and not adequately modelled in many research works. Thinner soft tissue would allow higher forces to be exerted on the hip compared to a thicker one [12,13] and inappropriately chosen impact mass and residual impact energy would adversely affect results of experiments to evaluate the efficacies of hip protectors [14]. Therefore, it is appropriate to ensure that the hip assembly response as close as possible to the real situation in the test condition to evaluate the effectiveness of a hip protector.

Hence, this research aims to systematically establish a baseline force, representative of fracture causing situation, before measuring the force attenuation capabilities of hip protectors. Since the soft tissue also attenuates some impact force, the optimum test condition should be between peak force, not less than 2500 N [15] and the impact attenuation ability of the soft tissue around 16 ± 7.8 (SD) % [16].

2. Methodology

2.1. Average hip surface geometry

The surface geometry of the hip was taken from the three-dimensional (3D) coordinates describing the average surface geometry of the hip, buttock and anterior thigh region of older women in a previous study where participants consisted of 15 Canadian women with a mean age of 77.5 years, mean body mass of 61.2 kg, mean height of 1.61 m, and mean body mass index of 23.6 kg/m^2 [11]. The 3D coordinates were plotted in SolidWorks 2016 (Figure 1) to create the surface geometry.

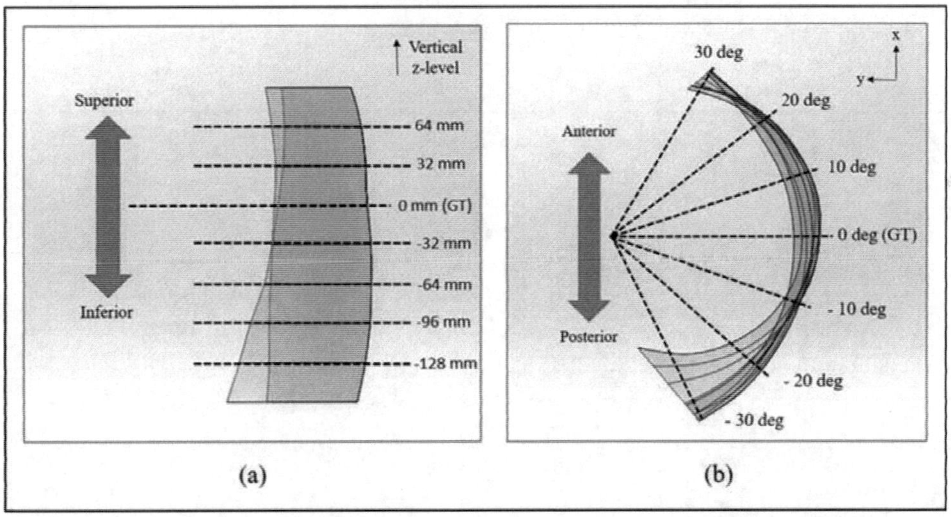

Figure 1. (a) Side view of hip surface geometry, (b) Top view of hip surface geometry used in creating the soft tissue.

The surrogate soft tissue was made using closed-cell polyethylene foams. A closed-cell polyethylene foams LD45 of density 45 kg/m^3 was used directly over the proximal femur and closed-cell copolymer foam Evazote EV50 of density 50 kg/m^3 was used over the regions anterior, posterior, and superior to the femur. The LD45 and EV50 polyethylene foam were glued together using a spray adhesive (Scotch Super 77) to form a single (220 × 150 × 90) mm^3 block. The block was then placed into a 3-axis CNC machine to machine the surface geometry of the human hip according to the SolidWorks model. The thickness of the foam above the greater trochanter was set to be 2.4 cm thick [11] (Figure 2).

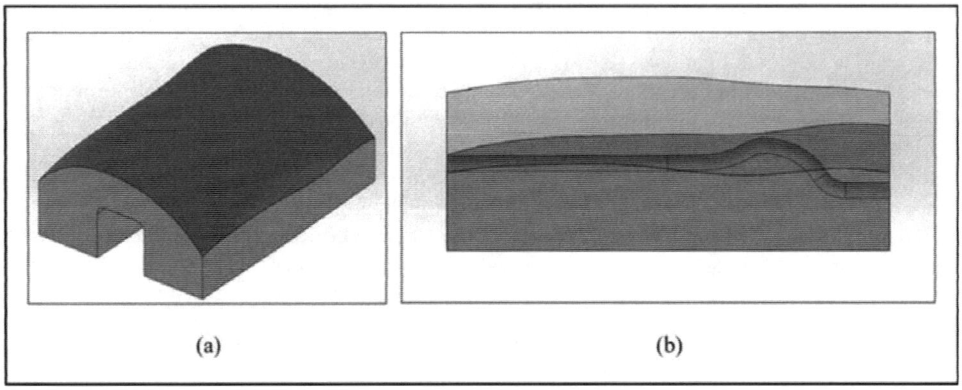

Figure 2. (a) the soft tissue model, (b) Transparent side view of the hip model.

2.2. *Simplified Femur*

Based on the geometry of femur employed by Keenan and Evans [15], a simplified model of the femur having the profile of the greater trochanter as traced from a Sawbones 4th generation computer-aided design (CAD) model in SolidWorks 2016 (Figure 3) was adopted for the testing. The Aluminium block (195 x 40 x 50) mm was machined by using a 3-axis computer numerical control (CNC) milling machine (Robodrill α-T21iFLb).

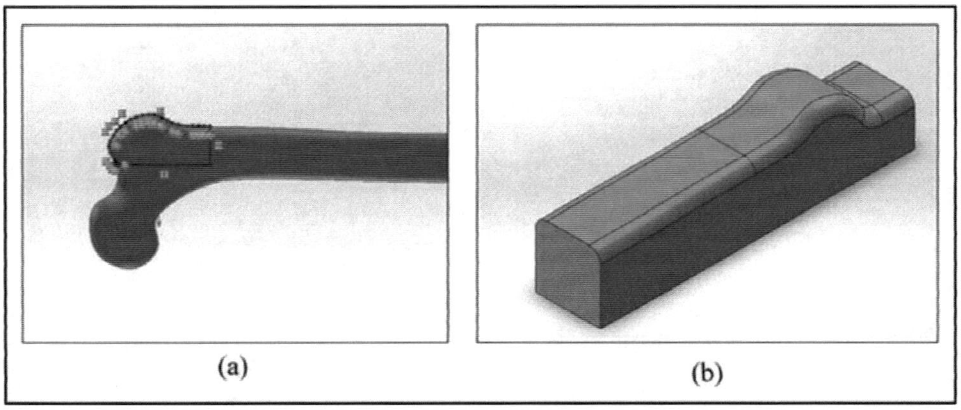

Figure 3. (a) Sawbones 3rd and 4th generation CAD model, (b) Simplified CAD femur bone.

2.3. *Experiment with impact simulator*

A test rig that closely adhered to the recommendation of the consensus from the International Hip Protector Research Group for biomechanical testing at the Copenhagen Consensus Conference [17] on the design of hip protector test rig was designed (Figure 4) but with the adoption of lower effective mass following the findings of researchers such as Mills [18] and Kannus et al. [5] and with no spring [19–22] having the intent of producing forces above the reported fracture threshold of 3.472 kN and observing the condition that mimics the impact attenuation provided by the human soft tissue [23,24].

Since it has been demonstrated that the simulated impact condition in a fall can be achieved by various combination of weights and heights [11], masses of 3.75 Kg, 4.75 kg and 5.52 Kg were dropped from a height of 100 mm, 300 mm, 500 mm, 600 mm, 700 mm, and 750 mm with 0.99 m/s, 1.72 m/s, 2.21 m/s, 2.43 m/s, 2.62 m/s and 2.71 m/s velocity respectively to ascertain the suitable drop impact height and femoral neck force for the hip arrangement to produce a representative fracture-causing impact in a surrogate hip arrangement containing an isolated femur. Hence, the impact energies were about 3.68 J, 11.04 J, 23.3 J, 27.96 J, 32.62 J, and 34.95 J, respectively. All tests were repeated three times for each impact condition.

The force transmitted to the bare femur was first ascertained at the femoral neck, followed by the force attenuated by the surrogate soft tissue. The force at the femoral neck when the soft tissue was not in place is the unprotected force F_u used in calculation and F_p is the impact recorded at the same point when the polyethylene foam protected the femur geometry. The percentage impact attenuation rate is defined as follows [25].

$$Impact\ Attenutation\ rate\ (\%) = \left(1 - \frac{F_p}{F_u}\right) \times 100\% \quad (1)$$

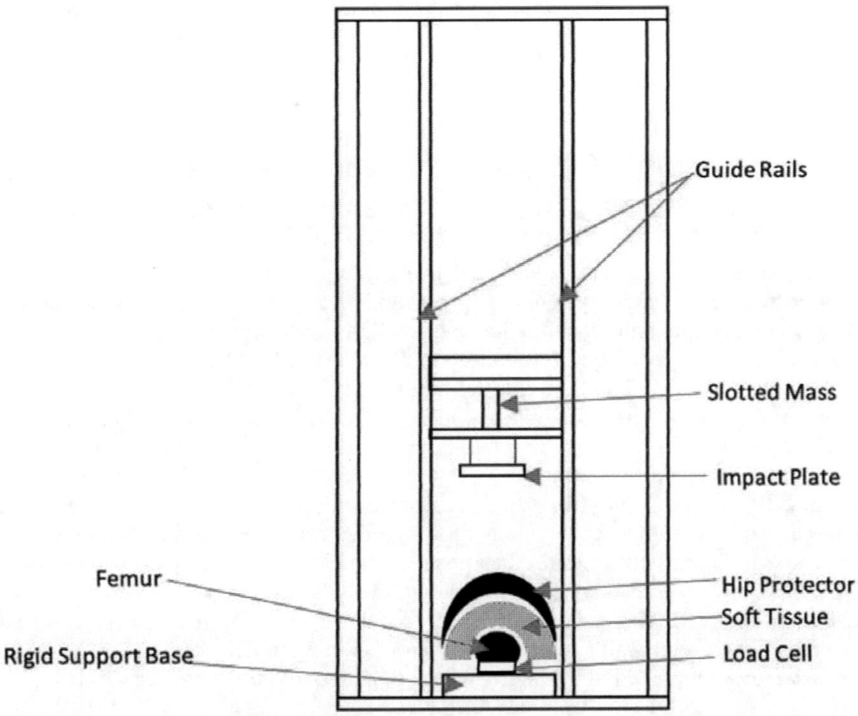

Figure 4. Graphical representation of the drop impact tester. In this case, simply justify the caption so that it is as the same width as the graphic.

3. Results

Considerable disparities were identified between the force attenuation provided by the soft tissue at different residual impact energy. The peak force was reduced up to 95.17% by the soft tissue when the impact weight was very low and up to 11.92% at very high impact energy. These differences in impact force attenuation demonstrate that the impact attenuation of the trochanteric soft tissue has to be correctly modelled when designing a compliant test system. The mean maximum peak force attenuation that is representative of impact force on the femoral neck in a sideways fall were observed at a residual impact energy of 32.5 J – 37.9 J that corresponds to 2.62 m/s – 3.71 m/s velocity of impact as shown in Table 1. These velocities correspond to the reported values of velocity in a typical fall [11,12]. The peak impact force attenuated is very high when lower impact mass and height of dropping the impact mass were low. Hence, these two factors were considerably increased one factor at a time to reach a representative test condition with minimal runs of the experiment.

Table 1. Drop Impact test results for polyethylene foam and aluminium femur

Height (cm)	Weight (kg)	Velocity (m/s)	Impact Energy (J)	Unprotected Force (kN)	Soft tissue protected force (kN)	Attenuation rate (%)
10	3.75	1.40	3.68	3.52 ± 00	0.17 ± 00	95.17%
30	3.75	2.43	11.04	3.56 ± 00	0.22 ± 00	93.82%
50	4.75	3.13	23.30	3.86 ± 01	2.00 ± 02	48.19%
60	4.75	3.43	27.96	6.40 ± 00	4.10 ± 00	35.94%
70	4.75	3.71	32.62	6.12 ± 03	4.54 ± 01	25.82%
75	4.75	3.84	34.95	6.88 ± 02	6.06 ± 00	11.92%
40	5.52	2.80	21.66	10.00 ± 08	2.53 ± 01	74.66%
50	5.52	3.13	27.08	11.52 ± 02	4.84 ± 00	57.99%
60	5.52	3.43	32.49	12.02 ± 02	7.01 ± 01	41.65%
70	5.52	3.71	37.91	12.85 ± 02	9.29 ± 08	27.74%

From the experiment conducted, the optimized arrangement for carrying out the impact attenuation test of a hip protector using our test system is gotten at an impact load of 4.75 kg at the height of 0.6 m - 0.7 m that produces an impact force within the range of 4.54 to 6.06 kN or with an impact mass of 5.52 kg at the height of 0.7 m for a more severe fall up to about 9.29 kN. At this height, the average force attenuated by the foam hip model and the aluminium femur is between 11% -27.74% which is close to the amount of force attenuated by the human tissue as previously reported [12,24,26]. Though other impact masses/impact heights variation can be employed [11] depending on the stiffness and other characteristics of the surrogate soft tissue employed, the force attenuation of the soft tissue used must first be ascertained to avoid the possibility of misrepresentation of the capability of a hip protector. Figure 5 shows the soft tissue reduces the impact force on the trochanter and increases the time to peak impact force. The experiment demonstrates the impact attenuation provided by the soft tissue could also influence hip protector testing result at different energy levels for falls to the sagittal plane.

This experiment reveals that the impact force in a sideways' fall and the percentage of attenuated force representative of an actual fall situation in human when the sagittal plane can be achieved despite the lower impact energy employed in this study when compared to previous results [8,16].

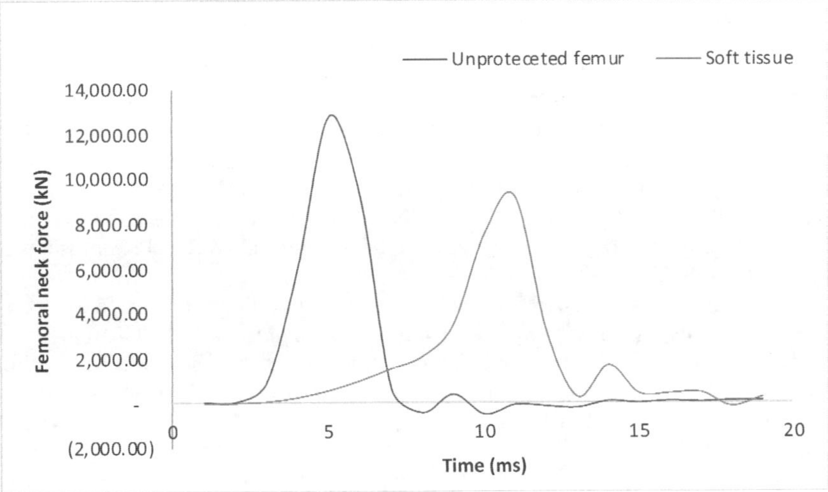

Figure 5. *A* Typical femoral neck reaction force versus time for impacts to the hip assembly with and without the soft tissue sample.

4. Conclusion

The conducted experiment shows that preliminary optimization of the test condition would ensure unrealistic impact energy or too severe femoral neck force will not be employed in evaluating the performance of a hip protector. The potential of overestimating or underestimating the capability of a hip protector is therefore eliminated, thereby ensuring only biomechanically effective hip protectors would be deployed for clinical trials or deployed for use by vulnerable individuals who depend on the intervention.

References

[1] Dhanwal D K, Dennison E M, Harvey N C and Cooper C 2011 Epidemiology of hip fracture: Worldwide geographic variation *Indian J. Orthop.* **45** 15–22

[2] Grisso J A, Kelsey J L, Strom B L, Ghiu G Y, Maislin G, O'Brien L A, Hoffman S and Kaplan F 1991 Risk Factors for Falls as a Cause of Hip Fracture in Women *N. Engl. J. Med.* **324** 1326–31

[3] Cowling P D 2004 Hip protectors and their role in hip fracture prevention: A review *McGill J. Med.* **8** 58–65

[4] Ekman A, Mallmin H, Michaëlsson K and Ljunghall S 1997 External hip protectors to prevent osteoporotic hip fractures *Lancet* **350** 563–4

[5] Kannus P and Parkkari J 2006 Prevention of hip fracture with hip protectors *Age Ageing* **35** 51–4

[6] Lauritzen J B and Askegaard V 1992 Protection against hip fractures by energy absorption *Dan. Med. Bull.* **39** 91–3

[7] Lauritzen J B 1996 Hip fractures: incidence, risk factors, energy absorption, and prevention *Bone* **18** S65–75

[8] Nabhani F and Bamford J 2002 Mechanical testing of hip protectors *J. Mater. Process. Technol.* **124** 311–8

[9] Lauritzen J B, Petersen M M and Lund B 1993 Effect of external hip protectors on hip fractures *Lancet* **341** 11–3

[10] Yahaya S A, Ripin Z M and Ridzwan M I Z 2019 Test systems for the biomechanical evaluation of hip protectors: a systematic review *Osteoporos. Int.* **31** 43–58

[11] Laing A C and Robinovitch S N 2008 The Force Attenuation Provided by Hip Protectors Depends on Impact Velocity, Pelvic Size, and Soft Tissue Stiffness *J. Biomech. Eng.* **130** 061005-1-061005–9

[12] van Schoor N M, van der Veen A J, Schaap L A, Smit T H and Lips P 2006 Biomechanical comparison of hard and soft hip protectors, and the influence of soft tissue *Bone* **39** 401–7

[13] Klingberg S, Mehlig K, Sundh V, Heitmann B L and Lissner L 2018 Lower risk of hip fractures among Swedish women with large hips? *Osteoporos. Int.* **29** 927–35

[14] Majumder S, Roychowdhury A and Pal S 2008 Effects of trochanteric soft tissue thickness and hip impact velocity on hip fracture in sideways fall through 3D finite element simulations *J. Biomech.* **41** 2834–42

[15] Keenan B E and Evans S L 2019 Biomechanical testing of hip protectors following the Canadian Standards Association express document *Osteoporos. Int.* 1205–14

[16] Robinovitch S N, Hayes W C and McMahon T A 1995 Energy-Shunting Hip Padding System Attenuates Femoral Impact Force in a Simulated Fall *J. Biomech. Eng.* **117** 409–13

[17] Robinovitch S N, Evans S L, Minns J, Laing A C, Kannus P, Cripton P A, Derler S, Birge S J, Plant D, Cameron I D, Kiel D P, Howland J, Khan K and Lauritzen J B 2009 Hip protectors: Recommendations for biomechanical testing-an international consensus statement (part I) *Osteoporos. Int.* **20** 1977–88

[18] Mills N J 1996 The biomechanics of hip protectors. *Proc. Inst. Mech. Eng. H.* **210** 259–66

[19] Mills N J 2007 Chapter 17 - Hip protector case study ed N J B T-P F H Mills *Polym. Foam. Handb.* 403–24

[20] Derler S, Spierings A B and Schmitt K-U 2005 Anatomical hip model for the mechanical testing of hip protectors *Med. Eng. Phys.* **27** 475–85

[21] Srewaradachpisal S, Tangtrakulwanich B, Dechwayukul C and Wongsiri S 2011 Hip Protectors : Comparative Study of FEM Simulation and Testing *The Second TSME International Conference on Mechanical Engineering* (Krabi: TSME) p BME09

[22] Minns J, Dodd C, Gardner R, Bamford J and Nabhani F 2004 Assessing the safety and effectiveness of hip protectors. *Nurs. Stand.* **18** 33–8

[23] Li N, Tsushima E and Tsushima H 2013 Comparison of impact force attenuation by various combinations of hip protector and flooring material using a simplified fall-impact simulation device *J. Biomech.* **46** 1140–6

[24] Melo J D D, Barbosa A S G and Guerra R O 2010 Mechanical Evaluation of Polymer Composite Hip Protectors *Int. J. Biomater.* **2010** 1–6

[25] Zhang T, Song Y and Gao S 2011 Research on the characteristics of buffer material used as hip protector *5th International Conference on Bioinformatics and Biomedical Engineering, iCBBE 2011* (IEEE) pp 1–4

[26] Robinovitch S N, McMahon T A and Hayes W C 1995 Force attenuation in trochanteric soft tissues during impact from a fall *J. Orthop. Res.* **13** 956–62

Reliability analysis on the reservoir dam spillway structure using fluid-structure interaction

Mohd Hafiz Zawawi[1], F C Ng[2], Aqil Azman[2], L H Tung[2], Ismail Abustan[3], Z Mohd Remy Rozainy[3] and Aizat Abas[2,*]

[1] Department of Civil Engineering, College of Engineering, Universiti Tenaga Nasional, 43000 Kajang, Selangor, Malaysia.
[2] School of Mechanical Engineering, Universiti Sains Malaysia, Engineering Campus, 14300 Nibong Tebal, Penang, Malaysia.
[3] School of Civil Engineering, Universiti Sains Malaysia, Engineering Campus, 14300 Nibong Tebal, Penang, Malaysia.

**Corresponding author*: aizatabas@usm.my

Abstract. This paper presents a combined reliability and flow analysis on the spillway of reservoir dam using the numerical fluid-structure interaction (FSI) approach. The spillway is an important structure to the operation of reservoir dam, as it regulates the water flow from upstream to downstream. Therefore, a rapid discharge of water from the reservoir along the spillway will causes possibly flooding at downstream and incurs high structural stress. Using the ANSYS software, the two-way FSI problem of a dam spillway is numerically simulated. Such FSI numerical study able to simultaneously simulate and obtain both the parameters on fluid flow and structural deformation. The current study shows that the downstream has the highest flow velocity; while the highest deformation of 10.7 mm being observed in the connection bridge section. As the general stress level of the overall spillway structure (max. 30.75 MPa) is less than the yield stress of corresponding materials, the spillway will not exhibit any structural failure upon the current discharge mode.

Keywords: Dam; Fluid-structure interaction (FSI); Reliability analysis; Spillway structure; Stress

1. Introduction

Dam is a crucial infrastructure to the everyday need and development of civilization. It collects and holds water in the reservoir, for which the water supply to be distribute over several regions. Dam also function as river mitigation, reduces the occurrence of flooding during the rainfall season. Moreover, dam frequently build with a hydroelectrical power station. The upstream water flow from the reservoir dam will have its potential energy converted into kinetic energy in the turbine, which finally converted to useful electricity [1, 2].

Spillway structure is built in the dam to regulate the water flow and thus to control the water level of dam reservoir. Therefore, the spillway is an important structure to prevent overtopping of dam reservoir which may leads to dam failure and the flooding at downstream area. Other factors that causes to dam break includes the induced vibration [3]. Currently, the issue on dam failure is of particular interests among the researchers. This is coincident with the catastrophically consequences shall the dam failed, as it detrimental to human lives, infrastructure and social economy [4, 5].

Reliability analyses on the dam structures are essential as a part of maintenance works to prevent any failures during normal operating or overtopping conditions. A regular-timed reliability analysis not

IOP Conf. Series: Materials Science and Engineering **920** (2020) 012030 doi:10.1088/1757-899X/920/1/012030

reduce the risks of failures but also flawless dam operations, thus saving the costs and eliminate the incurs losses [6, 7]. In the past literatures, there are various numerical studies on the dam reliability being conducted. Particularly by the means of fluid-structure interaction (FSI) numerical approach. FSI enables the instantaneous tracking of fluid flow and structure deformation, thus a more accurate depiction of actual phenomenon. There are few past literatures on the FSI study on the dam spillway structure [4, 5, 8]. The variation effect of sector gate opening were studied in Ref. [4, 5]; while the water flow along the spillway and downstream were numerically visualized in Ref. [8]. There is also a FSI reliability analysis being conducted to access the performance of dam banks operating at normal conditions [9].

This paper demonstrates the further application of FSI numerical scheme on the dam reliability analysis, particularly the spillway structure. In the present case, the free surface flow at the reservoir dam level is imposed. Lastly, the structural impact due to the water flow from reservoir was justified to determine for any possible cracking and failure.

2. Numerical simulation

Using the commercially available software, Ansys, the fluid-structure interaction (FSI) of dam spillway structure is numerically simulated. The associated primary modules are Fluent and Transient Structure, respectively for simulating the fluid flow and structural deformation [10, 11]. Both modules were later coupled and simultaneously solved by System Coupling module. For current study, two-way FSI is established by considering both the data transfers from fluid to structure and from structure to fluid. It is vital to consider a throughout two-way data transfer in the current reliability analysis to capture the realistic depiction of spillway flow problem by considering the relationship between water flow and structure deformation.

In the settings of Ansys Fluent, the three-dimensional water flow across the spillway is modelled to be incompressible, turbulence and unsteady. To track the instantaneous position of water flow, multiphase volume of fluid (VOF) model was employed with the implicit scheme. Meanwhile, the flow turbulence was modelled using the k-ε turbulence model to enable the capture the turbulence properties of spillway flow. Whereas for the settings in Ansys Transient Structure, standard deformation model was applied. The material assignment on the spillway structure is shown in Figure 1. The connection bridge sector is made up of structural steel while elsewhere of the spillway is made up of concrete.

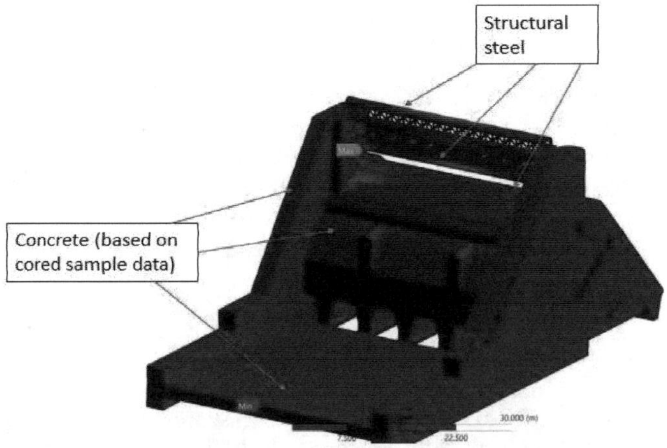

Figure 1. Material assignments on the spillway structure.

On the boundary conditions as shown in Figure 2, all the interfacial boundaries between fluid and structure are designated as no-slip wall and set to be fluid/structure interaction surfaces. At the upstream, the water is set to be at the ambient condition of atmospheric pressure equals to 0 Pa (gauge). The initial water level is set to be 10 feet above the ground level. The bottom part of the spillway structure is designated as fixed support. The modelling of gravity of magnitude 9.81 m/s^2 is also enabled.

The unstructural mesh of optimized mesh sizing were generated on both the fluid and structural domains. In the System Coupling, the data transfers were created between fluid and structural domains. First order implicit upwind and coupled schemes were adopted for the current simulation. Finally, optimum time size was used to execute the two-way FSI simulation.

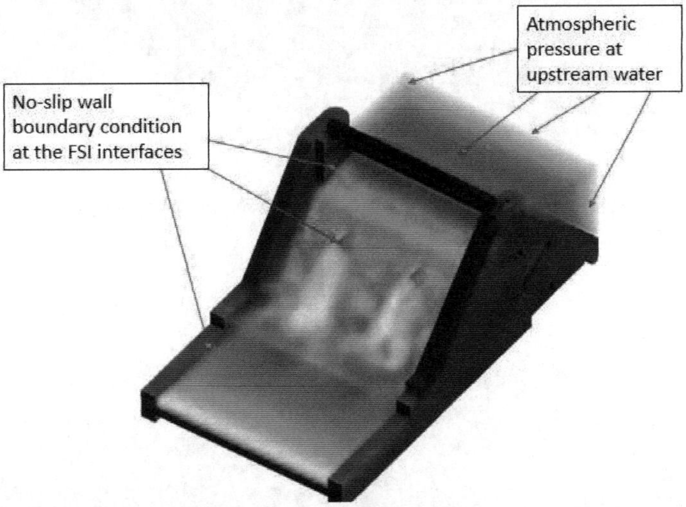

Figure 2. Boundary conditions applied for the fluid/structure interactions study of spillway.

3. Results and Discussion

This section depicts the FSI simulation findings on the reliability analysis of spillway structure. The high velocity region is observed at the spillway opening gap with a maximum value of 6.448 m/s and at the baffle block on the downstream spillway surface, as shown in Figure 3. Figure 4 shows the hydrostatic pressure across different upstream water levels. The deformation contour in Figure 5 shows that the connection bridge section sustained the largest deformation of 10.7 mm. However, such deformation appeared to be insignificant upon being experienced by the on-site personnel; thus negligible. Figure 6 depicts that the highest localized stress region at the sector gate being approximately 30.7 MPa. Meanwhile, the average nominal stress level on the spillway structure is within 10 – 20 MPa. As the maximal stress level of spillway structure is lower than the yield stress for structural steel of 370 MPa, no structural failure will be observed. Other sections of the spillway structure have stress level below 3.417 MPa, which is well below the yield stress of concrete of 20 MPa. Nonetheless, preventive care should be emphasized the high stress region, follows with regular maintenance to prevent the occurrence of cracking.

IOP Conf. Series: Materials Science and Engineering **920** (2020) 012030 doi:10.1088/1757-899X/920/1/012030

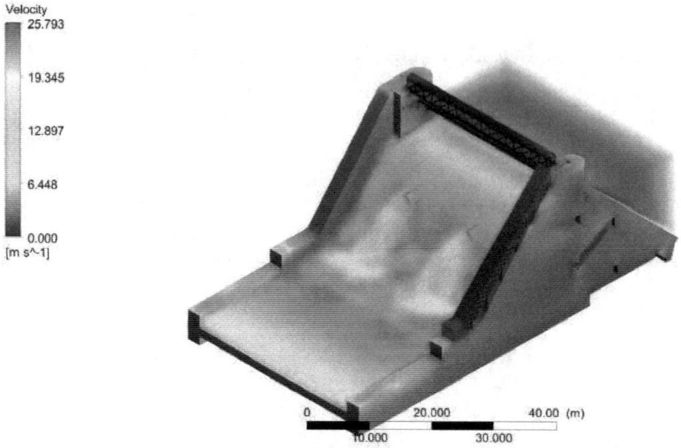

Figure 3. Velocity distribution of water flow along the spillway.

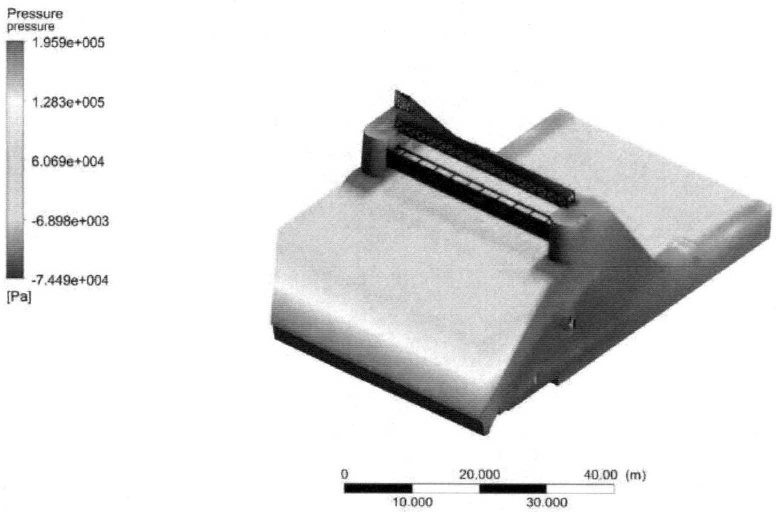

Figure 4. Pressure distribution of water along on the upstream reservoir.

Figure 5. Deformation distribution of spillway structure.

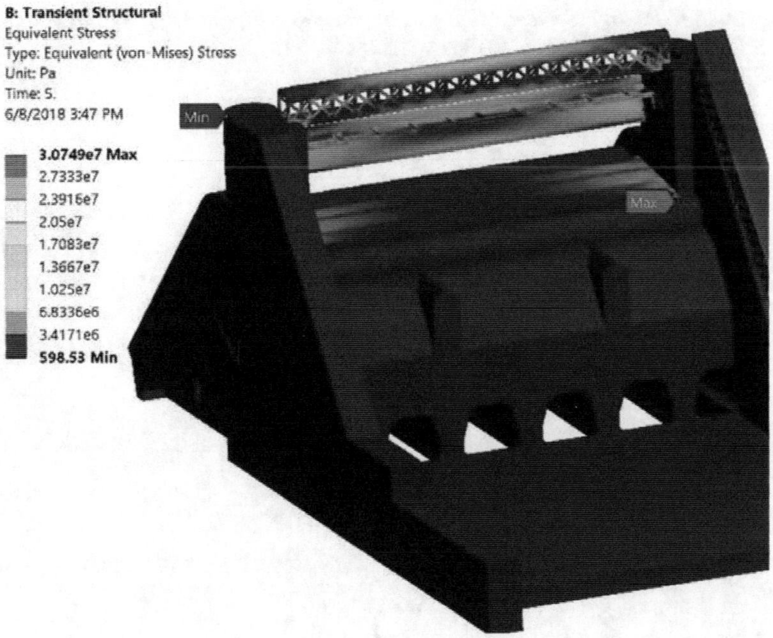

Figure 6. Stress distribution of the spillway structure.

4. Conclusions

In this paper, the fluid-structure interaction (FSI) numerical reliability analysis of dam spillway was successfully simulated using the commercial software ANSYS. The investigated spillway structure was subjected to upstream water flow with water height of 10 feet. The water flow with highest velocity was observed on the downstream. For the structural aspect, the connection bridge section sustains highest deformation of 10.7 mm and averagely has highest regional stress. As the overall stress level in the spillway structure is far less than the yield strengths of steel and concrete, no structural failure will be observed. Generally, this study would benefit the reservoir dam sector for the prediction of dam failure. Subsequently, suitable countermeasures can be taken accordingly based on the result of reliability analysis.

References:

[1] L M Sidek, F C Ros and N H A. Aziz 2011 *Student Conference on Research and Development*
[2] S Solava and N Delatte 2003 *Proceedings of the Third ASCE Forensics Congress*
[3] M H Zawawi, N A Aziz, M R M Radzi, N H Hassan, M Z Ramli, N M Zahari, M A Abbas, A Saleh, A Salwa, Z C Muda 2018 *AIP Conference Proceedings.* **2030** 020245
[4] F C Ng, A Abas, I Abustan, Z M R Rozainy, M. Z. Abdullah, A. B. Jamaludin and S. M. Kon 2018 *IOP Conf. Ser. Mater. Sci. Eng.* **370**(1) 012062
[5] F C Ng, A Abas, I Abustan, Z M R Rozainy, M. Z. Abdullah, A. B. Jamaludin, and S. M. Kon 2018 *IOP Conf. Ser. Mater. Sci. Eng.* **370**(1) 012063
[6] G L S Babu and A Srivastava 2010 Journal of Geotechnical and Geoenvironmental Engineering **136**(7)
[7] Francisco Ríos Bayona 2014 *KTH, Royal Institute of Technology*, Stockholm
[8] M N Nashurdin, A Abas, A Azman, F C Ng, M R M Radzi and A Hassani 2019 *AIP Conference Proceedings* **2129** 020044
[9] F C Ng, A Azman, A Abas, MH Zawawi, AZ Ahmad Mazlan, NM Zahari and Z Mohd Remy Rozainy 2020 *IOP Conf. Series: Materials Science and Engineering* **815** 012021
[10] ANSYS Fluent Documentation 2017 *ANSYS Inc.*
[11] ANSYS Mechanical Documentation 2017 *ANSYS Inc.*

IOP Conf. Series: Materials Science and Engineering **920** (2020) 012031 doi:10.1088/1757-899X/920/1/012031

Effect of the oil spill on the soil friction angle

Harris Ramli*, Hafizul Saat, Mastura Azmi and Fauziah Ahmad
School of Civil Engineering, Engineering Campus, Universiti Sains Malaysia, 14300 Nibong Tebal, Penang, Malaysia

*Corresponding author: cehmr@usm.my

Abstract. There are many factors governs the soil stability, and strength is one of the most important aspects. In the event of soil contamination by oil spills, the soil characteristic and its physical properties will change due to the intrusion of oil in the subsurface. This changes mainly related to the viscous properties of the oil in comparable to natural groundwater. As the soil friction angle is one of the main components of the soil shear strength, the viscosity of the oil will affect the soil friction angle. In this study, three types of typical sand were used to mix with water and oil, and the soil friction angle is determined using direct shear test on these samples under dried condition, 50% of soil saturation and at optimum moisture content. The three sand is classified as Sandy clay, Silt clay with sand and Silt with sand. It is found all samples followed soil compaction principal when the sample mixed with water or oil, except for Sandy clay soil. For sandy clay, as the soil friction angle supposed to increase when reaching the optimum moisture content (ϕ = 38.7°), the soil friction angle is continued to reduce (ϕ = 34.2°). This may due to less than 60% coarse grain of the grain size distribution in sandy clay. As coarse grain is the main factor for the grain angularity function that provides the interlocking between particles.

1. Introduction

Most oil spill cases happened in unexpected situation in the land and sea during transfer of the oil or leakage of storage tanks. However, there are oil spill cases happened intentionally as what occur in 1991 at Persian Gulf War [1], [2]. Quick action for remediation of the contaminated land is crucial as it will not just contaminate the ground but also affect the geotechnical properties of the soil. A better understanding of the effect of soil contaminated with hydrocarbon may be of help in assessing the recovery potential of a soil [3].

Soils might be contaminated by oil due to the leakage of the oil. Major tasks are required for reclamation and recovery of the contamination area [4]. Oil products such as heavy fuel oil, diesel oil or gasoline are the most common chemical spills occurred [5]. Other than that, in heavily industrial places, oil pollution is common where careless operators do not take consideration about the environmental issues. Such places include fuel depots and army polygons [6].

Benzene, toluene, ethylbenzene and xylene (BTEX) are one of the compositions in the fuel oil's contents which are hazardous materials subject to regulations in many parts of the world [7], [8]. Transportation of oil from one place to another by pipeline or vehicle is increasing the potential of an oil spill accident. Oil spill that occurred on land will affect the structure above ground because the soil structure and stability changed due to different physical and chemical characteristic of the oil.

The ground stability, mainly related to shear strength, is one of the most important key aspects to ensure the stability of civil engineering structure and infrastructure. As the soil friction angle is one of the main components of the soil shear strength, the viscosity of the oil will affect the soil friction angle. Therefore, this study is focused on the effect of the oil spill on the soil friction angle.

2. Method and Laboratory Tests

The focus of this research is to understand the effect of oil spill on the soil friction angle. Three typically sand used for construction at Penang, Malaysia was used in this research. The soil was classified before it mixed with the oil. The oil used was diesel and the properties are; density of 0.850 g/cm^3, kinematic viscosity of 1.5 – 5.8 mm^2/s at 40 °C, and vapor pressure of < 1hPa at 20 °C.

The soil samples are oven dried for 24 hours to disperse off all water from the samples. Samples need to be oven dried to control moisture content for each test. If the soil is partially dry, it will affect the result of the tests since it going to be mixed with diesel. Specific gravity is defined as the ratio of the unit weight of a given material to the unit weight of water. The small pycnometer method is used in this laboratory testing.

Sieve analysis test used to determine particle size distribution of larger diameter. To obtain the grain size distribution of soils, there are two separate procedures; mechanical sieve size for particle bigger than 63 μm and hydrometer analysis for smaller than 63 μm. The mass retained are recorded and the test is performed in accordance with BS 1377: Part 2: 1990 [9].

Liquid limit is the minimum water content at which the soil will flow under a small disturbing force. It is the boundary between the plastic and semi-solid states. In this experiment, liquid limit test was carried out by using cone penetrometer method as in BS 1377: Part 2: 1990: Clause 4.3.1. This method covers the determination of the liquid limit of a sample of soil in its natural state, or of a sample of soil from which material retained on a 425 μm test sieve has been removed.

Soil compaction Proctor test is carried out to determine the optimum moisture content and maximum dry density of a compacted specimen. In this standard Proctor test, the sample is compacted in a mould that has volume of 1000.98 cm³. The diameter and height of the mould are 105 mm and 115.6 mm respectively. 4-6 % of water is added, and the soil is mixed evenly. In this test, 4 % increment of water is being added to the soil. In this study, soil strength tested using direct shear test is intended to determine the shear strength of soil based on BS 1377: Part 7: 1990 [10]. Preparation of the samples is based on soil under clause 4.4.

3. Results and discussion

In this research, three typical sand used for construction in Penang, Malaysia was collected to be used to study the soil friction angle changes due to oil spill. Each sample has a different percentage of particle size. The percentage of gravel, sand, silt and clay can be determined from the particle size distribution. The ranges for the percentage of particle size are tabulated in Table 1. Sample A have higher composition of clay compared to sample B and sample C.

The soil Atterberg limits are determined at which the moisture content at the point of transition from semisolid to plastic state is the plastic limit (PL), and from plastic to liquid state is the liquid limit (LL). Liquid limit determined by the moisture content at 20 mm cone penetration. From experiments carried out liquid limit obtained from the graph of penetration resistance to moisture content for all three samples. Based on the graph, the liquid limit for sample A is 49.97%, sample B is 58.67% and sample C is 38.39%. Plastic index (PI) for all three samples were determined using the formula PI = LL – PL. Plastic index for sample A is 22%, sample B 29% and sample C 13%. The Atterberg limits result and soil classification is shown in Table 2.

Based on the results obtained, the soil can be classified according to the American Association of State Highway and Transportation Officials (AASTHO) Classification System. Sample A is classified as Sandy clay, sample B as Silt clay with sand and sample C as Silt with sand.

Based on Unified Soil Classification System (USCS). For sample A, the soil is classified as poorly graded sand with clay (SP-SC) because PI > 7 and plots above "A" line of Casagrande's Plasticity Chart [11]. However, for both sample B and sample C, the soil is classified as well graded sand with clay (SW-SC).

Table 1: Percentage of Gravel, Sand, Silt and Clay.

<table>
<tr><th>Particle</th><th colspan="2">Standard</th><th>Particle Name (Size, mm)</th><th>Sample A</th><th>Sample B</th><th>Sample C</th></tr>
<tr><td rowspan="4">Gravel</td><td>►</td><td>BS 1377</td><td>Fine Gravel (2 – 6 mm)</td><td rowspan="4">0.6</td><td rowspan="4">20.6</td><td rowspan="4">22.0</td></tr>
<tr><td>▲</td><td>AASHTO</td><td>Gravel (2 – 75 mm)</td></tr>
<tr><td>□</td><td>ASTM</td><td>Coarse Sand (2 – 4.75 mm)</td></tr>
<tr><td>○</td><td>USCS</td><td>Coarse Sand (2 – 4.75 mm)</td></tr>
<tr><td rowspan="4">Sand</td><td>►</td><td>BS 1377</td><td>Sand (0.060 – 2 mm)</td><td>57.2</td><td>50.2</td><td>57.6</td></tr>
<tr><td>▲</td><td>AASHTO</td><td>Sand (0.075 – 2 mm)</td><td rowspan="3">56.4</td><td rowspan="3">48.9</td><td rowspan="3">56.5</td></tr>
<tr><td>□</td><td>ASTM</td><td>Sand (0.075 – 2 mm)</td></tr>
<tr><td>○</td><td>USCS</td><td>Sand (0.075 – 2 mm)</td></tr>
<tr><td rowspan="4">Silt</td><td>►</td><td>BS 1377</td><td>Silt (0.002 – 0.060 mm)</td><td>9.2</td><td>20.4</td><td>20.4</td></tr>
<tr><td>▲</td><td>AASHTO</td><td>Silt (0.005 – 0.075 mm)</td><td rowspan="2">7.1</td><td rowspan="2">15.5</td><td rowspan="2">17.2</td></tr>
<tr><td>□</td><td>ASTM</td><td>Silt (0.005 – 0.075 mm)</td></tr>
<tr><td>○</td><td>USCS</td><td>Fines (< 0.075 mm)</td><td rowspan="2">43.0</td><td rowspan="2">30.5</td><td rowspan="2">21.5</td></tr>
<tr><td rowspan="4">Clay</td><td>○</td><td>USCS</td><td>Fines (< 0.075 mm)</td></tr>
<tr><td>►</td><td>BS 1377</td><td>Clay (< 0.002 mm)</td><td>33.0</td><td>8.8</td><td>0</td></tr>
<tr><td>▲</td><td>AASHTO</td><td>Clay (< 0.005 mm)</td><td rowspan="2">35.9</td><td rowspan="2">15</td><td rowspan="2">4.3</td></tr>
<tr><td>□</td><td>ASTM</td><td>Clay (< 0.005 mm)</td></tr>
</table>

Table 2: Soil Atterberg Limits and Soil Classification

	Sample A	Sample B	Sample C
Specific Gravity	2.67	2.65	2.63
Liquid Limit	49	58	38
Plastic Limit	27.2	29.8	25.6
Plastic Index	22	29	13
AASTHO Classification	Sandy clay (A-6, A-7-6)	Silt clay with sand (A-1-b)	Silt with sand (A-1-B,A-2-4, A-2-5, A-2-7)
USCS Classification	Poorly Graded Sand with Clay (SP-SC)	Well Graded Sand with Clay (SW-SC)	Well Graded Sand with Clay (SW-SC)

Proctor tests were performed on each sample to determine the optimum moisture content and maximum dry density of the soil. Optimum moisture content for Sandy clay, Silt clay with sand and Silt with sand are 19.7%, 15.3% and 13.4% respectively. The compaction curve of the proctor test for all samples is shown in Figure 1.

IOP Conf. Series: Materials Science and Engineering **920** (2020) 012031 doi:10.1088/1757-899X/920/1/012031

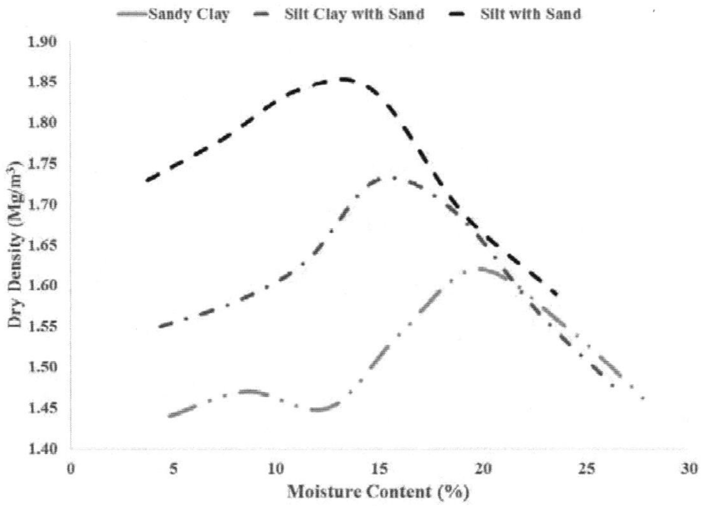

Figure 1. Compaction curve for all proctor test.

Direct shear test done was to determine the friction angle of soil and soil cohesion. Direct shear test is carried out on all three samples to get the friction angle of soil. Soil condition tasted in dried condition, 50% of soil saturation and optimum moisture content. The influence of moisture content on both shear strength parameters is very significant. As can be seen from the Figure 2, increases in the moisture content in the soil has caused a decrease in the internal friction angle but at optimum moisture content, the friction angle value increasing for all three samples is sandy clay, silt clay with sand and silt with sand.

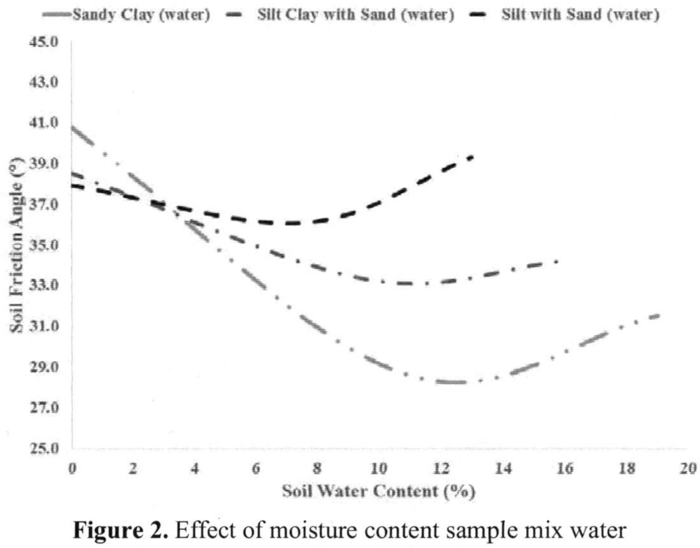

Figure 2. Effect of moisture content sample mix water

When the soil is in dry state, it is shown that the friction angle has highest value for all three samples. The value of friction angle, ϕ for sandy clay at dry condition is 40.8°, silt clay with sand is 38.5° and silt with sand is 37.9°. Similar condition found by sadek et. al (2011) when a dry sandy soil demonstrates a higher soil friction angle and shear strength compared to moist and wet sandy soil.

When the dry soil matrix in contact with water, water will start to fill the available void. Water also might replace the solid (soil) matrix volume into a water volume. When moisture content increased as a result of the existence of water, the distance between particles increased significantly. This cause a significant reduction of friction angle when compared to dry condition.

As the amount of water increases, water became as lubricant for soil particle to glide in between particle void and rearrange to obtain the most compact particle packing. In the process, the soil friction angle increased gradually. The most compact particle packing can be achieved when the soil is at optimum moisture content. As at optimum moisture content, the soil can also achieve the maximum dry density. This is also the condition where the highest soil strength can be achieved.

In considering for sandy soil, major component of soil shear strength is obtained mainly from the soil friction angle rather than soil cohesion. Thus, at maximum dry density, it is considered the highest soil friction angle value could be reached for the tested sandy soil. In accordance with compaction curve behaviour and principle, as the dry density would reduce if the soil is subjected to further moisture increment after the optimum moisture content, the soil friction angle should also be reduced.

Similar condition also occurs when the dry soil is in contact with oil. As shown in Figure 3, when the dry soil matrix in contact with oil, oil will start to fill the available void. With the increases in the moisture content of oil in the soil has caused a decrease the soil friction angle. When the oil content increase after the dry condition, both Silt clay with sand and Silt with sand exhibits a similar behaviour as the increasing of water content after the dry condition.

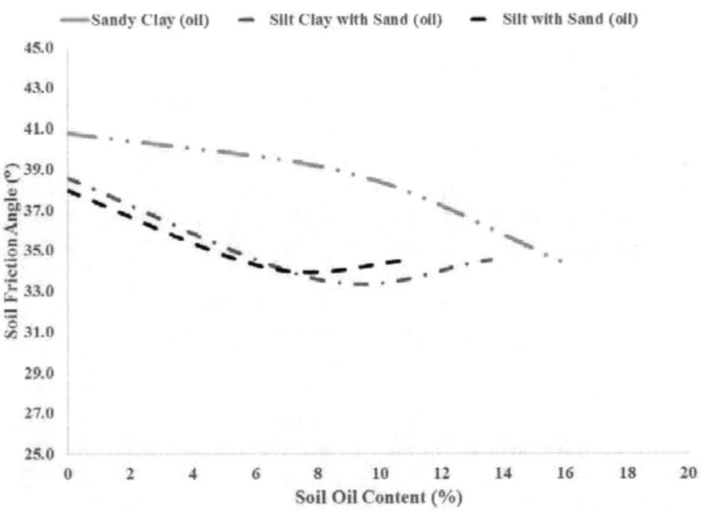

Figure 3. Effect of Moisture Content sample mix oil

Sandy clay soil, however, shown an opposite behaviour as the soil friction angle will be kept reduce even at the point of reaching the optimum moisture content. Main reason of the behaviour is since both Silt clay with sand and Silt with sand has the main grain size distribution from coarse grain (gravel and sand); about 70% to 80%. Unlike the sandy clay, less than 60% of the grain size distribution is from coarse grain. As coarse grain is the main factor for the grain angularity function that provides the interlocking between particle, the soil that contains abundance coarse grain may not be affected by the increase of oil content. In sandy clay soil, there is a possibility for the clay fraction instead it is acted as a filler inside a coarse grain particle, it may start to flocculate and becoming a larger clump when the clay fraction in contact with the oil. Such condition may cause the reduction of soil friction angle as the soil particle may not be able to rearrange its position to reach the maximum packing condition.

4. Conclusion

The soil friction angle is one of the most important components in soil strength. The effect of soil contaminated with diesel has been chosen for this study. Three types of typical sand have been chosen, and one of the sands is having slightly higher clay content. The soil friction angle is determined from the direct shear test. For all three samples, the soil friction angle is very high when it is in dry condition. Soil friction angle decrease when moisture content of water in the soil increase from the dry state; but friction angle increased at optimum moisture of water. A similar condition occurs when the oil content increase from the dry state towards the optimum moisture content for silt clay with sand and silt with sand. In contrast for sandy clay sample, the soil friction angle keeps reducing even when the oil content reaching the optimum moisture content value.

Acknowledgement

The authors wish to express sincere appreciation for the support and encouragement provided by the School of Civil Engineering, Universiti Sains Malaysia in making this project a success. The research work was funded by the Department of Higher Education, Ministry of Education Malaysia under the Fundamental Research Grant Scheme with grant number FRGS/1/2018/TK01/USM/03/4.

References

[1] A. H. Al-Rabeh, H. M. Cekirge, and N. Gunay, "Modeling the fate and transport of Al-Ahmadi oil spill," *Water, Air, Soil Pollut.*, vol. 65, no. 3–4, pp. 257–279, 1992.

[2] M. C. Bruckberger *et al.*, "Biodegradability of legacy crude oil contamination in Gulf War damaged groundwater wells in Northern Kuwait," *Biodegradation*, vol. 30, no. 1, pp. 71–85, 2019.

[3] V. Labud, C. Garcia, and T. Hernandez, "Effect of hydrocarbon pollution on the microbial properties of a sandy and a clay soil," *Chemosphere*, vol. 66, no. 10, pp. 1863–1871, 2007.

[4] M. Khamehchiyan, A. Hossein Charkhabi, and M. Tajik, "Effects of crude oil contamination on geotechnical properties of clayey and sandy soils," *Eng. Geol.*, vol. 89, no. 3–4, pp. 220–229, 2007.

[5] S. Halmemies, S. Gröndahl, M. Arffman, K. Nenonen, and T. Tuhkanen, "Vacuum extraction based response equipment for recovery of fresh fuel spills from soil," *J. Hazard. Mater.*, vol. 97, no. 1–3, pp. 127–143, 2003.

[6] J. Trzciński, D. J. Williams, and M. S. Zbik, "Can hydrocarbon contamination influence clay soil grain size composition?," *Appl. Clay Sci.*, vol. 109–110, pp. 49–54, 2015.

[7] A. Serrano, M. Gallego, and J. L. González, "Assessment of natural attenuation of volatile aromatic hydrocarbons in agricultural soil contaminated with diesel fuel," *Environ. Pollut.*, vol. 144, no. 1, pp. 203–209, 2006.

[8] B. Rajasekhar, I. M. Nambi, and S. K. Govindarajan, "Human health risk assessment for exposure to BTEXN in an urban aquifer using deterministic and probabilistic methods: A case study of Chennai city, India," *Environ. Pollut.*, vol. 265, 2020.

[9] BS 1377-2:1990, *Methods of test for soils for civil engineering purposes. Classification tests*. British Standards Institution, 1990.

[10] BS 1377-7:1990, *Methods of test for soils for civil engineering purposes. Shear strength tests (total stress)*. British Standards Institution, 1990.

[11] A. Casagrande, "Classification and identification of soils," *Trans. ASCE*, vol. 113, pp. 901–991, 1948.

[12] M. A. Sadek, Y. Chen, and J. Liu, "Simulating shear behavior of a sandy soil under different soil conditions," *J. Terramechanics*, vol. 48, no. 6, pp. 451–458, 2011.

IOP Conf. Series: Materials Science and Engineering **920** (2020) 012032 doi:10.1088/1757-899X/920/1/012032

Mechanical behaviour of Steel fibre reinforced concrete beams: a review

Al-Khwaja Aysar Fares Amin[1], Choong Kok Keong[1*] and Tan Geem-Eng[2]

[1]School of Civil Engineering, Unversiti Sains Malaysia
14300 Nibong Tebal, Pulau Pinang
[2] RIVO Precast SDN. BHD.
41050 Klang, Selangor, Darul, Ehsan, Malaysia

*E-mail: cekkc@usm.my

Abstract. Concrete is one of the most widely utilized material in the world. It is mainly used in construction due to its capability to be formed to any required shape. This because of the good workability of the fresh concrete. However, concrete is still a brittle material that makes it tends to fail under tension. Thus, a material with high tensile strength needs to be included in the mixture such as steel rebars and steel fibres (SF). This review paper summarizes previous studies that investigated the effect of SFs addition on the mechanical properties of reinforced concrete (RC) beams. Researchers in all previous studies agreed that the addition of fibres in RC beams can noticeably improve the mechanical properties, serviceability, and durability of RC beams. Generally, the effect of SFs on the RC beams was dependent on various factors such as aspect ratios, volume percentages, and type of SFs. It was confirmed that the optimum SFs content ranged from 0.75 to 1.5 V.%, depending on the strength of concrete. However, more studies are required to further understand the effect of fibres on the properties of concrete in beams with different cross-sectional shapes.

1. Introduction

One of the most valuable properties of the SFs is the excellent resistance to crack propagation and cracking. Commonly, SFs with reinforced concrete were used to monitor the cracks that happened due to both plastic and drying shrinkage. Due to the ability to arrest cracks, SFs tend to enhance the tensile strength of the concrete mixture only when the SF modulus of elasticity is larger than the mortar binder or concrete, also it can hold the concrete matrix even after wide cracking [16,20]. SFs not only improve the tensile strength, but also can enhance the extensibility of the concrete, at both ultimate and first crack, under flexural loading [4,16]. Another important factor that affect the concrete mix is the aspect ratio of SFs (length/diameter of SFs). Generally, the aspect ratio of SFs is proportional to the toughness and the flexural strength of concrete. However, sometimes too long SFs tend to form "ball" in the concrete mix, creating workability problems [20]. The inclusion of SFs not only improve the strength and toughness of concrete, but it can also lower the permeability of the RC. The reduced permeability of concrete can also lead to reduction in the water bleeding [19,20]. Several types of SFs generate larger shatter resistance, abrasion, and impact in concrete. The net results of the majority of studies that have been performed on the steel-fibres-reinforced concrete (SFRC) have agreed that inclusion of SFs have the ability to alter the character of the SFRC from brittle to ductile material. This changing will basically increase some characteristics of the SFRC. For instance, the energy absorption and its ability to resist repeatedly applied, impact or shock loading [1].

2. Effect of SFs on the mechanical properties of concrete

2.1. Effect on compressive strength

The inclusion of SFs has been shown to improve compressive strength of concrete. However, the improvement in the compressive strength of concrete is influenced by the volume content of the SFs. For example, Van Chanh el al. (2004) studied the inclusion of 1% SFs by volume which corresponds to 78.6 kg/m^3, and they indicated that this content slightly increased the compressive strength from 0 to 25 % [16]. Similarly, 20% improvement in the concrete compressive strength was observed by Mello et al. (2014) when 2.3% SF was used in the concrete mixture [10]. Figure 1 shows the effect of SFs content on compressive strength. It can be noticed from Figure 1 that the inclusion of 3% fibre content enhanced the compressive strength of plain concrete by around 15%. These findings with regards to the improvements in compressive strength by Mello et al. are in agreement with other investigations [16]. Moreover, Olivito and Zuccarello, (2010) studied the effect of SFs on the compressive strength. Results showed that the concrete cube specimens did not fail. They maintained their integrity until the end of the test. This is attributed to the fibres bridging effect on the specimens. They have further observed that the compressive failure of the SF reinforced samples was significantly altered from fragile to the ductile state [12]. In addition, Jang and Yun, (2018) have investigated the combined effects of coarse aggregate size and SFs on flexural and compressive strength of high-strength-concrete. Their results showed that both aggregate size and SF volume fraction have little effect on the modulus of elasticity and compressive strength of Steel-Fibre-Reinforced-Concrete. Furthermore, their results showed that the increase of fibre volume fraction can improve the toughness, post-peak behaviour, peak strain of Steel-Fibre-Reinforced-Concrete under compression [6].

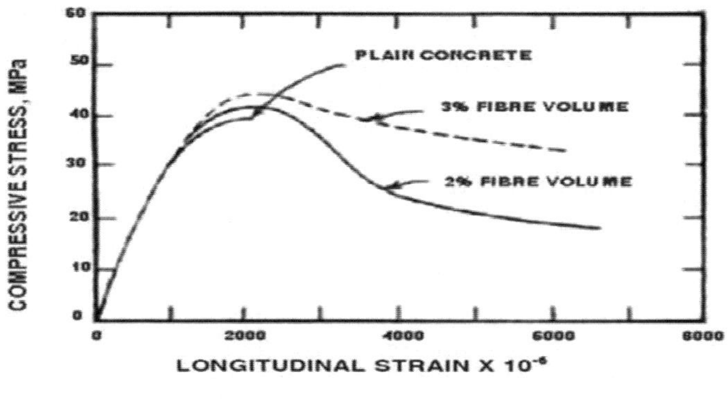

Figure 1. Stress-Strain curves in compression for SFRC [16].

The compressive behaviour of fibre reinforced concrete with end hooked SFs has been investigated [9]. Lee et al. (2015) performed uniaxial compression tests on 48-cylinder specimens with dimensions of 150 mm and 300 mm for diameter and height, respectively. The variables that they considered are length to diameter (fibre aspect ratio), fibre volumetric and concrete compressive strength. Their results revealed that the SF volume content and the SF aspect ratio can influence the compressive strength and the slump of SFRC; for instance, the compressive strength increased slightly when the fibre volume content increased for the aspect ratio of 45, while the compressive strength of the aspect ratio of 65 decreased when increasing the fibre content.

2.2. Effect on tensile strength

The effect of SFs addition on the tensile strength of concrete has been studied by several authors [10,12]. Generally, the investigations showed that the addition of SFs can significantly improve the tensile strength. However, the increment percentage was influenced by the volume content of the fibres ; for instance, Olivito and Zuccarello (2010) have investigated the tensile strength of Steel-Fibre-Reinforced-Concrete by performing a direct tensile test on eight different prismatic specimens which have dimensions of 30 × 80 × 350 mm with only one type of SFs but different lengths (22 mm and 30 mm) with only 1% of fibre content. The experimental study revealed that the maximum tensile strength for the 22 mm and 30 mm SFs is 3200 N and 5800 N, respectively. The results mean that the improvement of the maximum tensile strength was enhanced up to 44% of the maximum tensile strength for the 22 mm fibres specimens, whilst for 30 mm fibres specimens showed higher ultimate strain compared with 22 mm fibres specimens [12]. In addition, these results were in agreement with study conducted by Mello et al. (2014) where they indicated that the addition of SF content up to 3% resulted in a significant increase in the tensile strength of concrete from null to 121.5% [10].

2.3. Effect on flexural behaviour

Several studies have investigated the flexural behaviour of SFRC. Jang and Yun, (2018) investigated the flexural behaviour by performing a three-point bending test on three prismatic specimens (300×100×100 mm). Each specimen was fabricated using different SF percentages (0.5, 1.0, 1.5, and 2.0 V.%) and different aggregate sizes (8, 13, and 19 mm) in the concrete mixture by following American standards. The findings showed significant improvement in the equivalent flexural strength ratio ; for instance, the flexural load ranged between 50 – 55 kN with the use of 2% SFs, while the maximum flexural load for 0.5% SF ranged between 20 – 37 kN under conditions of different aggregate sizes. Similarly, the maximum flexural load ranged between 50 to 52 kN with the use of 1.5% SF under conditions of different aggregate sizes. The improvement that occurred to the flexural performance of SFRC is mainly related to the volume fraction of SFs in the tension zone of the prismatic specimen. Therefore, it was suggested to use 1.5% volume fraction of hooked SFs since it is more economical and efficient. On the other hand, the results showed that the use of small aggregate size has a small effect on mechanical properties compared to SFs volume fraction. But small aggregate size with a high percentage of SFs was found to help the SFs to be distributed homogeneously within the concrete mix [6].

Yoo et al. (2018) have studied the influence of employing SFs on the flexural behaviour of RC beams with low reinforcement ratio. In their study, they incorporated four different fibre volume content of 0.25%, 0.50%, 0.75%, and 1.00% into the concrete mixture. A control specimen without the addition of fibres was also cast. Furthermore, the SF influence on flexural behaviour of RC beams was also evaluated using different low reinforcement ratios of 44%, 66%, 78%, and 100% of ρ_{min}. The results indicated an improvement in the flexural performance of RC beams, including the cracking behaviour, post cracking flexural stiffness, deflection capacity and flexural strength with the increment of the reinforcement ratio up to minimum reinforcement ; for instance, the ultimate load reached 70 kN for 44% reinforcement ratio of ρ_{min} and 0.5% SF content. However, ultimate load up to around 120 kN was observed when using 1.0% SF content and 100% reinforcement ratio of ρ_{min} [19]. A similar study that investigated the effect of SF content on the flexural behaviour of concrete reported that the addition of 2.3% of SF improved the flexural strength up to 80.7%. However, a further increase in SF volume content to 3% led to a decrease in the flexural strength from 80.7 to 67.9 % (Mello et al., 2014) [10]. This clearly shows that the increase of SF content does not necessarily improve the flexural strength of concrete. Therefore, further studies are required to determine the optimum SFs percentage that can give maximum flexural strength. Additionally, the employment of SFs resulted in an improvement of the post cracking stiffness, cracking performance, yield loads and initial cracking. However, the effect on the ultimate load-carrying capacity was found to be low. The utilization of high fibre volume content with low reinforcement ratios has led to lower ductility indices. It was concluded that discontinuous SFs could not be used as a replacement of longitudinal steel reinforcements at moderate volume content (V_f= 1%) in terms of flexural strength, ductility and ultimate load carrying capacity margin[18] .

2.4. Effect on shear behaviour

Several authors investigated the effect of replacement of transverse reinforcement with SFs on shear behaviour. Dinh, (2009) studied the shear behaviour of RC beam with hooked-end SFs as a replacement for stirrups. Twenty-eight simply supported beams with a shear span-to-depth ratio of 3.5 were cast. The specimens were subjected to concentrated load. Two parameters were used in the investigation which include fibre types and fibre volume content between 0.75% to 1.5%. Three types of DRAMIX hooked-end SFs were used. The first two types had the same length of 30 mm with different aspect ratios of 55 and 80, while the third had a length of 55 mm with aspect ratio of 80. The results indicated that the employment of 0.75% of fibre volume content or greater led to a significant improvement in shear strength ; for example, the shear capacity for the 0.75% SF content reached 220.45 kN, while 270.63 kN shear capacity was observed when using 1.5% fibre content at the same fibre aspect ratio of 55. However, for fibre aspect ratio of 80, there was a slight difference of 17% higher than the 55 aspect ratio with 0.75% fibre content [4]. It can be concluded that there is a proportional relationship between the aspect ratio, fibre content, and shear capacity. Moreover, it was also observed by Dinh that hooked-end SFs can be employed as a minimum shear reinforcement in reinforced-concrete beams cast with normal strength concrete as well as within the considered depth range [4]. It is worth mentioning that these observations are in agreement with the ACI Code Section 11.4.6 [1], which states that the usage of hooked-end SFs in volume contents greater than or equal to 0.75% can replace the minimum shear reinforcement in normal-strength concrete beams as long as V_u lies down between the range of $0.5\varphi V_c \leq V_u \leq \varphi V_c$ (318-08, 2008) [1]. In addition to the abovementioned statements, the volume content of 0.75% has been proven to be the most recommended fibres ratio to be used as replacement for minimum reinforcement. Parra-Montesinos(2006) investigated the shear strength of 250 mm SFRC beams. The SFs (hooked end and deformed) replaced the minimum shear reinforcement in the beams completely. The obtained data showed that the addition of 0.75% of deformed or hooked SFs has shown to be the optimum volume fraction [13]. Moreover, it has been also indicated by Dinh that the utilization of fibres volume content had a greater effect on the shear strength of SFRC beams compared with effective beam depth and longitudinal reinforcement ratio. They further claimed that the increase in shear strength recuded when the employment of fibres volume fraction was more than 1%. The abovementioned studies have investigated the effect of hooked-end SFs with normal concrete on the shear behaviour. High strength concrete has also been employed and investigated in past studies. For example, a recent study claimed that the addition of 0.75% of hooked fibres as a replacement for the minimum shear reinforcement with high strength concrete resulted in an acceptable alternative [19].

Further studies have also investigated the effect of fibres on shear behaviour [2]. Cucchiara et al. (2004) studied the effect of hooked-end SFs on rectangular simply supported beams with and without stirrups. Fibres length with 30 mm with an equivalent diameter of 0.5 mm (aspect ratio 60) and nominal tensile strength of 1115 MPa were used. Fibre contents corresponding to 1% and 2% of the volume of concrete were included in the mix. The results indicated that the use of a sufficient percentage (1-2%) of SFs can change the brittle shear mechanism to a ductile flexural mechanism. Furthermore, the changing from brittle shear mechanism to a ductile flexural mechanism would allow large dissipation of energy. They also claimed that the use of SFs as an alternative to transverse reinforcements in simply supported beam produce a similar performance in traditional simply supported beam in terms of ultimate strength while using SFs and shear reinforcement together would be more suitable because stirrups allow a larger deformation capacity beyond the elastic limit.

2.5. Effect on crack behaviour

Addition of SFs have also been shown to positively affect the crack behaviour of concrete elements. Soulioti et al. (2011) have fabricated plain and fibre reinforced concrete specimens with 0.5, 1, and 1.5 V.% SF content. Their results showed that the fibre reinforced concrete samples noticeably retained the post cracking ability, unlike plain concrete samples which failed significantly by a single crack [14]. Several authors found that the cracking behaviour of RC beams has been improved by the addition of SFs. Olivito and Zuccarello(2010) stated that using higher fibre content of 1% as well as fibre length of

30 mm resulted in the increment of the ductility and tenacity of the SFRC[12]. Moreover, the study showed that this phenomenon occurs due to the high energy absorption and high deformability nature of SFRC. The study also showed that different fibre length can affect the post-cracking behaviour. Short fibres caused softening behaviour, while long fibres resulted in hardening behaviour. It is worth to mention that long fibres not only influence the post-cracking behaviour, but it can also increase the maximum load carrying capacity of the member. Furthermore, they observed that increase of fibre content resulted in increase in flexural strength, first crack strength, and ductility. The study also showed that SFRC has higher bending stiffness and different cracking style compared with those of normal concrete [12].

In addition, the effect of beam depth on the crack behaviour in the case of replacing the minimum transverse reinforcement with SFs content of 0.75% has also been investigated ; for example, Dinh(2009) concluded that wider spacing between inclined cracks when larger beam depths were utilized. Generally, beams with larger spacing between cracks exhibited lower shear strength compared to beams with small crack spacing. The influence of hooked-end SFs with different lengths on the crack behaviour has also been studied by Dinh(2009). As a result, fibre lengths of 60 mm allowed a larger inclined crack opening before the occurrence of complete failure compared to the type of failure observed with beams having a fibre length of 30 mm. However, the consolidation of long fibres was unfavourable. In addition, Dinh suggested the use of fibres with length less than the clear spacing between longitudinal bars to achieve reasonable consolidation [4]. As have been discussed before, SFs intend to change the character of the concrete from brittle to ductile as shown in Figure 2. Figure 2 shows the effect of using long and short straight SFs on the concrete cracks. Short straight SFs were used to bridge or links the microcracking but long straight SFs were used to close the macrocracks of the concrete [3].

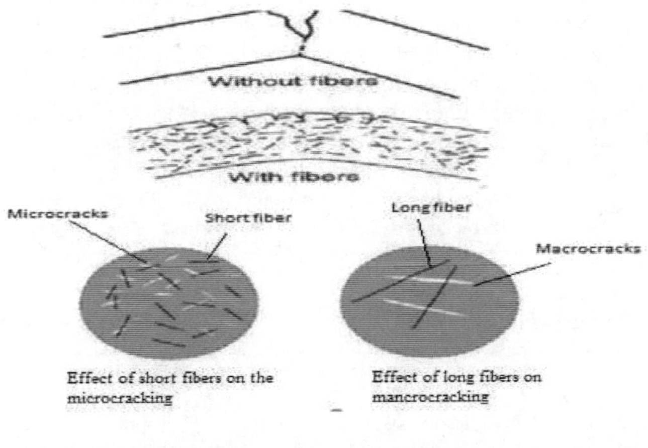

Figure 2. Effect of SFs on concrete characteristics and effect of long and short SFs on the cracks [3].

3. Orientation and distribution of SFs

SFs reinforcement can be employed as a three-dimensionally randomly distributed form in structural elements. This, as a result can enhance the shear resistance and crack control of the structural member. In addition, SFs can also act as a tensile layer that covers the steel reinforcement in the structural element. It is found that the efficiency is batter if two – dimensional orientation of SFs occurred in the structural element [16,20]. Several researchers intensively investigated the orientation and distribution behaviour of SFs in hardened concrete. It was stated in research studies [7,8] that the orientation of SFs resulted from a series of stages where these stages starts from the mixing of the concrete until the hardening of the concrete inside the formwork. In addition, the boundary conditions related to each stage of the fabrication process and the actions imposed, such as external vibration, gravity and wall effect can encourage many interaction stages where the orientation of SFs is modified. The orientation and distribution of SFs in SFRC specimens were also studied in [5,15] where the results obtained revealed that compaction process has a significant effect on the orientation of the SFs. During the compaction process, a phenomenon called wall effect occurred. They explained that this phenomenon of SFs orientation tends to make SFs orientated parallelly to the surface of the formwork. However, this phenomenon has shown to only affect the concrete that is in contact with the surface of the formwork. In the case of using long SFs in thin elements, the wall effect can be significant (see Figure 3).

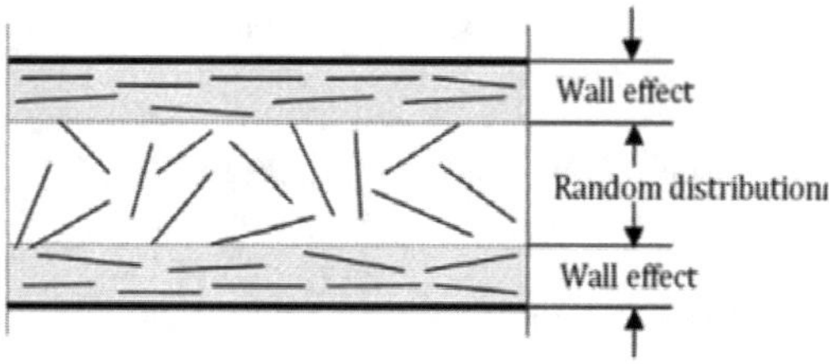

Figure 3. Illustration of wall effect inside formwork [15].

In the same studies, it was also reported that the vibration (180 Hz) of concrete can trigger the alignment and rotation of the SFs preferentially in a specific direction[5, 15]. The SFs tend to be orientated perpendicular to the direction of the external vibration. Studies have shown that the most significant factor affecting the SFs distribution and orientation is casting method, in particular for self-consolidation SFRC[17, 19]. According to the results, the required distribution and orientation of SFs in the construction site can be achieved by carefully controlling the casting method. However, Laranjeira et al, (2011) [7] showed that the existence of coarse aggregate does not allow the SFs to line up parallel to the direction of shear crack and random distribution and orientation of SFs is achieved instead. Yoo et al. (2016) have recommended the following methods to overcome this issue: (i) the inclusion of SFs slowly in the fresh concrete (dosage) and mixing again of fresh concrete and (ii) the casting of the fresh concrete parallel to the direction of the element length [17]. In another study, Michels et al. (2012) investigated the orientation of SFs in RC slab. The result revealed that increasing the slab thickness will reduce the orientation at cracked state [11]. A summary of the findings on the effect of SFs on mechanical properties of RC beam is shown in Table 1.

Table 1. Summary of findings on the effect of SFs on the mechanical properties of RC beams.

Studies behaviour	SF content (V.%)	Findings	Ref.
Compressive strength	2.3%	Addition of 2.3% SFs improved compressive strength up to 20% for plain concrete.	[10]
	3%	3% SF RC enhanced compressive strength to 16%.	[16]
Tensile strength	1%	The 30 mm fibres specimens showed higher ultimate tensile strength of 44% compared with 22 mm fibres specimens.	[12]
	3%	Tensile strength was improved up to 121% comparing with the control sample.	[10]
Flexural strength	0.5, 1.5, and 2%	The maximum flexural load for the 0.5, 1.5, and 2% ranged between 20-37, 50-52, 50-55 kN, respectively. The 1.5% found to be the optimum fibre content.	[6]
	0.5 and 1%	The ultimate load reached 70 kN for the reinforcement ratio of 44% of the ρ_{min} and 0.5% fibre content. While, ultimate load 120 kN for 1% fibre content and 100% reinforcement ratio of the ρ_{min}.	[18]

	2.3 and 3%	The flexural strength was increased up to 80.7% when using 2.3% fibre content. However, the improvement percentage reduced to 67.9% when using 3% fibre content.	[10]
Shear strength	0.75 and 1.5%	Shear stress reached 220.45 kN for the 0.75% fibre content. While, 270.63 kN shear stress was observed when using 1.5% fibre content, at the same aspect ratio (55). However, at aspect ratio of 88, a slight difference of 17% in the shear stress was observed when using 0.75% fibre content.	[4]
Crack behaviour	0.5, 1, and 1.5%	Results showed that the fibre RC samples retained the post cracking, unlike plain concrete, which failed significantly by a single crack.	[14]
	1%	Study showed improvement in the ductility and tenacity of the SFRC when increasing the fibre volume percentage and length.	[12]
	0.75%	Fibres having longer length (i.e. 60 mm) allowed larger inclined crack opening, compared with the shorter fibre length (30 mm). However, the consolidation of long fibres was unfavourable.	[4]
Orientation and distribution	-	Results showed that compaction and vibration methods can significantly influence the orientation and distribution of fibres.	[5,15]
	-	The study revealed that casting method can also affect the orientation and distribution of SFs in the concrete mixture.	[17,19]
	-	The existence of coarse aggregates will not allow the SFs to line up parallel to the direction of the shear cracks. However, it can lead to random distribution and orientation of fibres.	[7]

4. Conclusion

SFRC can be defined as a combination between normal concrete and thousands of small discontinuous discrete fibre distributed randomly in the concrete during mixing. SFs (SF) has become a promising material that can be employed in different structural elements to improve mechanical properties of concrete. Inclusion of fibres has shown to improve compressive strength, tensile strength, and shear behaviour. In addition, this material can resist repeatedly applied, impact or shock loading. The addition of SFs has shown to alter the characteristics of concrete from brittle to ductile as well as increase the ultimate strength of concrete. Further, it was found that the orientation and distribution of SFs are influenced by casting, vibration, and compaction methods. A review of the past studies have shown that inclusion of 0.75 V.% of SF in RC beams can fully replace shear links. Furthermore, it has been observed that hooked-end SF are the most suitable type of SF to be included with normal strength concrete, while straight SF resulted in better results if used with high strength concrete due to the high tensile strength of straight SF.

References

[1] ACI 318-08 Committee and International Organization for Standardization, 2008. Building code requirements for structural concrete and commentary. *ACI.*

[2] Cucchiara C, La Mendola L and Papia M 2004 Effectiveness of stirrups and steel fibres as shear reinforcement. *Cem Concr Compos* **26**, 777-786.

[3] Cuenca E 2014 Ph,D. thesis Universitat Politecnica de Valencia On shear behavior of structural elements made of steel fiber reinforced concrete *Springer*.

[4] Dinh H H 2009 Ph.D. thesis University of Michigan Shear Behavior of Steel Fiber Reinforced Concrete Beams without Stirrup Reinforcement.

[5] Gettu R, Gardner D, Saldivar H and Barragán B 2005 Study of the distribution and orientation of fibers in SFRC specimens. *Mater Struct* **38,** 31-37.

[6] Jang S J and Yun H D 2018 Combined effects of steel fiber and coarse aggregate size on the compressive and flexural toughness of high-strength concrete. *Compos. Struct* **185**, 203-211.

[7] Laranjeira F, Grünewald S, Walraven J, Blom C, Molins, C and Aguado A 2011 Characterization of the orientation profile of steel fiber reinforced concrete. *Mater Struct* **44,** 1093-1111.

[8] Laranjeira F, Molins C and Aguado A J 2010 Predicting the pullout response of inclined hooked steel fibers. *Cem Concr Res,* **40,** 1471-1487.

[9] Lee S C, Oh J H and Cho J Y 2015 Compressive behavior of fiber-reinforced concrete with end-hooked steel fibers. *Materials,* **8,** 1442-1458.

[10] Mello E, Ribellato C and Mohamedelhassan E 2014 Improving concrete properties with fibers addition. *Int J Civ Eng* **8,** 249-254.

[11] Michels J, Waldmann D, Maas S and Zürbes A 2012 Steel fibers as only reinforcement for flat slab construction–Experimental investigation and design. *Constr Build Mater* **26**, 145-155.

[12] Olivito R and Zuccarello F 2010 An experimental study on the tensile strength of steel fiber reinforced concrete. *Compos Part B-ENG,* **41,** 246-255.

[13] Parra-Montesinos G J 2006 Shear strength of beams with deformed steel fibers. *Conc. Eng. Int* **28**, 57-66.

[14] Soulioti D, Barkoula N, Paipetis A and Matikas T 2011. Effects of fibre geometry and volume fraction on the flexural behaviour of steel-fibre reinforced concrete. *Strain,* **47,** e535-e541.

[15] Torrents J M, Blanco A, Pujadas P, Aguado A, Juan-García P and Sánchez-Moragues M Á 2012 Inductive method for assessing the amount and orientation of steel fibers in concrete. *Mater Struct*, **45**, 1577-1592.

[16] Van Chanh N 2004 Steel fiber reinforced concrete. Faculty of Civil Engineering Ho chi minh City university of Technology. *Seminar Material*. 108-116.

[17] Yoo D Y, Banthia N, Kang S T and Yoon Y S 2016 Effect of fiber orientation on the rate-dependent flexural behavior of ultra-high-performance fiber-reinforced concrete. *Compos. Struct.* **157,** 62-70.

[18] Yoo D Y and Moon D Y 2018 Effect of steel fibers on the flexural behavior of RC beams with very low reinforcement ratios. *Constr Build Mater* **188**, 237-254.

[19] Yoo DY, Yuan T, Yang J M and Yoon, Y S 2017 Feasibility of replacing minimum shear reinforcement with steel fibers for sustainable high-strength concrete beams. *Eng. Strut* **147**, 207-222.

[20] Zandi Y, Husem M and Pul S 2011 Effect of distribution and orientation of steel fiber reinforced concrete. Proceedings of the 4th WSEAS international conference on Energy and development-environment-biomedicine. World Scientific and Engineering Academy and Society (WSEAS), 260-264.

Effect of hydrogen embrittlement towards thermal and mechanical behavior of NiTi shape memory alloy

C W Ng[1], A S Mahmud[1], M F Razali[1*] and M N Ahmad[2]

[1]Nanofabrication and Functional Materials Research Group, School of Mechanical Engineering, Engineering Campus, Universiti Sains Malaysia, 14300 Nibong Tebal, Pulau Pinang, Malaysia

[2]Mechanical Section, Universiti Kuala Lumpur Malaysian Spanish Institute, Kulim Hi-Tech Park, 09000 Kulim, Kedah, Malaysia

*E-mail: mefauzinizam@usm.my

Abstract. NiTi arch wires are susceptible to hydrogen embrittlement upon contact with dental brackets in oral cavity during orthodontic treatment. This study investigated the effect of hydrogen absorption and diffusion over time towards the thermal and mechanical properties of NiTi shape memory wire. The hydrogen absorption process was carried out via electrolytic charging at constant current density for 16 hours in 1.0 weight percent (wt.%) sodium sulphate solution. The hydrogen charged wires were aged at room temperature in air for different durations to allow further inward diffusion of the hydrogen into the specimens. The results show that after hydrogen charging, the latent heat of forward and reverse martensitic phase transformation of the NiTi wire changed from 19.96 to 11.98 J/g, and 19.21 to 13.42 J/g, respectively. Further suppression and disappearance of thermal transformation peaks were observed as the charged specimen aged for 7 days. The transformation stress level on tensile deformation increased by almost 90 MPa after hydrogen charging, and exhibited non-flat stress plateau after further aging.

1. Introduction

Nickel titanium (NiTi) is a popular biomaterial in orthodontic application for its ability to deliver constant recovery force after deformation owing to its shape memory behavior, in addition to its favorable biocompatibility. NiTi orthodontic arch wire and stainless steel dental bracket are one of the most commonly used wire-bracket combinations in the orthodontic treatment. The treatment of teeth malocclusion has been benefited from the almost constant recovery force generated by the deformed NiTi orthodontic wire to initiate and promote the tooth movement. However, the contact between the two different alloys in the oral cavity with the presence of electrolyte, for instance the saliva and mouthwashes will give rise to the galvanic corrosion [1, 2]. The galvanic corrosion has caused the degradation of NiTi arch wire in oral cavity over a long treatment duration, such as loss of ductility and even fracture [3].

The loss of functionality and fracture of NiTi orthodontic arch wire during treatment has been claimed to be caused by hydrogen absorption and subsequent embrittlement effect [3–5]. The hydrogen embrittlement has caused the fracture of the charged NiTi wire in tensile deformation test to occur before and during the stress-induced martensite phase transformation [3, 6, 7]. On top of that, the cross-sectional views of the failure surfaces have shown that the fracture of the wire happened in a brittle manner, without necking and reduction of area as observed in the as-received wire specimen [3, 7]. In addition to the hydrogen charging, the aging of the charged wire specimens in air under room temperature has also shown the similar effect towards the NiTi alloy [7].

This paper studied the effect of hydrogen charging and aging toward thermal martensite transformation and tensile deformation behavior of NiTi shape memory alloy. The set-up for hydrogen charging in this study used stainless steel and NiTi alloy as electrodes in order to resemble the actual condition in clinical orthodontic treatment. The galvanic current measured between the NiTi wire and stainless steel bracket combination is only in the range of nano-amperes (nA) [8, 9]. However, in order to accelerate the charging process and to achieve a significant amount of hydrogen concentration in a short time, the higher current density was used in this study. Both the hydrogen charging and aging have shown to affect the thermal phase transformation behavior of the NiTi alloy, while the aging has more significant effect towards the tensile deformation behavior as compared to hydrogen charging.

2. Experimental procedure

NiTi wire with composition of 50.6 atomic percent (at.%) Ni and diameter of 0.5 mm was used in this study. The as-received wire specimens were cut into length of 50 mm, followed by heat treatment at temperature of 600 °C for 30 minutes. The heat-treated wires were polished using SiC abrasive paper, followed by cleaning using acetone and distilled water. The hydrogen charging of the NiTi wires was achieved through electrolytic charging using a direct current (DC) power generator. The NiTi wire specimens were immersed into the 1.0 wt.% sodium sulphate (Na_2SO_4) solution and cathodically charged, together with 304 austenitic stainless steel anode. The schematic representation of the experimental setup is shown in figure 1 below. The current density used in the electrolytic charging process was set at 1000 A/m^2. A multimeter was connected in series to the electrical circuit to measure and monitor the electrical current. The hydrogen charging was carried out for 16 hours. The aging of the charged wire specimens was carried out under room temperature in air for 1, 4 and 7 days, respectively.

The changes in the thermal phase transformation behavior of the hydrogen charged and aged wire specimens were characterized using TA Q20 differential scanning calorimeter (DSC). The wire specimens were initially heated to a temperature of 150 °C, followed by cooling to a temperature of -100 °C and subsequently heated to a temperature of 150 °C, at a constant cooling and heating rate of 10 °C/min, respectively. The effect of hydrogen charging and aging towards the tensile deformation behavior of NiTi wire was evaluated via tensile deformation test. The tensile test was conducted at room temperature using Instron 3367 universal tensile machine equipped with a 30 kN load cell. The gauge length of the wire specimens was set at 20 mm. The wire specimens were deformed up to 10% strain followed by unloading. The loading and unloading rates were set at 1 mm/min, respectively.

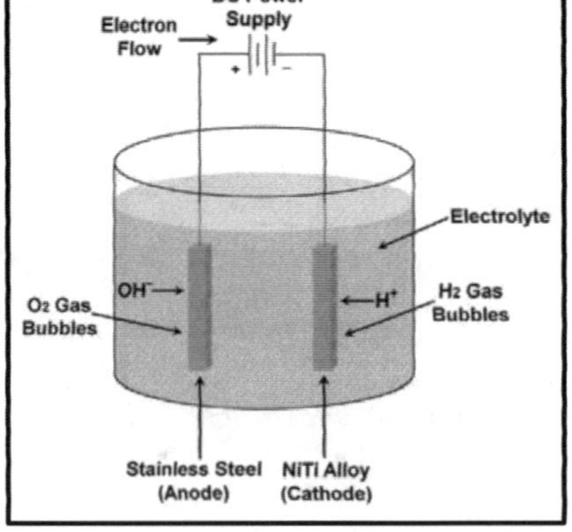

Figure 1. Schematic diagram of the experimental setup for hydrogen charging of NiTi wire.

3. Results and discussion

Figure 2 shows the DSC curves of NiTi specimens after hydrogen charging for 16 hours and subsequently aged at different durations. It is seen that the as-annealed wire specimen possesses a sharp and high thermal transformation peak during cooling and heating. The martensite start (M_s) and austenite finish (A_f) temperatures of the annealed specimen occur at -6.06°C and 20.97°C, respectively. The hydrogen charging and aging for 1 day have not affect the reverse transformation temperatures of the wire specimens during heating, while the aging for 4 and 7 days have shift the A_f temperatures from 20.97°C to 27.14°C and 27.29°C, respectively. It is observed that the enthalpies of forward and reverse thermal martensite phase transformation have change from 19.96 to 11.98 J/g, and 19.21 to 13.42 J/g, respectively after hydrogen charging. The aging of the charged specimens in room temperature further suppresses the phase transformation until the near complete disappearance of both forward and reverse transformation peaks, as seen from the wire specimen aged for 7 days. The charging and aging process has more effect on the forward transformation (A→ M) as compared to the reverse transformation (M→ A), which is in agreement with the previous studies [10, 11]. In addition to the reduction of the enthalpy of phase transformation, several new transformation peaks are observed from the specimens aged for 4 and 7 days. The suppression of thermal martensite phase transformation is caused by the hydrogen absorption and diffusion over time from the sub-surface into the core of the wire specimens. The diffusion of hydrogen reduces the volume of the hydrogen-free NiTi matrix, while the presence of hydrogen in the NiTi matrix creates additional resistance for the reverse transformation, thus causing it to occur at several stages.

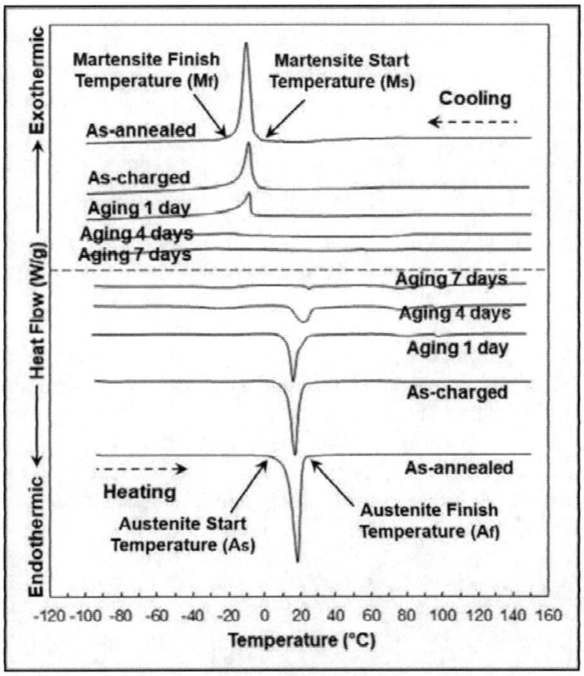

Figure 2. DSC curves of NiTi wire specimens charged at 1000 A/m^2 for 16 hours and aged at room temperature.

Figure 3 shows the tensile deformation behavior of the hydrogen charged and aged wire specimens. The stress-strain curves of the as-annealed and hydrogen charged wire specimens manifest the shape memory effect and martensite reorientation of the NiTi alloy, as the testing temperature is just above its A_f temperature. The as-annealed wire specimen has the transformation stress level of around 245 MPa, which is the lowest among all the tested specimens. The hydrogen charging causes the transformation stress levels to increase from 245 MPa to 332 MPa, with the increment of ~90 MPa as compared to the as-annealed specimen. The aging at room temperature for 1 to 7 days has increases the transformation stress as much as ~150 MPa from 245 MPa to 398 MPa, while causing the transformation stress plateaus to propagate in a non-flat manner. The increase of the critical stress levels at the onset of the stress plateaus as seen from the aged specimens suggests that the diffusion and presence of hydrogen hinder the martensite reorientation, causing more energy needed to initiate the transformation process [12]. Additionally, the residual strain of each charged and aged wire specimen after complete unloading during the tensile deformation is lower than the as-annealed specimen.

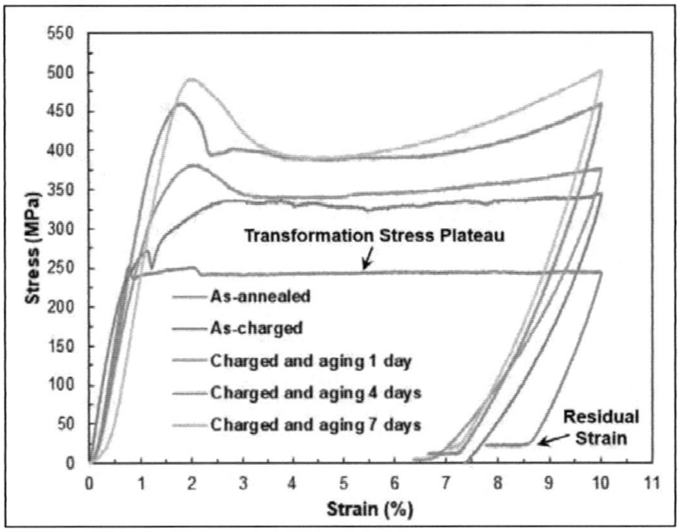

Figure 3. Tensile deformation behavior of NiTi wire specimens charged at 1000 A/m^2 for 16 hours and aged at room temperature for 1 to 7 days.

4. Conclusions

The hydrogen absorption and diffusion were found to suppress the thermal phase transformation of NiTi alloy. The enthalpies of forward and reverse phase transformation were decreased from 19.96 to 11.98 J/g, and 19.21 to 13.42 J/g, respectively, in addition to the disappearance of transformation peaks. On the other hand, the hydrogen charging caused the martensite reorientation to occur at a higher transformation stress level, with increment of the transformation stress level as much as 90 MPa. The diffusion of hydrogen into the NiTi alloy matrix caused the stress-induced martensite phase transformation to exhibit non-flat stress plateau behavior.

References

[1] Asaoka K, Yokoyama K, and Nagumo M 2002 *Metall. Mater. Trans.* A **33**(3) 495–501
[2] Schiff N, Boinet M, Morgon L, Lissac M, Dalard F, and Grosgogeat B 2006 *Eur. J. Orthodont.* **28** 298–304
[3] Yokoyama K, Hamada K, Moriyama K, and Asaoka K 2001 *Biomaterials* **22** 2257–62
[4] Ogawa T, Oda T, Maruoka K, and Sakai J 2015 *Int. J. Mech. Mater. Eng.* **10**(1) 1–9
[5] Sarraj R, Hassine T, and Gamaoun F 2019 *Mater. Res. Express* **5**(1) 015704
[6] Gamaoun F, Hassine T, and Bouraoui T 2014 *Phil. Mag. Lett.* **94**(1) 30–36
[7] Gamaoun F, Ltaief M, Bouraoui T, and Zineb T B 2011 *J. Intel. Mat. Syst. Str.* **22**(17) 2053–59
[8] Bakhtari A, Bradley T G, Lobb W K, and Berzins D W 2011 *Am. J. Orthod. Dentofac.* **140**(1) 25–31

[9] Iijima M, Endo K, Yuasa T, Ohno H, Hayashi K, Kakizaki M, and Mizoguchi I 2006 *Angle Orthod.* **76**(4) 705–11
[10] Runciman A, Chen K C, Pelton A R, and Trépanier C 2006 *SMST–2006: Proc. Int.Conf. on Shape Memory and Superelastic Technology (California)* ed B Berg *et al* (Ohio: ASM International) pp 185–96
[11] Snir Y, Carl M, Ley N A, and Young M L 2017 *Shape Mem. Superelast.* **3**(4) 443–56
[12] Rokbani M, Saint-Sulpice L, Chirani S A, and Bouraoui T 2018 *J. Int. Mat. Syst. Str.* **29**(16) 3242–53

Acknowledgements
The authors are grateful for the financial support provided by Universiti Sains Malaysia under the research grant FRGS 203/PMEKANIK/6071405.

Development of a Flow Visualization Technique in Wind Tunnel for Hydrokinetic Turbine Application

Abdelrahman T. Ismail and Noorfazreena M. Kamaruddin*

School of Aerospace Engineering, Universiti Sains Malaysia, 14300 Nibong Tebal, Penang, Malaysia

*Corresponding author: fazreena@usm.my

Abstract. A smoke wire technique was developed in a wind tunnel to visualize a drag-driven turbine flow structure for hydrokinetic application. A Nichrome wire with a diameter of 0.5 mm was initially heated and then coated with Safex oil to illustrate the streamlines that represent the turbine flow characteristics. The developed smoke wire oil coating technique was compared at airflow speeds of 3 m/s to 10 m/s to represent equivalent water flow speed corresponding to Reynolds numbers (Re) of 29, 563 to Re= 98, 543, respectively. Several parameters that influence the turbine flow patterns, such as the diameter of wire and oil coating technique, were tested in the wind tunnel to investigate the effect of these parameters to visualize the stream lines at low and high speed case. The qualitative result based on these various configurations was analyzed and discussed in this paper to identify the most practical flow visualization technique in a wind tunnel that could effectively be captured to represent the turbine flow structure for hydrokinetic application.

Keywords: Flow visualization technique, hydrokinetic, turbine, wind tunnel

1. Introduction

The flow visualization technique is commonly used in qualitative measurement. It consists of graphical details that describes the flow structure, as opposed to a quantitative measurement that consists of data and values [1]. It helps to understand the flow behaviours surrounding a model that can be used to improve the performance [2]. There are various techniques established in the literature to visualize flow. One of the most popular technique is to provide a thin and discrete smoke filament by using a smoke wire [3]. It was initially developed in 1950s by Raspet and Moore to measure a velocity profile of their model and was later used to study the flow characteristics [4]. It is a simple flow visualization technique that is widely used to capture laminar flow at a Reynolds number of below 25,000 [5, 6].

To illustrate flow motion, smoke is produced by heating a wire coated with an oil. The heated wire causes the oil to vaporize and form streak lines that represent the flow's characteristics [7]. Wire strength, resistivity, diameter and size are important parameters that influence the capability to illustrate the flow dynamics effectively. The diameter of the wire determines the smoke intensity and thickness of the stream lines [8]. The most common materials used in this wire are stainless steel, Nichrome and Tungsten. Experimental work performed by FrisPost et al. [9] using Tungsten wire managed to capture and maintain a clear and smooth streak lines for a considerably high speed of 18 m/s. In addition, the wire can be twisted as a single or double coil to maintain sustainable streak lines for a longer duration [10].

Although the diameter and the type of the wire are some of the important aspects that contribute to a good visualization of the flow, the white smoke produced through this method is usually difficult to see with the naked eye, especially in a wind tunnel experiment. To overcome this issue, the test section of a wind tunnel is usually covered with a black cloth to minimize any reflection and to provide some contrast with the white smoke lines [11-12]. Sieverding and Van Den Bosche used coloured smoke instead of the typical white smoke by coating the wire with Waxoline dye, also known as Sudan dye [13].

The development of the smoke wire techniques has revealed various results and accuracy. One of the most challenging results to obtain in a wind tunnel experiment is to capture the flow motion at a high speed (for flow visualization), particularly beyond 3 m/s. Most smoke wire techniques are only effective at a very low speed [14]. In order to overcome this limitation, this paper aims to develop a flow visualization technique that capable to illustrate the flow motion and maintain a constant visible streak lines at a low and high-speed ranging from 3 m/s to 10 m/s, respectively. A turbine model was used to illustrate the flow structure for hydrokinetic application, and the results will be focused on evaluating the visibility of the streak lines to represent the effectiveness of the developed flow visualization technique.

2. Methodology

2.1. Wind Tunnel

The wind tunnel used for this experiment was a closed-circuit wind tunnel located at the Science and Engineering Research Centre (SERC), Aerospace Wind Tunnel Laboratory, Universiti Sains Malaysia (USM) as shown in Fig. 1. The airflow inside the wind tunnel was driven by an axial fan motor to the test section with a dimension of 1.0 m (width), 0.8 m (height), 1.8 m (length). The test section had a contraction ratio of 10:1 and a turbulence level of 0.1%, for a flow speed of up to 80 m/s.

Fig. 1: Wind Tunnel

2.2 Hydrokinetic Turbine Model

A three-bladed vertical axis drag-driven hydrokinetic turbine was used as a model to illustrate the effectiveness of the developed flow visualization technique that represents the flow behavior in the wind tunnel. It has a simple design that consists of a shaft with the blades attached to the bottom section of the shaft as shown in Fig. 2. The blade has a height of 0.085 m and a diameter of 0.14 m, with an aspect ratio of 0.61. The shaft was attached to a bearing connected to a pulley that was placed at the top of the wind tunnel test section. This has allowed the turbine to hang freely where the driven flow stream to the test section that interacts with the blades, eventually causing the turbine to rotate.

Fig. 2: The 3-bladed turbine model

2.3 Oil Coating Techniques

The smoke wire oil coating technique requires meticulous effort, as any excessive oil applied on the wire affects the smoke clarity and intensity. In the current work, a manual technique requiring the usage of a syringe from the top of the wind tunnel test section was applied to a 0.5 mm Nichrome wire that had initially heated and coated with Safex oil, as shown in Fig. 3.

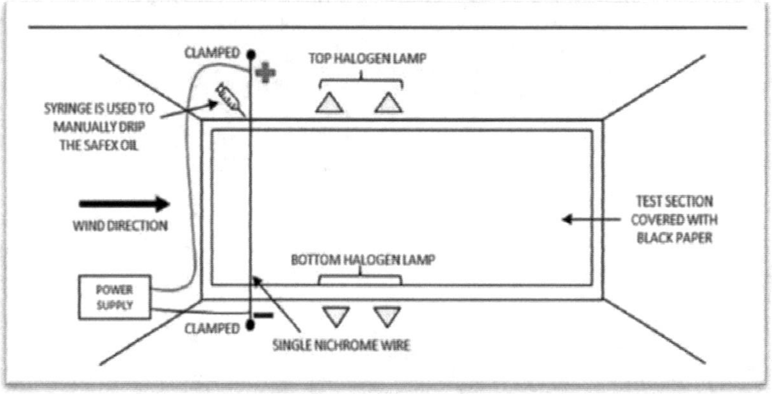

Fig. 3: Manual oil coating technique

2.4 Experimental Procedure

In this experiment, the smoke wire oil coating technique was tested at different flow speeds of 3 m/s to 10 m/s, corresponding to Reynolds number of Re=29,563 to Re=98,543, respectively. At the speed of 3 m/s, the test model was positioned at a static state and as the speed increasing to 10 m/s, the test model was rotating and can be considered at a dynamic state. This was performed to evaluate the flow streak lines clarity at high speed. The test section was covered with a black cloth to minimize any reflection that could affect the results. Several halogen light bulbs were installed at the top and bottom of the wind tunnel test section to highlight the smoke flow patterns.

3. Results and Discussion

3.1 Effect of smoke wire technique at low speed

The results revealed that the streak lines were thin, clear, uniform and straight at a low speed of 3 m/s, indicating that laminar flow before it started to impinge on the turbine as shown in Fig. 4. As the airflow started to interact with the turbine blade, the flow has curled around the blade and some small circulation can be seen behind the blade. This eventually leads to a turbulence flow structure as the streak lines were no longer straight. The smoke has managed to capture the formation of some small wake vortices. The manual technique was effective at low speed using a Nichrome wire with a diameter of 0.5 mm, as clear and visible streak lines were captured which represent the turbine flow characteristics.

Fig. 4: Manual technique at low speed

3.2 Effect of smoke wire technique at high speed

The flow characteristics at a high speed of 10 m/s is shown in Fig. 5. The smoke streak lines were still visible and uniform, but started to fade as the turbine started to rotate. The usage of a Nichrome wire with a diameter of 0.5 mm managed to illustrate the flow pattern at this speed range. However, the manual method was very challenging, as the oil dripping rate could not be controlled properly, leading to rapid rate of dispersed smoke lines. Therefore, the thickness of the oil coated on the wire eventually varied and resulting in inconsistent smoke intensity. Table 1 has summarized the results at low and high speed for brief comparison.

Fig. 5: Manual technique at high speed

Table 1: Case studies and corresponding flow regimes

Case study	Flow regime	Flow description
1	Low speed (3 m/s)	Clear smoke intensity and large wake Thick streaklines
2	High speed (10 m/s)	Smoke intensity decrease significantly Fade and thin streaklines

3.3 Future semi-automated flow visualization technique

A new semi-automated flow visualization technique is proposed to overcome the inconsistent smoke intensity observed with the current manual using syringe technique that is capable to capture the flow structure motion precisely at high speed as shown in Fig. 6. A Nichrome wire with a diameter of 0.5 mm is connected to a handle winch at the top of the wind tunnel test section, while a weight is hanged at the bottom to provide sufficient tension to pull the wire down as the handle winch is released. The wire is coated with Safex oil (filled in a box) to ensure consistency during the coating process.

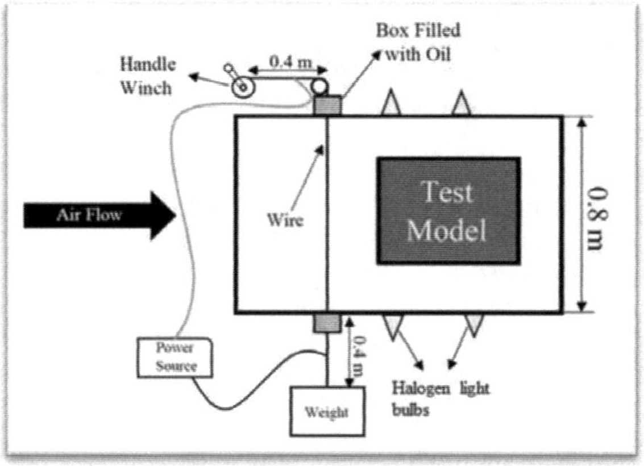

Fig. 6: Semi-automated smoke wire oil coating technique

4. Conclusion

A smoke wire flow visualization technique has been developed in a wind tunnel to visualize a drag-driven turbine flow structure. The results showed that the usage of a Nichrome wire with a diameter of 0.5 mm has proven to be effective as compared to 0.36 mm (used previously) as it has managed to capture the streak lines clearly at low speed of 3 m/s and even at high speed of 10 m/s. However, the manual technique using the syringe could not capture the flow structure precisely due to poor dripping oil rate and inconsistent smoke intensity.

5. Acknowledgement

The authors would like to acknowledge the financial support from the Ministry of Education Malaysia under the Fundamental Research Grant Scheme (FRGS): 203.PAERO.6071433.

6. References

[1] M. Atkins and M. Boer, "Flow Visualization," in Application of Thermo Fluidic Measurement Techniques, pp. 15-59 (2016).

[2] J. B. Barlow, W. H. Rae Jr and A. Pope, "Low speed wind tunnel testing." INCAS Bulletin, 7 (1), 133 (2015).

[3] T. J. Mueller and F. Brown, "Flow visualization by direct injection," in Fluid mechanics measurements: Routledge, pp. 367-450, (2017).

[4] R. C. Nelson and B. M. Babie, "Flow Visualization by Direct Injection Technique." Encyclopedia of Aerospace Engineering, (2010).

[5] D. J. Szarko, "Smoke-wire visualization of an oscillating flow in a gas spring," Massachusetts Institute of Technology, 1993.

[6] A. J. Smits, Flow visualization: techniques and examples, World Scientific. (2012).

[7] S. M. Batill and T. J. Mueller, "Visualization of transition in the flow over an airfoil using the smoke-wire technique." AIAA Journal, 19 (3), 340-345 (1981).

[8] S. Yarusevych, P. E. Sullivan and J. G. Kawall, "Smoke-wire flow visualization in separated flows at relatively high velocities." AIAA Journal 47 (6), 1592-1595 (2009).

[9] N. Gao and X. Liu, "An improved smoke-wire flow visualization technique using capacitor as power source." Theoretical and Applied Mechanics Letters, 8 (6), 378-383 (2018).

[10] S. S. Dol, M. A. M. Nor and M. Kamaruzaman, "An improved smoke-wire flow visualization technique," in Proceedings of the 4th WSEAS International Conference on Fluid Mechanics and Aerodynamics, Crete Island, Greece, Published, pp. 21-23, (2006).

[11] T. Corke, D. Koga, R. Drubka and H. Nagib, "A New Technique for Introducing Controlled Sheets of Smoke Streaklines in Wind Tunnels ", (1977).

[12] Salleh, M.B., Kamaruddin, N., and Mohamed-Kassim, Z., "A qualitative study of vortex trapping capability for lift enhancement on unconventional wing," in IOP Conference Series: Materials Science and Engineering, Vol. 370, No. 1: IOP Publishing, p. 012054, (2018).

[13] C. Sieverding and P. Van Den Bosche, "The use of colored smoke to visualize secondary flows in a turbine-blade cascade." Journal of Fluid Mechanics, 134, 85-89 (1983).

[14] Shen Xi, T., Kamaruddin, N., and Mohamed-Kassim Z., "Flow Visualization Study of Wing-Tip Vortex for a Wing with Cavity." Journal of Engineering and Applied Sciences, 12 (15), 3813-3817 (2017).

Evaporation rate of molten copper in argon and nitrogen gas flow and characteristics of condensed particles

S Yokoyama[1*], K Nakano[2] and J Sasano[1]

[1] Department of Mechanical Engineering, Toyohashi University of Technology, Tempaku-cho, Toyohashi, Aichi, 441-8580, Japan
[2] Department of Mechanical Engineering, Graduate School of Toyohashi University of Technology,Tempaku-cho, Toyohashi, Aichi, 441-8580, Japan

*Email: yokoyama@me.tut.ac.jp

Abstract. Copper has been levitation-melted in room temperature Ar and N_2 gas flows to investigate the evaporation rate and characteristics of condensed particles. The evaporation measured experimentally increased with the increases of the temperature of the levitated Cu, and the gas flow rate. The site, the concentration and temperature at the vapor condensation in the gas film around the levitated Cu was analysed using the temperature distribution and the critical supersaturation concentration of Cu vapor. There was a linear relationship between the calculated and the experimentally obtained evaporation rates. The calculated evaporation rate was larger than the experimentally measured rate. The condensed particles were metallic Cu. When the temperature at the vapor condensation site was lower than the melting point of Cu, the vapor condensed as a solid phase, and the solid particles did not grow with collision-coalescence of the particles. Therefore, the size of them was not influenced by the temperature of levitated Cu and the gas flow rate. When the temperature at the condensation site was higher than the melting point, the size of them was increased by the distance which was the difference in the position at the melting point and the vapor condensation.

1. Introduction

One of the methods for preparing ultrafine particles of metals and alloys is a low pressure gas evaporation method [1]. In this method, a metal is evaporated and metal vapor condenses under reduced pressure. It has been already known that the size of the particles obtained with this method increases with an increase in atmospheric pressure. However, there are scarecely any research in which particles are produced at atmospheric pressure. On the other hand, studies of the evaporation rate in a gas stream are mostly performed in a high-temperature gas stream, and there are few studies in a gas stream at lower temperature than that of the evaporated substance. Evaporation in gas proceeds through a continuous process of evaporation, vapor diffusion, condensation and particle growth. Thus, these four processes for the evaporation in a low temperature gas stream completes in the gas film on the surface of substance [2]. Particle size is influenced by the vapor condensation site and so on in the gas film.

Recently, high temperature process such as a plasma process has become easily available. It is also important to investigate the evaporation phenomenon at high temperatures.

The evaporation rates and characteristics of the condensed particles of molten Al and Fe in a stream of argon gas at room temperature were investigated in previous works [3-7]. Temperature ranges for aluminum and iron were 1873 K to 2273 K, and 2173 K to 2273 K respectively. The tmperature of the iron vapor condensation was below the melting point of iron. Whereas condensed particle size of iron was unchanged with temperature of levitation-melted iron, gas flow rate and so on, the particle size of aluminum was influenced by them. Then, as one of a series of fundamental studies of evaporation of metal in room temperature gas stream, evaporation of copper of which melting temperature was between Al and Fe was investigated in this study.

2. Experimental

Figure 1 depicts the main part of the experimental apparatus [3,5]. The levitation melting apparatus was a high frequency induction furnace (130 kHz, 60 kW, Hitachi Kokusai Electric). The temperature of the levitated specimen was measured through the prism by a two-colour thermometer (Chino, IR-FL2), and was manually controlled at less than plus or minus 10 K. The flow rate of the gas was adjusted by a mass flow controller. The purities of an Ar and a N_2 gases used were 99.99 vol% and 99.9995 vol% respectively. These gases were passed through the two columns (Nikka Seiko) to eliminate water vapour and oxygen gas included as impurities. Those contents were less than 2 ppb. The sample of Cu was cut from a cylinder rod with purity of 99.9 wt% by a fine cutting device. Cut sample was ground by emery paper to remove oxide layer on the sample. Finally, sample weight was adjusted to be 2.20 ± 0.05 g by cutting with a nipper.

The Cu sample was levitation-melted in Ar or N_2 gas stream. The levitated sample melted 60 s later after turning on the power. Ten seconds later, evaporation from the sample was observed. This time was set as the starting time. The rise in temperature was kept constant to measure the evaporation weight from the sample weight. After the temperature of the levitated copper was kept constant for a predetermined time, the sample was dropped onto a water-cooled copper mold which was inserted from the end of the reaction tube.

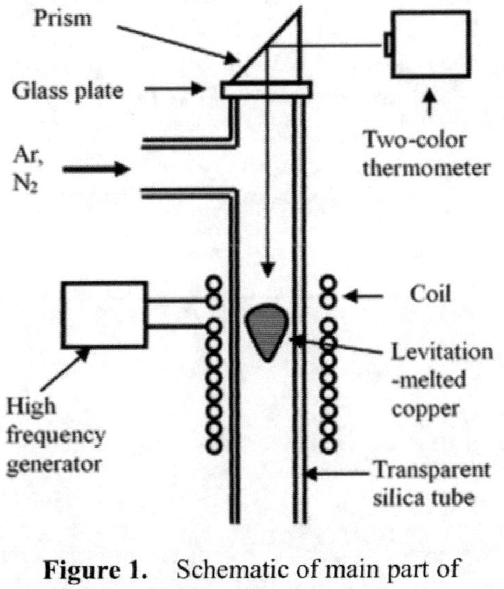

Figure 1. Schematic of main part of experimental apparatus.

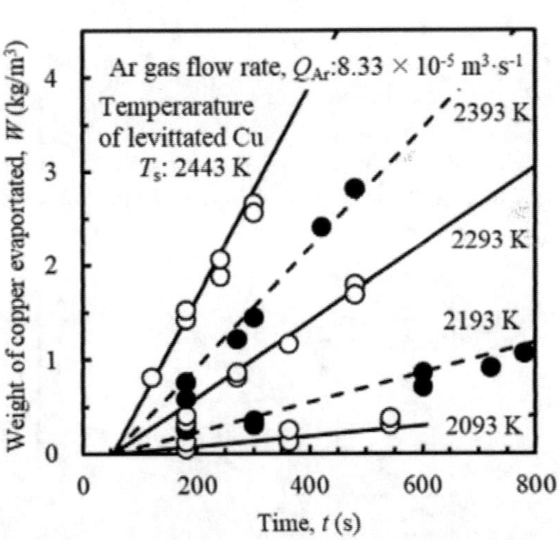

Figure 2. Change in weight of evaporated Cu with time. T_s is the temperature of levitated Cu.

The particles for the observation of a TEM (transmission electron microscope, Hitachi, H800) with an electron diffraction device were collected at the end of the reaction tube by placing a collodion film bound copper mesh in the exhausted gas mixture for a moment. The particles for measuring the contents of nitrogen and oxygen and analysing the crystal structure were collected by adhesion on a filter for a vacuum cleaner. Both the particles for a TEM and for measuring content and structure were collected in a gas stream. The time that the former particles are in contact with air was sorter than that of the latter.

The size of the particles, which were photographed with TEM, was measured by the image analyzer (Nireco, LUZEXIII). More than 500 Particles were investigated. The particles that were attached to each other were classified as isolated particles. The content of nitrogen in the particles was measured by extracting nitrogen in a He gas (Horiba, EMGA-1300). Crystal structure was analysed using XRD (X-ray diffractometry, Rigaku, RINT-2200).
The experiments were carried out to investigate the effect of temperature (1993 K – 2493 K) of the levitated copper while keeping the gas flowrate (1.67~25.0 $\times 10^{-5} m^3 s^{-1}$ (NTP (normal temperature and pressure))) on both the evaporation rate and the characteristics of the condensed particles.

3. Results

3.1 Evaporation rate

Figure 2 showed the change in the evaporated weight of levitation-melted sample with time. In this figure, the evaporated weight was given by dividing the weight loss with initial surface area of the levitated Cu. The initial surface areas of the levitated Cu for 1993 K, 2093 K, 2193 K, 2293 K, 2393 K, 2443 K, and 2493 K were 1.94×10^{-4} m^2, 1.96×10^{-4} m^2, 1.98×10^{-4} m^2, 2.01×10^{-4} m^2, 2.04×10^{-4} m^2, and 2.05×10^{-4} m^2 respectively. However, the surface area decreased by approximately 13 % when the weight loss by evaporation was maximum. The evaporated weight linearly increased with time at each temperature. The straight lines did not pass through the origin of this figure. This indicated that it took time for the sample to reach the predetermined temperature after the evaporation was castle observed visually. The evaporated weight in the N_2 gas flow also increased linearly with time. The gradients of theses straight lines expressed the evaporation rates respectively. The evaporation rate increased with the temperature rise. Here, the evaporation rate was the production rate of condensed Cu particle.

Figure 3 shows the influence of gas flow rate on the evaporation rate. Generally, the experimental evaporation rates, $\dot{W}_{obs}$, increased with an increase in the gas flow rate, Q. In addition, the evaporation rate in N_2 gas flow was larger than that in Ar gas flow, and its dependency of gas flow rate was larger than that in Ar gas flow rate.

The comparison of the evaporation rates in the Ar and the N_2 gas flow at the same temperature and gas flow rate was shown in figure 4. The evaporation rate in the N_2 gas stream was approximately 1.27 times larger than that in the Ar gas stream.

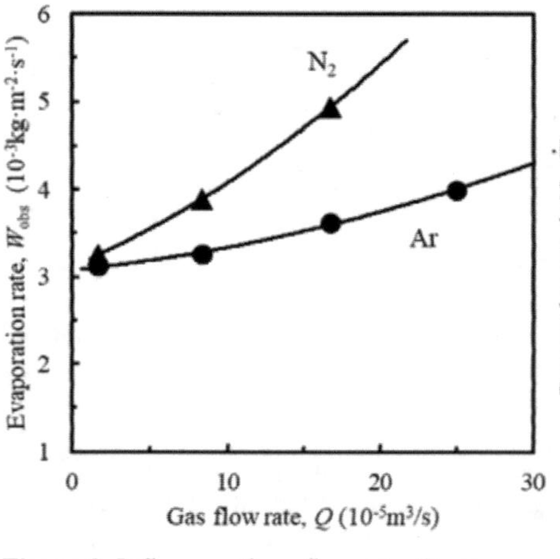

Figure 3. Influence of gas flow rate, Q, on evaporation rate obtained experimentally, $\dot{W}_{obs}$.

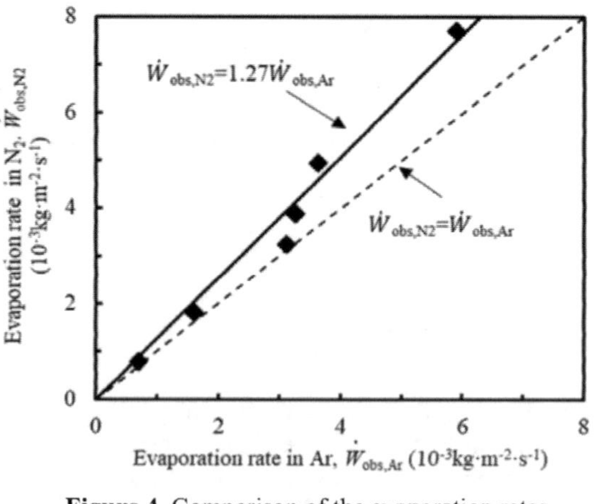

Figure 4. Comparison of the evaporation rates in Ar, $\dot{W}_{obs,Ar}$, and N_2 gas flow, $\dot{W}_{obs,N2}$.

Figure 2 showed the change in the evaporated weight of levitation-melted sample with time. In this figure, the evaporated weight was given by dividing the weight loss with initial surface area of the levitated Cu. The initial surface areas of the levitated Cu for 1993 K, 2093 K, 2193 K, 2293 K, 2393 K, 2443 K, and 2493 K were 1.94×10^{-4} m^2, 1.96×10^{-4} m^2, 1.98×10^{-4} m^2, 2.01×10^{-4} m^2, 2.04×10^{-4} m^2, and 2.05×10^{-4} m^2 respectively. However, the surface area decreased by approximately 13 % when the weight loss by evaporation was maximum. The evaporated weight linearly increased with time at each temperature. The straight lines did not pass through the origin of this figure. This indicated that it took time for the sample to reach the predetermined temperature after the evaporation was castle observed visually. The evaporated weight in the N_2 gas flow also increased linearly with time. The gradients of theses straight lines expressed the evaporation rates respectively. The evaporation rate increased with the temperature rise. Here, the evaporation rate was the production rate of condensed Cu particle.

Figure 3 shows the influence of gas flow rate on the evaporation rate. Generally, the experimental evaporation rates, $\dot{W}_{obs}$, increased with an increase in the gas flow rate, Q. In addition, the evaporation rate in N_2 gas flow was larger than that in Ar gas flow, and its dependency of gas flow rate was larger than that in Ar gas flow rate.

The comparison of the evaporation rates in the Ar and the N_2 gas flow at the same temperature and gas flow rate was shown in figure 4. The evaporation rate in the N_2 gas stream was approximately 1.27 times larger than that in the Ar gas stream.

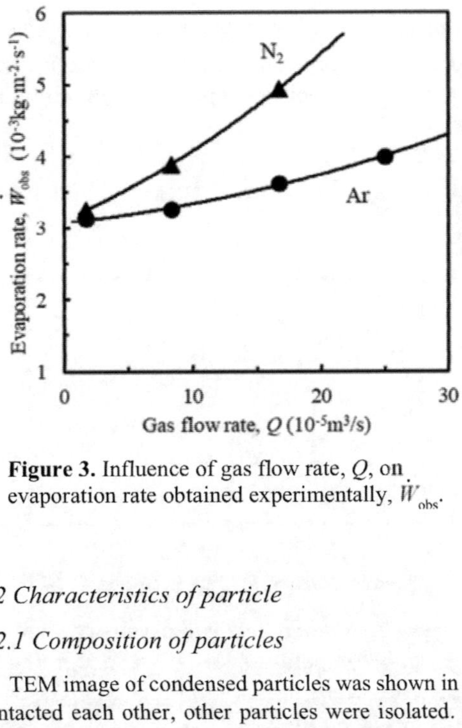

Figure 3. Influence of gas flow rate, Q, on evaporation rate obtained experimentally, $\dot{W}_{obs}$.

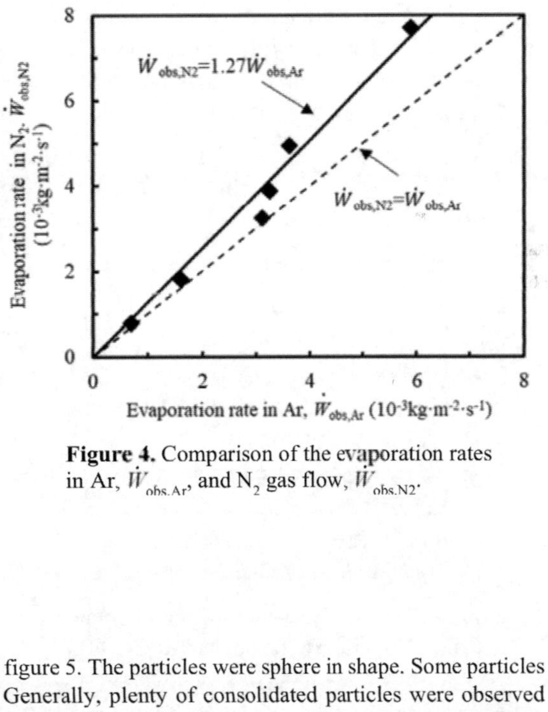

Figure 4. Comparison of the evaporation rates in Ar, $\dot{W}_{obs,Ar}$, and N_2 gas flow, $\dot{W}_{obs,N2}$.

3.2 Characteristics of particle

3.2.1 Composition of particles

TEM image of condensed particles was shown in figure 5. The particles were sphere in shape. Some particles contacted each other, other particles were isolated. Generally, plenty of consolidated particles were observed when the temperature of the levitation-melted copper was relatively low, or when the gas flow rate was relatively high. The particles did not have metallic luster and had reddish colour.

IOP Conf. Series: Materials Science and Engineering **920** (2020) 012035 doi:10.1088/1757-899X/920/1/012035

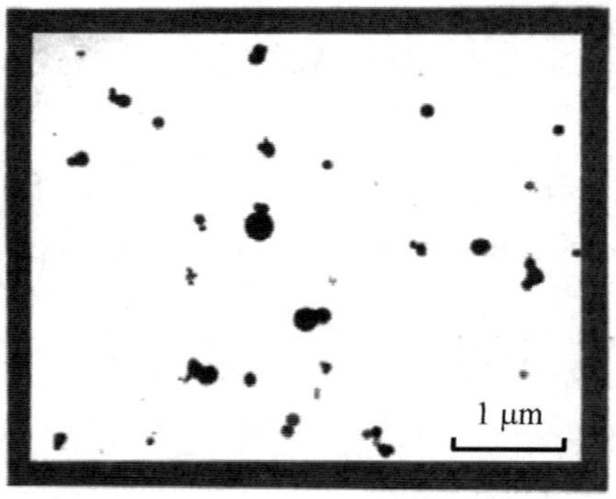

Figure 5. TEM image of particles. The particles were prepared in Ar gas flow rate, Q_{Ar}, of 8.33×10^{-5} m^3/s at 2273K of the levitated Cu.

Figure 6 shows the relation between the nitrogen content, C_N, in the particles and the temperature of levitated copper. Generally, the nitrogen content in the particles decreased with increasing temperature irrespective of gas species. The nitrogen content in the particles was less than approximately one-tenth of the nitrogen content in Cu_3N (6.84 wt%). The nitrogen content of the particles prepared in argon stream was smaller than that in nitrogen gas. Because the Ar gas did not include N_2 gas, the copper particles could not react with N_2 in the evaporation and condensation process. As described earlier, the particles were exposed to the air when they are collected. At that time, the particles were considered to have reacted with N_2.

This figure also depicted the nitrogen content of the levitated copper. The nitrogen content of the copper melted in the Ar gas flow could not be detected. The nitrogen content of the levitation-melted copper in the N_2 gas stream indicated the solubility of N_2 gas in the molten copper. The solubility for 1 atm of N_2 gas pressure is expressed as the following equation [8]:

$$\ln[\text{mass\%N}] = -(7406/T) - 5.863 \quad (1)$$

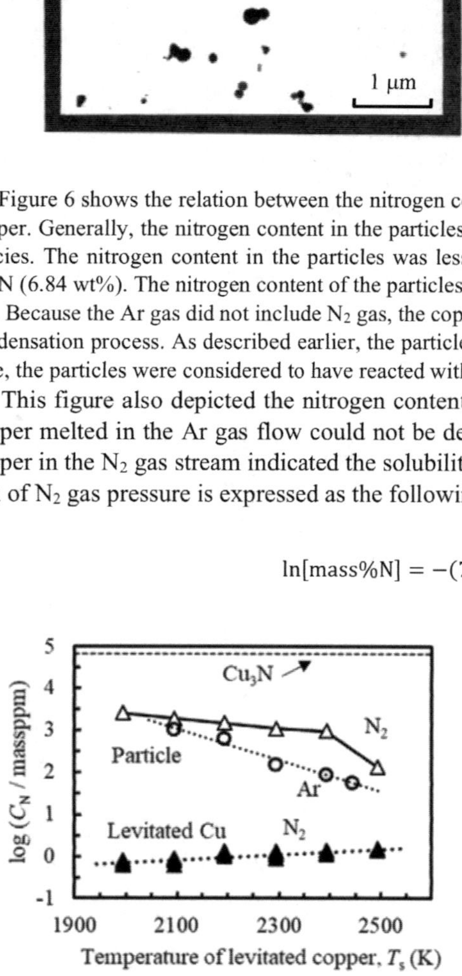

Figure 6. Relation between nitrogen content, C_N, and temperature of levitated Cu. The particles were prepared in $8.3\times10^{-5}m^3/s$ of the gas flow rate.

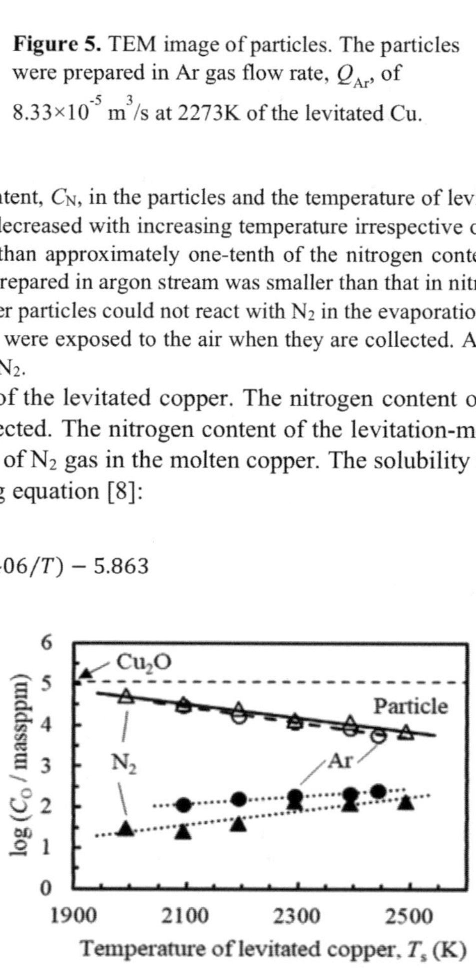

Figure 7. Relation between oxygen content, C_O, and temperature of levitated Cu. The particles were prepared in $8.3\times10^{-5}m^3/s$ of the gas flow rate.

Figure 7 shows the relation between the oxygen content, C_O, in the particles and the temperature of levitated copper. Generally, the oxygen contents in the particles decreased with increasing temperature irrespective of gas species. The oxygen content of the particles prepared in argon stream was approximately 10% smaller than that in nitrogen gas. In addition, the oxygen content of the particles was less than approximately one-half of the content in the Cu_2O (11.2 wt%). The oxygen content in the levitated Cu increased with increasing the temperature.

As shown in figure 6 and figure 7, the nitrogen content in the particles and the levitated Cu increased with increasing the oxygen content in them respectively. The temperature dependency of the relationship in the oxygen and nitrogen contents in the particles were different from the dependency in the levitated Cu. As described earlier, no oxide and nitride were not detected from the XRD and the electron diffraction detected no nitride and oxide. The oxygen and nitrogen seems to be forcibly dissolved in copper.

3.2.2 Mean particle diameter

The particle size distribution is shown in the figure 8. The value of the vertical axis of this figure represents the number-based ratio in percent. The distribution showed a distribution with one mode. Therefore, the mean particle size becomes a representative value of the distribution.

Figure 9 and Figure 10 show the influence of the temperature of the levitated Cu and the gas flow rate on the mean particles size respectively. Generally, the mean size increased with the temperature rise, and decreased with the increase in the gas flow rate. As shown in figure 2 and figure 3, the evaporation rate increased with the temperature rise and with increasing the gas flow rate. Therefore, the temperature and gas flow rate on the horizontal axis in figure 9 and figure 10 correspond qualitatively to the evaporation rate. Accordingly, there was no correlation between the mean size and the evaporation rate.

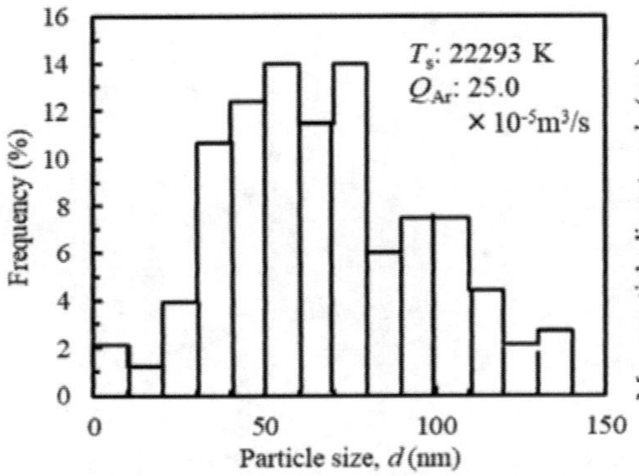

Figure 8. Histogram of particle size distribution.

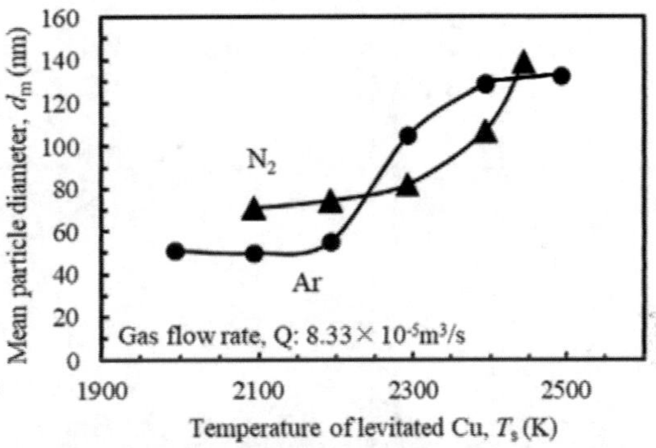

Figure 9. *Influence of temperature of levitated* Cu on mean particle size.

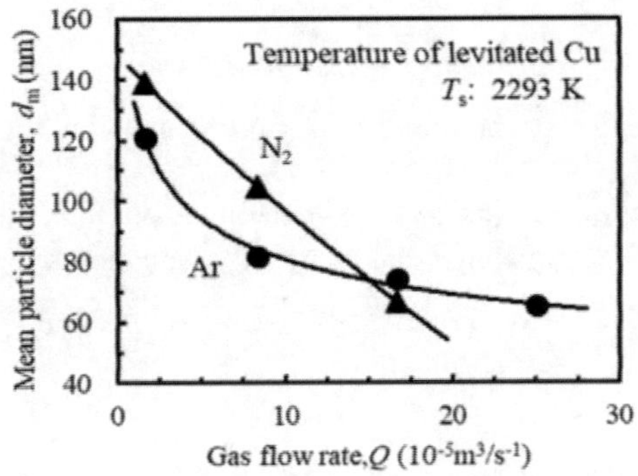

Figure 10. Influence of gas flow rate on mean particle size.

IOP Conf. Series: Materials Science and Engineering **920** (2020) 012035 doi:10.1088/1757-899X/920/1/012035

4. Discussion

4.1 Evaporation rate

Because the details of analysis of the evaporation rate have been described in previus work, they will be briefly described here. The outline of the model is as follows: According to the film theory for fluid-solid system [12], a stagnant gas film exists around a body in forced fluid flow. Vapor condenses in the gas film [8,10].

Assuming that the levitation-melted sample is sphere, the steady state evaporation rate of the sphere in forced convection with equimolar interdiffusion is expreesed as:

$$\dot{W} = M_{\mathrm{Cu}} D_{\mathrm{Cu}} \left(1 + \frac{r_n - r_s}{r_s}\right) \frac{C_s - C_n}{r_n - r_s} \tag{2}$$

where $\dot{W}$ is the evaporation rate [kg· mol^{-1}· s^{-1}], M_{Cu} molar atomic weight of Cu (kg/mol), D_{cu} the interdiffusion coefficient of Cu vapor and the gas system (m^2/s), C the concentration of Cu vapor (mol/m^3), and r the radial distance from the center of levitated Cu (m). Subscript s is the surface of the levitated Cu, and n the vapor condensation. In order to calculate the evaporation rate, we must know the concentration, position, and the temperature at vapor condensation.

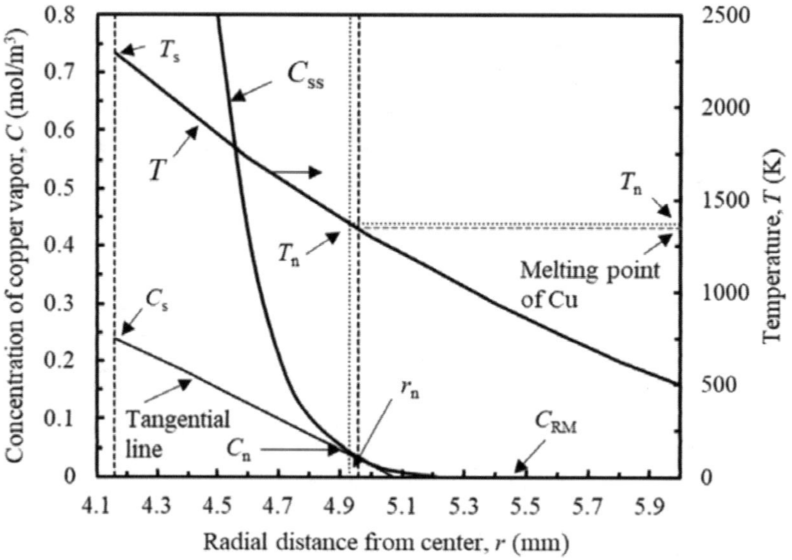

Figure 11. Distributions of temperature and concentration of Cu vapor in gas film.

The temperature distribution around the levitated sample is determined as follows. First, the temperature film thickness is determined using the following Rantz and Marshall's equation which represents heat transfer around the sphere in forced convection [9].

$$Nu = 2.0 + 0.6 Re^{1/2} Pr^{1/3} \tag{3}$$

where Nu, Re, and Pr are dimensionless numbers, and Nusselt, Reynolds and Prandtl numbers respectively.

Next, the temperature distribution in the gas film was determined from the following equation using the temperature of the levitated Cu.

$$\frac{T - T_b}{T_s - T_b} = \frac{(1/r) - (1/r_b)}{(1/r_s) - (1/r_b)} \tag{4}$$

where subscript b is the gas bulk. The temperature distribution of the main part is shown in figure 11. The flim thickness of temperature is given by subtracting r_s from r_b.

The nucleation rate, I ($m^{-3}s^{-1}$), given by Becker, Dörning, and Zeldovich [10] is as:

$$I = n_1 \frac{\alpha p_{Cu}}{\rho_{Cu} k_B T} \cdot \left(\frac{2\sigma_{Cu} M_{Cu}}{\pi N_A}\right)^{1/2} \cdot \exp\left\{-\frac{16\pi\sigma_{Cu}^3 M_{Cu}^2 N_A}{3R^3T^3\rho_{Cu}(\ln\alpha)^2}\right\} \tag{5}$$

where n_1 is the number concentraton of monomar vapor (m^{-3}), p the vapor pressure (Pa), M the molecular weight (kg/mol), N_A the Avogadro constant (mol^{-1}), k_B the Boltzmann constant (J/K), α saturation ratio (-), ρ the density (kg/m^3), and σ the surface tension (N/m). When the temperature is once set, the physical property values such as the surface tension are determined. I/n_1 becomes a function of the supersaturation ratio, α. I/n_1 is calculated at each temperature. Figure 12 shows the relation between I/n_1 and the supersaturation ratio at each temperature. Generally, after I/n_1 gradually increased with increasing the supersaturation ratio, it steeply increased with the ratio. The supersaturation ratio at 0.1 of I/n_1 at each temperature, this point almost corresponded to the point where I/n_1 increased steeply, was defined as the critical supersaturation ratio above which nucleation occured. This point corresponds to the point where I/n_1 starts up rapidly.

Using the critical supersaturation ratio and vapor pressure at each temperature, the critical supersaturation concentration, C_{ss}, is calculated, and shown in figure 11. The tangent line is drawn from the suface concentration, C_s, toward the curve, C_{ss}. It is assumed that Cu vapor was condensed at the concentration, C_n, and position, r_n, of the contact point.

Finally, the inter-diffusion coefficient at mean temperature between the surface and the vapor condensation site is calculated from Chapman-Enskog's first approximation [11].

The relationship between the calculated and the experimentally obtained evaporation rates is shown in figure 13. There was a good liner relation between the calculated and experimental evaporation rates irrespective of the temperature, the gas flow rate, and the gas species. The calculated evaporation rate was approximately two times larger than the exprerimental evaporation rate. In the case of evaporation of Al, the calculated evaporation rate of Al in a He or an Ar gas stream was approximately 1.1 and 1.8 times larger than the experimental evaporation rate respectively [3]. In the case of Fe evaporation, the calculated and the experimental evaporation rates in an Ar gas stream agreed well each other [5]. Therefore, it can be said that this model roughly explains the evaporation phenomena.

Thickness of the film for mass transfer is given from Rantz and Marshall's equation which represents mass transfer around the sphere in forced convection [9]. This equation is as:

$$Sh = 2.0 + 0.6Re^{1/2}Sc^{1/3} \tag{6}$$

where Sh and Sc are the Sherwood and Schmidt numbers respectively, and are dimensionless numbers. In figure 11, the point, C_{RM}, represents the position of the film thickness. In addition, generally, the thickness of the gas film for the mass transfer is smaller than that for the heat transfer. Assuming that the Cu vapor condensed at C_{RM}, that is, Cu vapor does nod concodense in the gas film, the calculated evaporation rate was approximately one third of the experimental evaporation rate. Therefore, it can be said that Cu vapor condensed in the gas film.

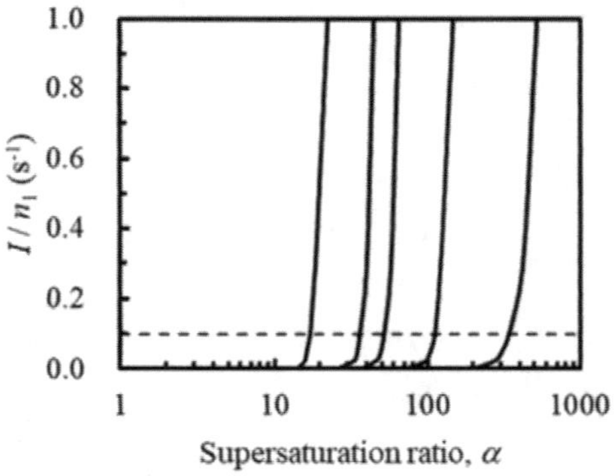

Figure 12. Relation between I/n_1 and supersaturation ratio.

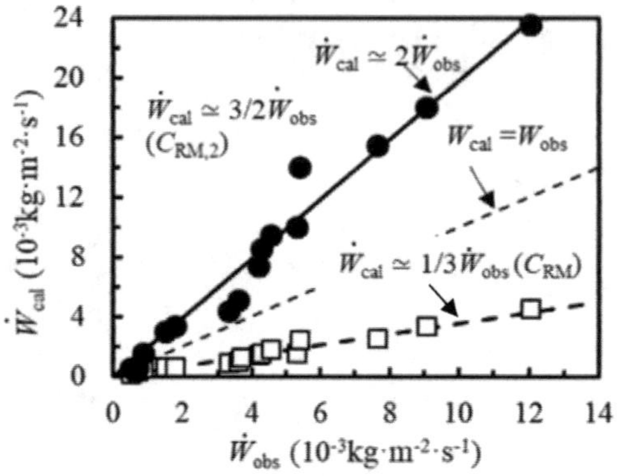

Figure 13. Comparison of experimentally obtained and calculated evaporation rates, $\dot{W}_{obs}$ and $\dot{W}_{cal}$.

4.2 Particle size

Figure 14 shows the relation between the mean particle size and the temperature at the vapor condensation. When the temperature at the condensation was below the melting point of Cu, the mean particle size was unchanged with the temperature. When the temperature was above the melting point, the mean particle size increased with the temperature rise. The condensed particles are liquid when the vapor condenses above the melting point. Therefore, the particles are easy to grow by collision-coalescence.

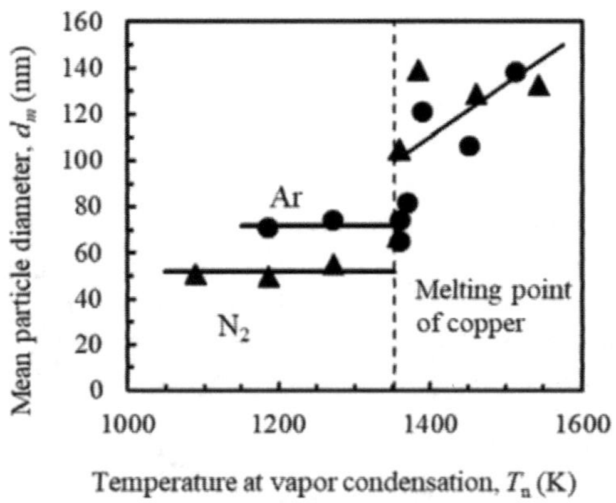

Figure 14. Influence of temperature of Cu vapor condensation on mean particle diameter.

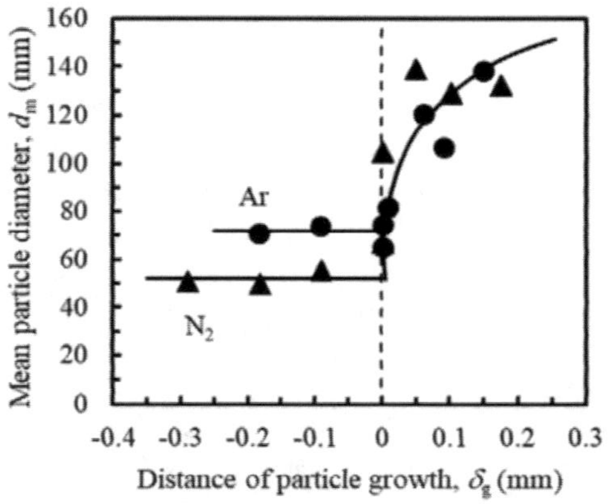

Figure 15. Relation between distance of particle growth and mean particle diameter.

As shown in figure 11, the difference from the position of the melting point of Cu to the position of the vapor condensation was defined as the distance of the particle growth, δ_g. Figure 15 shows the relationship between the particle growth distance and the average particle diameter is shown in the figure. The negative growth distance is due to the presence of the condensation position where the radius is smaller than the melting point position. The negative growth distance is due to the vapor condensing where the radius is larger than the melting point position. The correlation between the distance and the mean particle size was better than the relationship between the melting point and the mean particle size. In practice, the size and frequency of the condensed particles seem to have an effect, so further study is needed for grain growth.

In the case of Fe evaporation, because the temperature of the iron vapor condensation was below the melting point of Fe, the particle size was constant. In the case of evaporation of Al, because the temperature of the Al vapor condensation was higher than the melting point of Al, the particle size increased with an increase in the growth distance. The condensation temperature of Cu vapor became higher or lower than the melting point depending on the experimental conditions. The results of this study were consistent with previous results.

In addition, the distribution of the particles were expressed by the Weibull distribution with the shape parameter of 1.15 to 3.06. Generally, this parameter relates the range of the particle distribution. The parameter shows the tendencies to decrease with the temperature rise, and to increase with the increase in the gas flow rate. Therefore, the range became wide with the temperature rise, and narrow with the increase in the gas flow rate.

5. Conclusion

Copper was levitation-melted in an argon or nitrogen gas stream to study experimentally the evaporation rate and the characteristics of condensed particles. Both of them were analysed with using the model where the Cu vapor condenses in the gas film. The results and discussions were summarized as follows:

The evaporation rate increased with increasing the temperature of the levitated Cu and the gas flow rate. The evaporation rate in N_2 gas flow was larger than that in Ar gas flow. The rate calculated from this model was approximately two times larger than the experimental rate irrespective of the temperature of the levitated Cu, the gas flow rate, and gas species.

The condensed particles were spherical metallic copper. When the temperature at the condensation of vapor was below the melting point of Cu, the mean particles size was almost constant. When the temperature was above the melting point, the mean size increased with the increase in the distance of particle growth.

References

[1] Uyeda T, Hayashi C and Tasaki A 1996 *Ultra-Fine Particles,* (London: Noyes Publications)
[2] Turkdogan E T 1980 Physical chemistry of high temperature technology (San Diego: Academic Press)
[3] Yokoyama S, Saitoh K, Itoh N and Kawakami M 1995 *J. Japan Inst. Met. Mater.* **59** 814
[4] Yokoyama S, Saitoh K, Itoh N and Kawakami M 1995 *J.Japan Inst.Met.Mater.* **59** 1030
[5] Yokoyama S, Kokubo S, Itoh N and Kawakami M 1993 *J.Japan Inst.Met.Mater.* **57** 54
[6] Yokoyama S, Kokubo S, Itoh N and Kawakami M 1993 *J.Japan Inst.Met.Mater.* **57** 282
[7] Yokotama S, Saito K 2007 *Mater. Sci. Forum* **561-563** 1047
[8] Muizz B M N A, Nik H M Nor, Yokoyama S 2017 *High Temp. Mater. Proc.* doi.org /10.1515/
[9] Ranz W. E., Marshall W. R, 1952 Chem. Engng. Prog. **48** 141, 173
[10] Abraham F F, 1974 *Homogeneous nucleation theory*, (NY and London: Academic Press), pp.88
[11] Reid R C, Prausnitz J M, Sherwood T K, (Translater M Hirata), 1985 *The properties of Gases and Liquids 3rd* ed. (Japanese ver.), (Tokyo: Mcgrow-Hill), pp.546, pp.676

Acknowledgements

We acknowledge that we have used the TEM and the XRD used in this study in the Cooperative Research Facility Center at Toyohashi University of Technology.

IOP Conf. Series: Materials Science and Engineering **920** (2020) 012036 doi:10.1088/1757-899X/920/1/012036

Numerical Study of Droplet Behavior on an Oscillating Plate Using OpenFOAM

T Suzuki[1,*] and A Mitsuishi[2]

1 Department of Mechanical Engineering, Toyohashi University of Technology, Toyohashi 441-8580, Japan
2 Department of Mechanical System Engineering, Tokyo University of Agriculture and Technology, Tokyo 184-8588, Japan

*Corresponding author: takashi@me.tut.ac.jp

Abstract. The interfacial behaviour of a liquid droplet placed on an oscillating plate was simulated numerically by VOF method for the better understanding of the transition process of the surface fluctuation from the axisymmetric mode to the complicated mode which is leading to the interface burst and spray formation. The target of analysis was a water droplet of 100μL placed on a flat plate. In the place of the plate oscillation, a sinusoidal acceleration-oscillation was applied with the gravitational acceleration. The frequency of the acceleration-oscillation was ranging from 300 Hz to 900 Hz. The range of the acceleration-oscillation amplitude was from 3g to 60g. The computational domain was set to include a quarter of the droplet. In the numerical simulation, axisymmetric concentric waves and superposed azimuthal waves appeared on the droplet surface. The surface fluctuation grew in magnitude with an increase in the acceleration-oscillation amplitude, and the complexity of the interface increased due to the coalescences of the crests of the surface fluctuation.

1. Introduction

The vibration-induced liquid atomization has been applied widely in many industrial processes [1]. A unique cooling method of microelectronic packages was also proposed that utilized a two-phase mist flow generated by the vibration induced atomization [2]. Concerning to these applications, experimental investigations had been made on the interfacial behaviour of a liquid droplet placed on a vertically oscillating plate. It was revealed that an axisymmetric wave appeared on the droplet surface at relatively small oscillation amplitudes, the surface fluctuation grew in magnitude and complexity with an increase in the amplitude, and fine droplets were ejected from the crests of the fluctuation at larger oscillation amplitudes [3]. The experimental investigations had been also made on the critical oscillation amplitude for the fine-droplet ejection (onset of atomization) and the spray characteristics of ejected fine-droplets [4,5]. However, the transition process of the surface fluctuation from the axisymmetric mode to the complicated mode has not been understood well. Further investigation has been required on the surface fluctuation of the oscillated liquid droplet. Numerical simulations should be suitable for the detailed investigation of the surface fluctuation.

In the present study, for the better understanding of the transition process, the interfacial behaviour of a water droplet on an oscillating plate was simulated numerically by VOF method with dynamic adaptive mesh refinement. In the place of the plate oscillation, a sinusoidal acceleration-oscillation with relatively high frequency was added to the gravitational acceleration term. The interDyMFoam of OpenFOAM [6] was employed for solving the flow equations after the modification to make the acceleration term a function of time. The shape of the fluctuating surface, which was obtained numerically, resembled to the previous experimental observation [3]. It was considered that the flow field and the interface motion were well simulated by the employed computational method. The flow field and the interface motion were simulated numerically in several cases. The features of the surface fluctuations were revealed and the transition process of the surface fluctuation was discussed from the results.

2. Computational Method

The problem to be investigated in this study involves a water droplet of 100μL volume located on a vertically oscillating flat plate in atmospheric air, which is schematically shown in Figure 1 (a). The radius of a volume-equivalent half-sphere to the droplet is 3.63mm. The fluid flow, as well as the motion of the air-water interface, was simulated numerically by the volume of fluid (VOF) method. The computational domain is shown in Figure 1 (b). To reduce the computational cost, the computational domain was chosen as a cubic of 5mm which contained a quarter of the droplet, assuming the fluid motion and the interface shape were symmetrical with respect to the orthogonal cutting planes. In the place of the plate oscillation, a sinusoidal acceleration-oscillation was applied to the field with the gravitational acceleration in the simulation. The contact angle of the gas-liquid interface onto the plate was assumed to be constant. Test fluids, air and water, were both assumed to be incompressible and immiscible.

The governing equations of the incompressible fluid flow are as follows:

$$\nabla \bullet \mathbf{u} = 0 \tag{1}$$

$$\frac{\partial \rho \mathbf{u}}{\partial t} + \left[\nabla \bullet (\rho \mathbf{u}\mathbf{u})\right] + \left[\nabla \bullet \boldsymbol{\tau}\right] = -\nabla p + \rho \mathbf{G} + \mathbf{F_s} \tag{2}$$

where t is the time, $\mathbf{u}$ denotes the velocity vector, $\boldsymbol{\tau}$ is the viscos tensor, p is the pressure, $\mathbf{G}$ is the acceleration, and $\mathbf{F_s}$ is the surface tension force which is modelled by CSF method and acts in the interface region only. The transport properties of mixed fluid density, ρ, and viscosity, μ, are evaluated from the volume-fraction weighted properties of two fluids:

$$\rho = \alpha\, \rho_1 + (1-\alpha)\,\rho_2, \quad \mu = \alpha\, \mu_1 + (1-\alpha)\,\mu_2 \tag{3}$$

where ρ_1 and ρ_2 are the densities of water and air respectively, and μ_1 and μ_2 are the viscosities of water and air respectively. The void fraction, α, has the value of 0 for air and 1 for water, and the plane of α=0.5 represents the air-water interface. The transport equations of α can be written as:

$$\frac{\partial \alpha}{\partial t} + \nabla \bullet (\mathbf{u}\alpha) + \nabla \bullet \left[\mathbf{u_r}\alpha(1-\alpha)\right] = 0 \tag{4}$$

where $\mathbf{u_r}$ is the artificial interface compression velocity. The last term is introduced and optimized in order to keep the free-surface tracking accurate. The acceleration in z-direction was oscillated sinusoidally around the gravitational acceleration, $-g$.

$$\mathbf{G} = \left(0 \quad 0 \quad -g + A\sin(2\pi f t)\right)^T \tag{5}$$

where A and f are the amplitude and frequency of the acceleration oscillation respectively.

Equations (1)-(5) are solved numerically using interDyMFoam solver of OpenFOAM version 4.0 [6] after the modification to make the acceleration term in equation (2) a function of time. Specific settings of the boundary conditions are listed in Table 1. A hexagonal uniform grid of mesh-width 0.1mm was employed. Local adaptive mesh refinement (AMR) was used to ensure sufficient grid-resolution around the air-water interface. Figure 2 shows typical example of the locally refined grid created by the interDyMFoam. The minimum mesh width in the interface region was 0.025mm.

IOP Conf. Series: Materials Science and Engineering **920** (2020) 012036 doi:10.1088/1757-899X/920/1/012036

At the start of the simulation, the water droplet with the shape of one-eighth sphere was placed at the corner of the cuboid among the bottom wall and the symmetry planes. The numerical calculation was advanced for 2 seconds with zero oscillation amplitude and enhanced water viscosity until the droplet shape reached a steady state at which the restoring force by surface-tension balanced with the gravitational deforming force. The calculation was advanced with linearly increasing the acceleration-oscillation amplitude, A, to the settling value, A_s, within the duration of $10T$, then the calculation was proceeded further with the acceleration-oscillation amplitude of A_s. The time variation of the acceleration-oscillation amplitude during the start-up process is shown in Figure 1 (c), where T is the period of the acceleration-oscillation and is equal to $1/f$.

The numerical simulation was performed within the range of the oscillation frequency, f, from 300 Hz to 900 Hz, which was much higher than the free-oscillation frequency of a sessile drop [7].

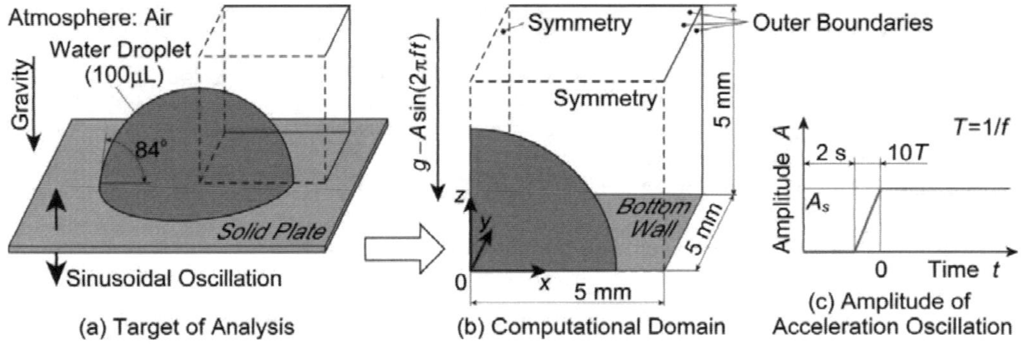

Figure 1. Problem definition and computational domain.

Table 1. Boundary conditions.

Boundaries	Velocity	Pressure	VOF(α)
Bottom Wall	noSlip	fixedFluxPressure	constantAlphaContactAngle (theta0 84)
Outer Boundaries	pressureInletOutletVelocity (uniform (0 0 0))	totalPressure (uniform 0)	inletOutlet (uniform 0)
Planes of Symmetry		symmetryPlane	

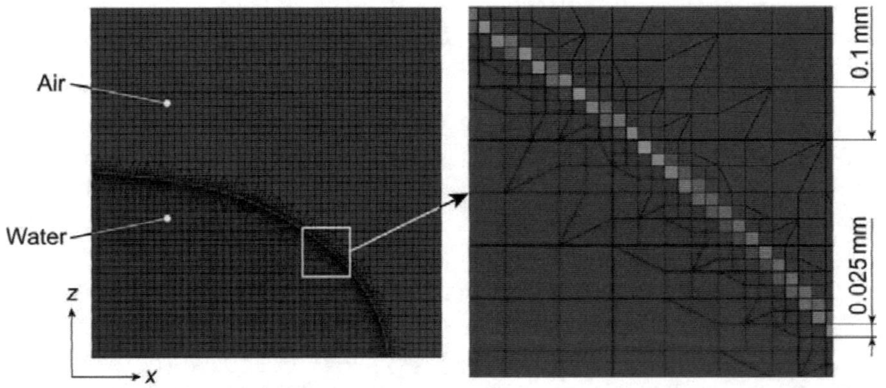

Figure 2. Typical example of local adaptive mesh refinement around air-water interface.

Table 2. Transport Properties of air and water.

Properties	Air	Water

Density ρ kg/m^3	1.21	998
Kinematic viscosity μ/ρ m^2/s	1.51 x 10^{-6}	1.12 x 10^{-6}
Surface Tension σ N/m	0.0714	

The range of the settling value, A_s, of the acceleration-oscillation amplitude was from $3g$ to $60g$. The volume of water within the computational domain was 25μL. A static contact angle of 84° was employed after the photographic observation of James et al. [3]. The fluid properties used in the simulations are listed in Table 2.

3. Results and discussion

Figure 3 shows typical examples of numerical results. Each image in Figure 3 is top view of the instantaneous shape of the air-water interface. When the acceleration-oscillation amplitude is relatively small, the droplet interface is relatively calm and almost axisymmetric as shown in (a) of the figure. Concentric axisymmetric waves are also observed faintly on the interface. At slightly large oscillation amplitude, azimuthal waves can be observed on the contact-line of droplet as shown in (b). The azimuthal waves are also observed subtly on the crests of the axisymmetric wave inside the contact-line. At larger oscillation amplitude, as the azimuthal waves grow in magnitude, the interfacial fluctuation becomes complicated one and the axial symmetry of the droplet shape is disappeared as shown in (c). The surface fluctuation should transit to the droplet-ejection mode at further large amplitudes. As the shape of fluctuating surface resembled to those of previous experimental observation [3], the flow field and the interface motion were considered to be well simulated by the employed computational method.

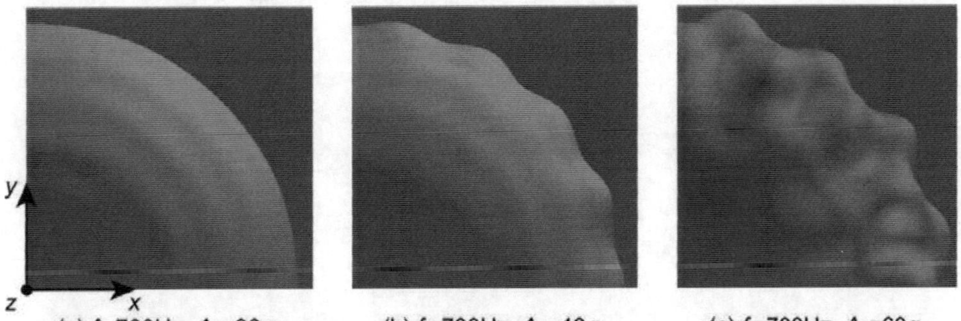

Figure 3. Top-views of the instantaneous interface shape of typical numerical simulations.

3.1. Concentric wave (Case 1)

A detailed investigation was made on the interfacial fluctuation appeared in the case of relatively small oscillation amplitude. Figure 4 shows the radial distribution of the time-averaged interface height and the radial variations of the vertical interface displacement. The droplet resembles the overturned bowl in shape as shown in the figure. There is a small-amplitude fluctuation on the interface. As the radial position of the fluctuation crests do not move, the interfacial fluctuation should be the standing wave. The fluctuation amplitude is large near the central axis of droplet and becomes small with an increase of the radial position, except near the circumference of the droplet where the interface is inclined and the vertical displacement is magnified. Figure 5 shows the time variation of the interface height at three points 1mm apart from the central axis (identified by up-arrow in Figure 4). The azimuths of these points are 22.5° apart as illustrated in the right-side of the figure. The fluctuation waveforms are sinusoidal and in the same phase, although the amplitudes of fluctuation are slightly different one another. The period, T, of the fluctuation corresponds to the reciprocal of the acceleration-oscillation frequency, f. That is, the fluctuating frequency of the interface is identical to the acceleration-oscillation.

These features of the surface fluctuation are similar to the surface wave observed during the surface-tension measurement by the capillary-wave method [8]. The surface fluctuation should be the concentric capillary standing wave. The fluctuation should not be able to grow greatly because the frequency of the acceleration-oscillation is much higher than the free-oscillation of droplet.

Figure 6 shows the distributions of the vertical displacement of interface from the time-averaged height at several representative moments. The almost axisymmetric concentric waves can be clearly seen. The azimuthal waves of minute amplitude are also observed. The azimuthal waves are appeared to be superposed on the concentric waves. Because of the azimuthal waves, the amplitudes of the interface height fluctuations of Figure 5 should be slightly different depending on the azimuths. As the displacement distribution at $t=t_0$ shown in (a) is very similar to that at at $t=t_0+T$ shown in (c), the azimuthal waves and the concentric waves should fluctuate in the same frequency, in this case.

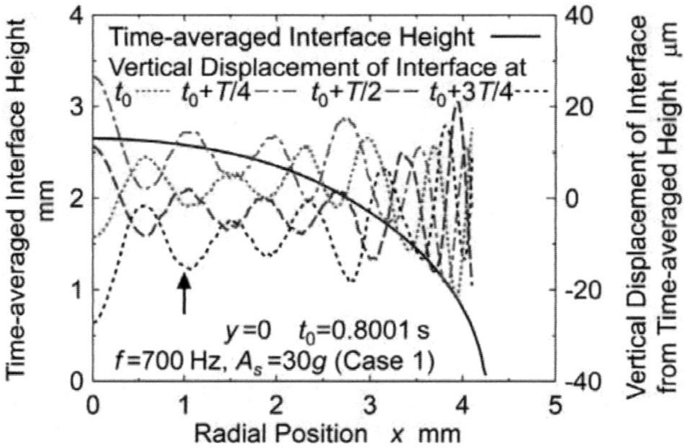

Figure 4. Radial distributions of time-averaged interface height and vertical displacements of interface.

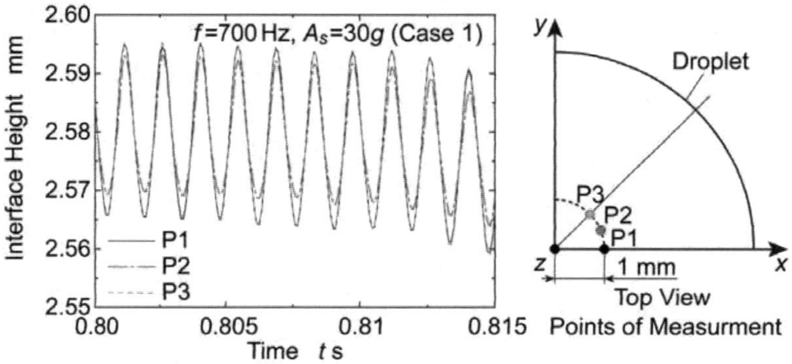

Figure 5. Time-variation of interface-height at representative points 1mm apart from central axis.

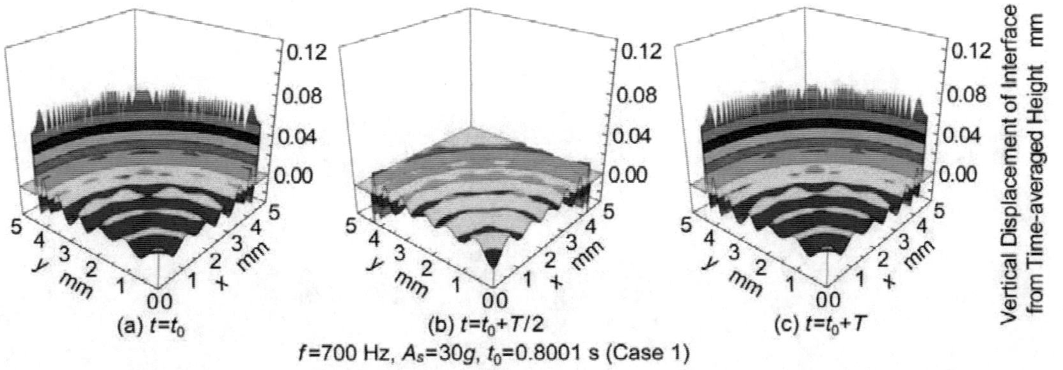

Figure 6. Vertical displacement of interface from time-averaged interface height at several moments.

3.2. Concentric wave with azimuthal wave (Case 2)

A detailed investigation was made next on the interfacial wave appeared in the case of the intermediate oscillating amplitude. Figure 7 shows the distributions of the vertical interfacial displacement from the time-averaged interface height at several representative moments. The azimuthal waves, as well as the concentric waves, can be observed on the droplet interface. The azimuthal waves are appeared to be superposed on the concentric waves. The amplitude of the azimuthal waves is large near the circumference of droplet and is not so large neat the central axis. The azimuthal waves at $t=t_0$ shown in (a) of the figure are in the same phase to those at $t=t_0+2T$ shown in (e) but in the opposite phase to those at $t=t_0+T$ shown in (c). The crests of the azimuthal waves do not move with time. The azimuthal waves should be the standing wave, but the fluctuating frequency of the azimuthal waves may be different from the concentric waves which are synchronized with the acceleration-oscillation.

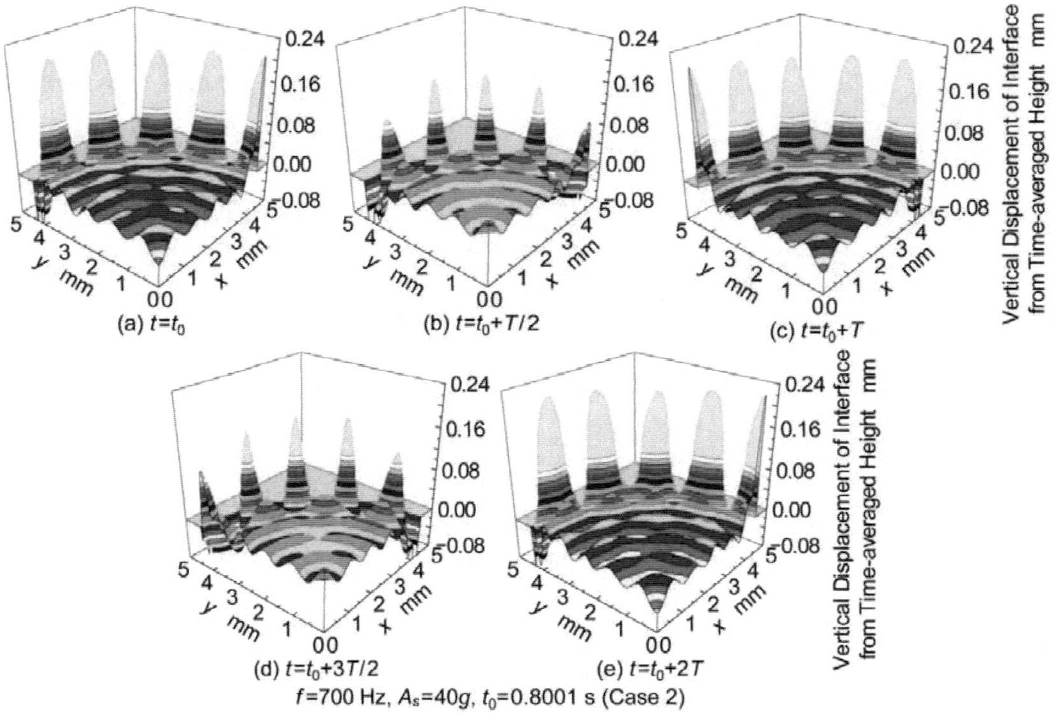

Figure 7. Vertical displacement of interface from time-averaged interface height at several moments.

The time-variations of the droplet radius on y=0 and the results of FFT analysis are respectively shown in (a) and (b) of Figure 8. Although the waveform of the droplet-radius distribution is sinusoidal and only a peak can be found on the FFT result at the acceleration-oscillation frequency in the case of relatively small oscillation amplitude (Case 1) as shown in (I) of the figure, the waveform of the droplet-radius distribution is distorted and two peaks can be found on the FFT result at the acceleration-oscillation frequency and half of the frequency in this case as shown in (II). These results suggest that the large-amplitude azimuthal waves near the circumference of droplet should be the subharmonic waves of the order 1/2.

The characteristics of the azimuthal waves were investigated in more detail. Table 3 lists the number of waves within a quarter of droplet, the wavelength, λ_a, and the fluctuating frequency, f_a, of the azimuthal wave superposed on each concentric wave. The estimated wave-velocities, $f_a\lambda_a$, and the theoretical velocity, c, of the surface-tension wave with wavelength of λ_a, which is obtained from equation (6), are also listed in the table. The wave-velocities were estimated assuming that each azimuthal wave was a standing wave and composed of right-handed and left-handed waves.

$$c = \sqrt{\frac{2\pi\sigma}{\lambda_a \rho_l}} \tag{6}$$

The azimuthal wave on the 1st concentric wave fluctuates at the frequency of the acceleration-oscillation, which is identical to the fluctuating frequency of the concentric wave. However, the azimuthal waves on other concentric waves fluctuate at half of the acceleration-oscillation frequency. The estimated wave-velocities of all azimuthal waves almost agree with the velocity of the surface-tension wave. The azimuthal waves should be the standing wave of the surface-tension wave whose integer multiple of the half-wavelength matches the circumferential crest-length of each concentric wave.

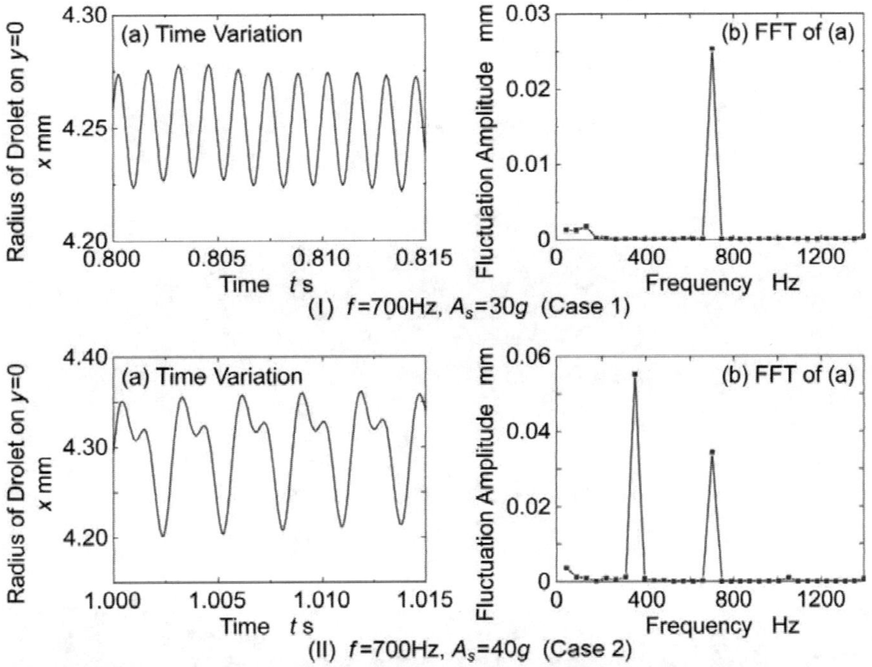

Figure 8. Time variations of droplet radius and results of frequency analysis.

Table 3. Characteristics of azimuthal wave superposed on each concentric wave.

ID of Concentric Waves	Radius, r mm	Num. of Waves in Domain	Wave-length, λ_a mm	Freq. of Wave, f_a Hz	Vel. of Wave, $f_a\lambda_a$ m/s	Vel. of Surface-tension Wave, c m/s
#1	0.60	1.0	0.94	700	0.660	0.691
#2	1.53	1.5	1.60	350	0.559	0.531
#3	2.38	2.5	1.49	350	0.522	0.549
#4	3.28	3.5	1.47	350	0.514	0.553
#5	3.83	4.5	1.34	350	0.467	0.580
#6	4.2	4.5	1.47	350	0.513	0.54

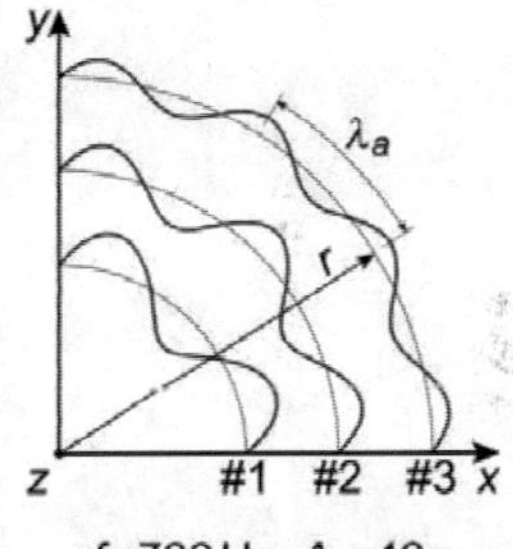

f=700 Hz, A_s=40g (Case 2)

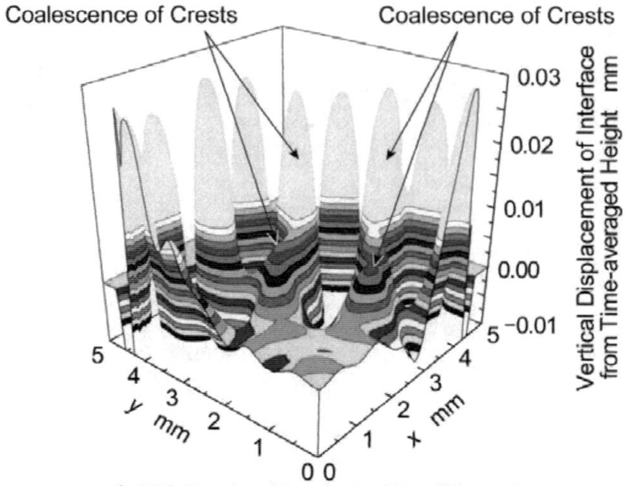

Figure 9. Vertical displacement of interface from time-averaged interface height showing coalescence of fluctuation crests.

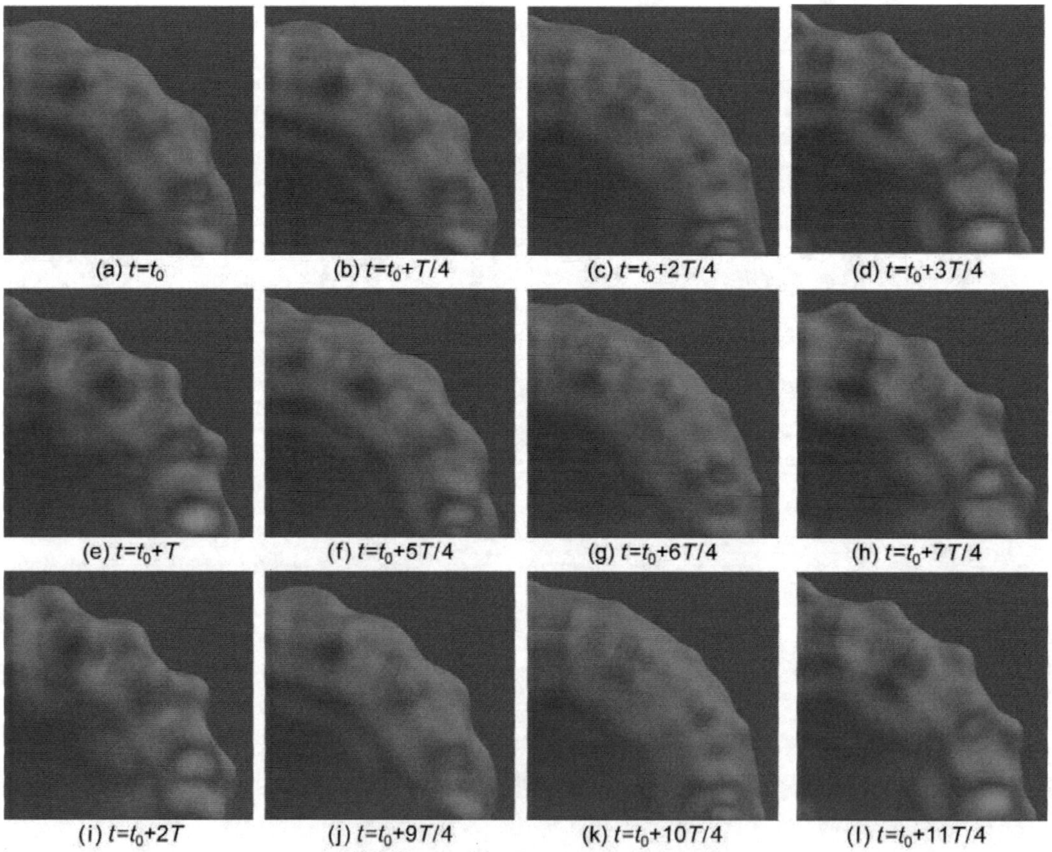

Figure 10. Change of droplet interface in shape over time.

3.3. Complicated wave (Case 3)

The interfacial fluctuation at larger acceleration-oscillation amplitude was investigated next. Figure 9 shows the instantaneous distribution of the vertical interfacial displacement from the time-averaged interface height. The amplitude of the azimuthal waves is larger than that of Case 2 as shown in the figure. The amplitude of the azimuthal waves is large not only near the circumference of droplet but also in the range of intermediate radius. However, the amplitude is not so large near the central axis of the droplet. Due to the large amplitude of the azimuthal waves, the crests of the interface fluctuation become high and the bases of the crests spreads in the radial direction at intermediate radius. As the results, the adjacent crests of the azimuthal waves coalesced and merged into one, as indicated by the arrows in the figure. This should be the reason why the interfacial fluctuation becomes complicated one at relatively large acceleration-oscillation amplitudes.

Figure 10 shows the change of the droplet interface in shape over time. The azimuthal wave can be observed obviously on the contact-line of droplet. The interface fluctuates in complex manner especially in the range of relatively large radius. Although the asymmetry of the interface shape disappears, the periodicity of the fluctuation still remains in this case, as can be seen from the figure.

3.4. Map of fluctuation modes

The surface fluctuations were simulated in several cases. Figure 11 shows the map of observed fluctuation modes. The acceleration threshold for the droplet ejection of Vukasinovic et al. [4] is also shown in the figure. Within the range of relatively small acceleration-oscillation amplitude, almost concentric waves were observed. The azimuthal waves superposed on the concentric waves become obvious in the range of acceleration-oscillation amplitude larger than the dashed line in the figure. The surface fluctuation is appeared to transit to the complicated one at the acceleration-oscillation amplitudes above the dotted line in the figure. It is considered from these results that the fine-droplets would be ejected from the crests of the fluctuating interface at further large acceleration-oscillation amplitudes.

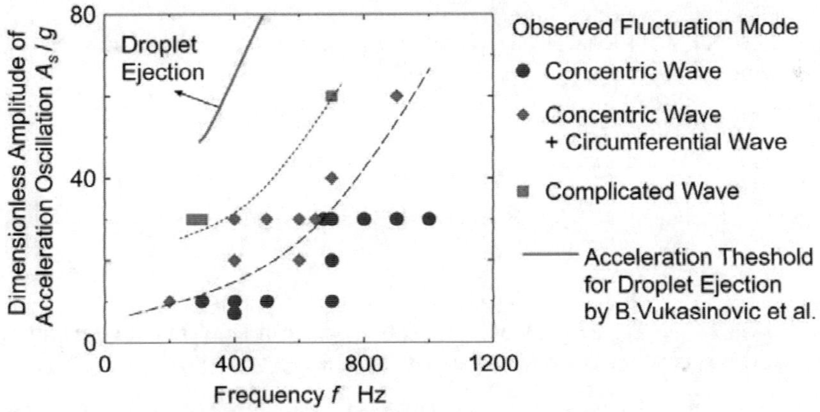

Figure 11. Map of interfacial fluctuation modes.

4. Conclusions

The interfacial behavior of water droplet on an oscillating plate was simulated numerically by VOF method with dynamic adaptive mesh refinement for the better understanding of the transition process of the surface fluctuation from the axisymmetric mode to the complicated mode. In the place of the plate oscillation, a sinusoidal acceleration-oscillation at relatively high frequency was applied with the gravitational acceleration. The interDyMFoam of OpenFOAM was employed for solving the flow equations after the modification to make the acceleration term a function of time. Following conclusions were deduced:

- Numerically obtained shape and motion of the droplet surface resembled to those of previous experimental observation. It was considered that the flow field and the interface motion were well simulated by the employed computational method.
- At relatively small acceleration-oscillation amplitude, almost axisymmetric concentric waves appeared on the droplet interface. The concentric waves were standing waves and the frequency of surface fluctuation was identical to the acceleration-oscillation.
- At slightly large acceleration-oscillation amplitude, azimuthal waves also appeared on the interface. The azimuthal waves were standing waves and superposed on the axisymmetric concentric waves. Near the droplet circumference, the azimuthal waves had large amplitude and showed subharmonic oscillation of the 1/2 order. The azimuthal waves were the standing wave of the surface-tension wave whose integer multiple of the half-wavelength matched the circumferential crest-length of each concentric wave.
- At larger acceleration-oscillation amplitude, the azimuthal waves became obvious not only near the circumference but also in the range of intermediate radius. As the result, the axial symmetry of the droplet shape is disappeared, and the adjacent crests of the azimuthal waves coalesced into one. This should be the reason why the surface shape becomes complicated one at relatively large acceleration-oscillation amplitudes.

5. References

[1] Nasr G G, Yule A J and Bendig L 2002 *Industrial Sprays and Atomization* (London: Springer-Verlag)

[2] Heffington S N, Black W Z and Glezer A 2002 Vibration-induced droplet atomization heat transfer cell for high-heat flux applications *Thermal Challenges in Next Generation Electronic Systems* (Rotterdam: Milpress) eds Joshi Y K and Garimella S V pp 123-132

[3] James A J, Vukasinovic B, Smith M K and Glezer A 2003 Vibration-induced drop atomization and bursting *J. Fluid Mech.* 476 pp 1-28

[4] B.Vukasinovic B, Smith M K and Glezer A 2007 Mechanisms of free-surface breakup in vibration-induced liquid atomization *Phys. of Fluids* 19 012104

[5] Vukasinovic B, Smith M K and Glezer A 2004 Spray characterization during vibration-induced drop atomization *Phys. of Fluids* 16 pp.306-316

[6] Greenshields C J 2016 OpenFOAM User Guide version 4.0 (The OpenFOAM Foundation) https://openfoam.org/version/4-0/

[7] Hoon J H, Kang B H and Kim H Y 2006 The lowest oscillation mode of a pendant drop *Phys. of Fluids* 18 021702

[8] Iino M, Suzuki M, Ikushima A J and Okuda Y 1984 Surface Waves on Liquid Helium *Japanese J. of appl. Phys.* 23 Suppl. 23-1 pp 54-56

Acknowledgments

This study was assisted by graduate school students of Toyohashi University of Technology. Authors would like to express special thanks to Miss Kumada S for her great support.

Failure analysis of various fiberglass cross-arm designs under multi-axial loading

Daud Mohamad[1], Salmia Beddu[1], Agusril Syamsir[2], Nazirul Mubin Zahari[1], Muhammad Fauzinizam Razali[3,*], Sareh Aiman Hilmi Abu Seman[3], Aizat Abas[3] and Fei Chong Ng[3]

[1]Department of Civil Engineering, Universiti Tenaga Nasional, 43000 Kajang, Malaysia.
[2]Institute of Energy Infrastructures (IEI), Universiti Tenaga Nasional, 43000 Kajang, Malaysia.
[3]School of Mechanical Engineering, Universiti Sains Malaysia, Engineering Campus, 14300, Nibong Tebal, Penang, Malaysia.

*E-mail: mefauzinizam@usm.my

Abstract. This study investigated the type of failure modes of different cross-arm designs under multi-axial static loading. The failure modes were numerically predicted from real-scale finite-element models of cross-arm integrated with a Hashin damage subroutine. Three finite-element models of cross-arm were considered; a model with standard cross-arm design, a standard design engaged with braces, and a standard design engaged with sleeves. The failure analysis of the composite cross-arm was focused on the location and type of failure of the structure upon its application. This investigation revealed that under the applied load, the cross-arm with installed sleeves exhibited the lowest total deflection of 0.21 m. Every cross-arm design exhibited fiber failure due to the tension and compression mode, regardless of the support system installed. Additionally, the installation of sleeves and braces on the cross-arm structure successfully reduced the number of areas associated to failure.

Keywords: Fiberglass Cross-arm, Hashin Failure, Multi-axial Loading.

1. Introduction

After the event of serious weather, recent damage to electrical installations in a number of areas has generated a growing interest in upgrading the design of these substations and energy tower buildings structures situated at those with natural risks [1-3]. The structural reliability is the chance of the structure not reaching its limits in a certain period (e.g. failure state). Therefore, the development of a plan to improve system resilience needs a knowledge of how these loads affect grid failure initiation and advancement. Consequently, the efficacy of distinct models can be recognized to guarantee that the grid is more solid and resilience. Installations of sleeve and braces are the most commonly used methods for resisting horizontal forces in a frame structure, as well as being highly efficient and economic [4, 5]. Even though the effect of sleeve and braces on structural performance has been studied previously [5-8], the knowledge on the type of failures experienced by the composite structure upon the use of sleeve and braces supports is still lacking.

This paper aims to predict the modes of failure of different cross-arm structure design under extreme multi-axial loading condition. Detail finite-element models of the composite cross-arm structure were developed in Abaqus/Explicit FEM code. The Hashin's failure criterion [9] was implemented using a VUMAT subroutine to predict the location and failure modes of the loaded structure. The methodology examined three distinct cross-arm constructions with and without sleeve and braces. The magnitude of displacement and the type of failure induced were compared in between designs to access the effectiveness of sleeves and braces installations on the structural reliability.

2. Development of the cross-arm model

2.1 Geometry and boundary condition

Three numerical models of composite cross-arm were developed in this study by using a finite-element analysis software program Abaqus/CAE. In the overall, each model consisted of two upper arms, two lower arms and two steel connectors. The cross-section of the arms is square (0.14 x 0.14 m), with a thickness of 0.01 m and 4.9 m length. There are 7 stacked plies of glass laminate throughout the thickness, with each ply thickness of 0.0014 m. The stacking sequence of the square arm is [45/ -45/ 0/ 90/ 0/ 90/ 0].

Figure 1 presents the considered designs of the cross-arm model. Figure 1(a) denotes the standard frame design of cross-arm, which consisted of four arms connected at one end and fixed at the other end. The cross-arm model was developed from a total of 73,472 linear hexahedral element with reduced integration (C3D8R). The steel connector was developed from a rigid element (R3D4). As the simulation involves severe bending deformation, multiple elements were set throughout the beam thickness to avoid the issue of hourglassing and severe element distortion [10].

One end of the cross-arm was fixed in every direction, and the other end was tie constraint towards the steel connector. Three loads of magnitude 6524 N, 84,992 N and 46,416 N were applied in longitudinal (x-axis), vertical (y-axis) and transverse (z-axis) directions, respectively. These load magnitudes were selected after considering the extreme exposure of loading condition with factor of safety of 4, such reported in [11]. These loads were applied at the reference point (RP) coupled at the steel connector.

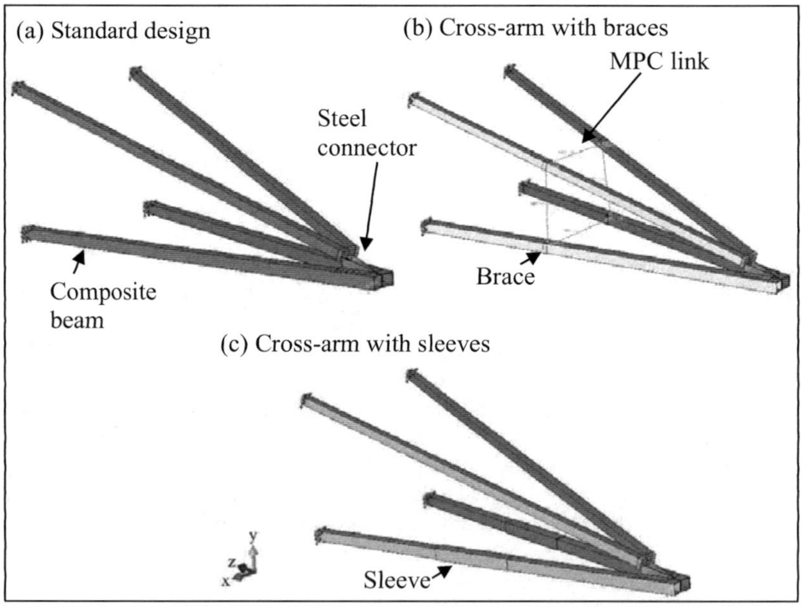

Figure 1. Geometry and boundary conditions defined on the cross-arm models: (a) standard design, (b) with braces and (c) with sleeves.

Figure 1(b) denotes the standard design of the cross-arm, with additional braces installed and connected at the center of each arm. This design concept was achieved by creating and linking multiple-point constraint (MPC) link between the center of the arms. The main idea is to fix the distance between the linked arms so that flexural deformation can be transferred from the center to the other region of the cross-arm. This is to replicate the case of cross-arm installed with a steel braces, with the purpose of reducing the bending deformation at the center region of the cross-arm (see Figure 1 (b)).

Figure 1(c) denotes the standard cross-arm design installed with a pair of 1-meter long fiberglass sleeve at the center length of the lower beam. The sleeves have the same stacking sequences and thickness with the beams of the cross-arm. A surface-to-surface contact was selected upon assigning the interaction between the sleeve and the lower arm surface, with the normal behaviour was set to "hard' contact and the tangential behaviour set to penalty formulation. These sleeves act as an alternative support system against the brace system proposed in the previous paragraph.

The analysis was considered as quasi-static and carried out with dynamic explicit solver. For the post-processing output, the user-dependent state variables SDV 1 to SDV 4 was requested from the field output manager. The Hashin subroutine characterized these variables onto the four failure modes of composite; fiber breakage in tension (SDV 1), fiber buckling in compression (SDV 2), matrix cracking in tension (SDV 3) and matrix crushing in compression (SDV 4). Additionally, the total deflection magnitude of the whole cross-arm structure was also requested for comparison.

2.2 Mesh convergence test

A mesh convergence test was performed to identify the optimal number of elements that satisfactorily balance numerical result and computational cost. The convergence test was performed by increasing the number of elements of the arms, by varying the element size from 0.06 mm (coarse) to 0.01 mm (very fine) by increments of 0.01 mm. In overall, five numerical runs were considered with the total deflection of the cross-arm set as the convergence criteria. Table 1 summarizes the total deflection of the whole model, and it is recognized that the deflection results converged as the mesh density increased. Thus, the mesh of 165,984 elements was selected for the model as it has relatively enough accuracy and lower computational cost than other meshes.

Table 1. Summary of mesh refinement model.

Run	Element size (mm)	Element number	Total deflection (m)	Computational period (h)
1	0.05	33,264	0.122	0.9
2	0.04	41,328	0.193	2.2
3	0.03	73,472	0.257	11.4
4	0.02	165,984	0.262	22.3
5	0.01	662,592	0.265	40.6

2.3 Laminate properties

The composite arms were made from fiber glass. Table 2 summarizes the mechanical properties of the composite arm, which were obtained from the manufacturer. The material data were assigned by means of the *USER MATERIAL instruction in Abaqus framework, to directly link the properties with the VUMAT/Hashin subroutine.

Table 2. Mechanical properties of the glass-fiber laminate.

No.	Parameters	Value (unit)
1	Density	1.8 g/cm^3
2	Young's modulus, E_1	17000 MPa
3	Young's modulus, E_2	5000 MPa
4	Young's modulus, E_3	1500 MPa
5	Poisson's ratio (V_{12}=V_{13}=V_{23})	0.28
6	Shear modulus (G_{12}=G_{13}=G_{23})	4000 MPa
7	Ultimate tensile stress, X_{1T}	320 MPa
8	Ultimate compressive stress, X_{1C}	150 MPa
9	Ultimate tensile stress, X_{2T}	80 MPa
10	Ultimate compressive stress, X_{2C}	60 MPa
11	Ultimate shear stress, S_{12}	80 MPa
12	Ultimate shear stress, (S_{13} = S_{23})	50 MPa

2.4 Hashin failure criteria

A VUMAT/Hashin subroutine was integrated into the developed finite-element model to predict the failure modes of the cross-arm. In general, the Hashin failure criterion is based on the work of Hashin [9]. The advantage of Hashin criterion over other stress-based criteria is it identifies four distinct modes of failure for composite material, in specific, fiber breakage in tension, fiber buckling in compression, matrix cracking in tension and matrix crushing in compression. These failure modes of fiber and matrix occur whenever the following equations are satisfied (equivalent or greater than 1).

If $\sigma_{11} \geq 0$, the tensile fiber failure criterion is:

$$\left(\frac{\sigma_{11}}{X_{1T}}\right)^2 + \left(\frac{\sigma_{12}}{S_{12}}\right)^2 + \left(\frac{\sigma_{13}}{F_{13}}\right)^2 \geq 1 \quad (1)$$

If $\sigma_{11} \leq 0$, the compressive fiber failure criterion is:

$$\left(\frac{\sigma_{11}}{X_{1C}}\right)^2 \geq 1 \quad (2)$$

If $\sigma_{22} \geq 0$, the tensile matrix failure criterion is:

$$\frac{(\sigma_{22}+\sigma_{33})^2}{X_{2t}^2} + \frac{\sigma_{23}-\sigma_{22}\sigma_{33}}{S_{23}^2} + \frac{\sigma_{12}^2+\sigma_{13}^2}{S_{12}^2} \geq 1 \quad (3)$$

If $\sigma_{22} \leq 0$, the compressive matrix failure criterion is:

$$\left[\left(\frac{X_{2c}}{2S_{23}}\right) - 1\right]\frac{(\sigma_{22}+\sigma_{33})}{X_{2c}} + \frac{(\sigma_{22}+\sigma_{33})^2}{4S_{23}^2} + \frac{(\sigma_{23}^2-\sigma_{22}\sigma_{33})}{S_{23}^2}] + \frac{\sigma_{12}^2+\sigma_{13}^2}{S_{12}^2} \geq 1 \quad (4)$$

Where σ, X and S is the components of applied stress, the ultimate stress and the shear stress, respectively.

3. Results and discussion

This paper deals with the failure prediction of different cross-arm designs under severe flexural loading. Three cross-arm designs were considered, which are the standard cross-arm design and a standard design with installed sleeves and braces for the supports. The installation of the supports at the middle region of the lower arm was motivated by our previous work [12], which reported the critical bending deformation about the middle region of the cross-arm. The numerical approached considered in this study served as a cost and time saving method to access the effectiveness of the supports towards the reliability and integrity of the cross-arm structure.

Table 3 summarizes the total deflection, and the failure modes exhibited by the different design of composite cross-arm upon loading. It is seen that the largest total deflection of 0.262 m was recorded both for the cross-arm with standard design and with installed braces. The installation of the 1-meter sleeves on the lower arm successfully increased the stiffness of the structure, thus considerably reduced the total deflection to about 0.21 m.

For the failure analysis, every design ends up with a failure of the fiber in tension (SDV 1) and compression (SDV 2). This finding shows that the installation of the support system such as sleeve and braces at the lower part of the cross-arm is not effective in preventing the occurrence of the failures. Interestingly, the installation of braces on the cross-arm has led to the existence of matrix cracking in tension (SDV 3); a failure which was not appeared in standard design case. This type of failure was induced due to the stiffening of the structure, as the middle regions of the cross-arm were constrained from deformation by the braces.

Table 3. Deflection and failure modes of different cross-arm designs.

No	Parameters	Cross-arm designs		
		Standard	Braces	Sleeves
1	Total deflection (m)	0.26	0.26	0.21
2	Fiber breakage in tension (SDV 1)	1	1	1
3	Fiber buckling in compression (SDV 2)	1	1	1
4	Matrix cracking in tension (SDV 3)	0	1	0
5	Matrix crushing in compression (SDV 4)	0	0	0

Figure 2 shows the locations of fiber breakage in tension (SDV 1) for different cross-arm designs upon loading. The 0.0 and 1.0 magnitude denote the non-failure and failure region with respect to the deformation of the composite structure. For every design, the fiber breakage occurred about the end where the arms were pinned to the steel connector. This type of failure appeared mostly on the upper part of the cross-arm, signifying that this section suffered in tension the most during the loading. Between the designs, it was observed that the design with installed sleeves and braces suffered more in terms of fiber failure, such indicated by the greater area of red contours at the upper part of the cross-arm. This observation highlighted that the stiffening of the lower arm (due to the installed support) had caused the upper arm to bear most with the applied loads and experienced greater deformation. Essentially, it is now clear that the installation of the supports on the composite trusses structure in transverse direction is not that effective in preventing structural failures as in the building structure case [4, 5].

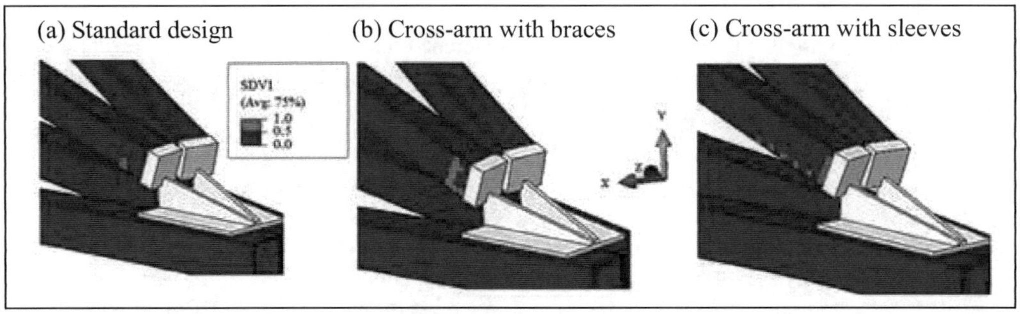

Figure 2. Failure locations of cross-arm structure due to fiber breakage in tension (SDV 1): (a) standard design, (b) with braces and (c) with sleeves.

Figure 3 shows the locations of fiber buckling in compression (SDV 2) for different cross-arm designs upon loading. It is seen that the large bending deformation appeared somewhere near the pinned area, and the location of failure is approximately identical to the one observed on the onsite failed cross-arm (unpublished work). It is interesting to note that the total deflection reported in previous section was referring to this region of lower part of the cross-arm. Since there was no support installed for the standard design, the large deflection of the structure caused the failure to occur at two locations; one about the middle region of the lower arm and the other one occurs about the fixed ends. On the other hand, the installation of supports such as braces and sleeves successfully reduced the number of failure location into one, which localized at the region near the pinned location.

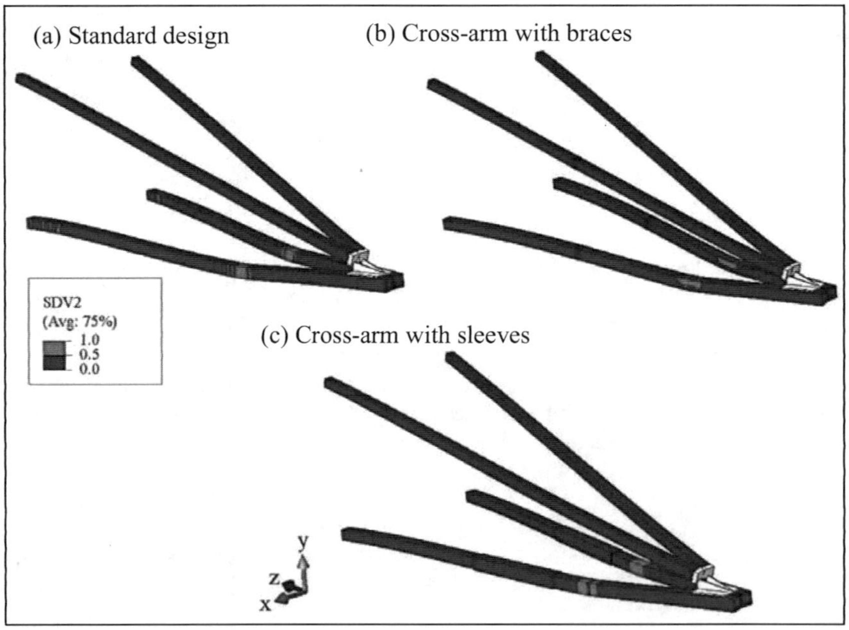

Figure 3. Failure locations of cross-arm structure due to fiber buckling in compression (SDV 2): (a) standard design, (b) with braces and (c) with sleeves.

4. Conclusions

Every cross-arm design considered in this study exhibited failures in terms of fiber breakage and buckling due to tension and compression load, respectively. The cross-arm installed with braces presented extra failure of matrix cracking in tension load. In overall, the installation of additional supports for the existing cross-arm material is not effective in preventing the failure of the structure for the current loading condition. It is suggested for the manufacturer to improve the integrity of the current cross-arm structure by strengthening the properties and increasing the number of layers of the composite laminates.

References

[1] A. L. López, L. E. P. Rocha, D. L. Escobedo, and J. S. Sesma 2009 *Proceedings of the 11th Americas Conference on Wind Engineering* (San Juan: Puerto Rico) p 22.
[2] K. D. Pitilakis et al. 2007 *Proceedings and Monographs in Engineering, Water and Earth Sciences* p 583.
[3] A. Nadhirah et al. 2017 *Int. J. Appl. Eng. Res.* **12** 15228.
[4] J. H. Ling, A. B. A. Rahman, I. S. Ibrahim, and Z. A. Hamid 2017 *Int. J. Concr. Struct. Mater.* **11** 525.
[5] M. A. Haque, M. A. Masum, M. M. Ratul, and Z. Tafheem 2018 *Proceedings of the 4th International Conference on Civil Engineering for Sustainable Development* p 4905.
[6] M. Xu, Z. Xu, L. Zhang, and S. Liu 2018 *Aust. J. Mech. Eng.* **16** 68.
[7] D. Mohamad et al. 2019 *IOP Conf. Ser. Mater. Sci. Eng.* **530** 12028.
[8] D. Mohamad et al. 2019 *IOP Conf. Ser. Mater. Sci. Eng.* **530** 12027.
[9] Z. Hashin 1980 *J. Appl. Mech.* **47** 329.
[10] "Abaqus Analysis User's Manual," 2015.
[11] M. Selvaraj ,S. Kulkarni, & R. R. Babu 2013 *Composite Structures* **96** 1.
[12] D. Mohamad et al. 2019 *IOP Conf. Ser. Mater. Sci. Eng.* **530** 12029.

Acknowledgements

The authors acknowledge Tenaga Nasional Berhad (TNB), UNITEN R&D, and Institute of Energy Infrastructure (IEI) for the lab facilities and financial support (TNB Seeding Fund: U-TS-RD-19-03). Special thanks to those who contributed to this project directly or indirectly

AUTHOR INDEX

Institute of Physics
Dirac House, Temple Back
Bristol BS1 6BE UK

ISSN: 1757-8981
ISBN 978-1-7138-3189-1

CURRAN ASSOCIATES INC.
proceedings.com